Liming and Agriculture in the Central Pennines

The use of lime in land improvement from the late thirteenth century to c. 1900

David S. Johnson

BAR British Series 525
2010

Published in 2016 by
BAR Publishing, Oxford

BAR British Series 525

Liming and Agriculture in the Central Pennines

ISBN 978 1 4073 0738 1

BAR Publishing is the trading name of British Archaeological Reports (Oxford) Ltd. British Archaeological Reports was first incorporated in 1974 to publish the BAR Series, International and British. In 1992 Hadrian Books Ltd became part of the BAR group. This volume was originally published by Archaeopress in conjunction with British Archaeological Reports (Oxford) Ltd / Hadrian Books Ltd, the Series principal publisher, in 2010. This present volume is published by BAR Publishing, 2016.

Printed in England

BAR
PUBLISHING

BAR titles are available from:

BAR Publishing
122 Banbury Rd, Oxford, OX2 7BP, UK
EMAIL info@barpublishing.com
PHONE +44 (0)1865 310431
FAX +44 (0)1865 316916
www.barpublishing.com

Contents

List of Figures

List of Tables

Acknowledgements

No field-based study such as this could have been undertaken without the willing co-operation and consent of a multitude of landowners and tenant farmers, and my thanks are extended to them all. Similarly, I owe a debt of gratitude to the archivists, librarians and staff of the many record offices and libraries visited during the archival research process. It may be invidious to single any out but I must draw special attention to the staff of the North Yorkshire County Record Office in Northallerton where I spent many a day. There are also those many individuals who directed me in certain directions or drew my attention to something they had identified in their own research: they will know who they are and I thank them, too.

Undertaking doctoral research can be akin to ploughing a lonely furrow but I was fortunate in having Professor Ian D Whyte, of Lancaster University, as my supervisor. A noted historical geographer and student of landscape himself, with a keen interest in enclosure and liming, he was the ideal person to be advised and prodded by. My thanks to him are profound and enduring.

I would also like to express my gratitude to my examiners, Professor Tom Williamson and Dr Mike Winstanley, for giving me such a stimulating and, I have to say, thoroughly enjoyable final experience, and for suggesting how I might tweak the thesis for publication. I am also very grateful to Dr Kathleen Whyte and Mrs Jill Sykes for their helpful comments on the text. However, I must include the disclaimer that any errors or misinterpretations are entirely my own.

Summary

This book focuses on the historic use of lime as a soil additive, and sets liming in the context of agricultural land improvement alongside draining, paring, marling and the use of other soil conditioners. It is concerned with the Central Pennines, centred on the Yorkshire Dales National Park, and adjacent lowlands.

Previous work has tended to concentrate on kilns as industrial monuments or on archaeological investigations of lime burning and quarrying sites. No other detailed regional study of the history of agricultural liming has been identified in the literature.

The book examines the time frame during which liming is known to have been practised, and investigates whether or not widespread liming was largely a phenomenon of the era of parliamentary enclosure; it looks into the spatial extent within which land was limed, whether dominantly moorlands and upland wastes or lowlands as well; and investigates the possibility that liming was not just undertaken where limestone bedrock occurs but also more widely; and considers the possibility that the benefits of liming may have been appreciated at all levels of the farming hierarchy rather than just by landowners and their stewards and agents.

An empirical, evidence-based approach was adopted with data having been obtained by both field- and desk-based methods. Archaeological excavation was employed, alongside the field surveying techniques of landscape archaeology, to identify the spatial distribution of liming and to isolate examples of late medieval or early modern lime burning. Extensive archival research concentrated on estate, manorial and farm records and on contemporary written sources.

The key conclusions are that liming in the study area has been recognised from the late sixteenth century with tentative earlier dates, and was extensively practised beyond the end of the time frame set, *ie* 1900; that lime was applied in lowland fringes and dale bottoms in addition to higher ground; that liming was an established practice in areas far from the nearest source of limestone bedrock; and that the benefits of applying lime were appreciated by tenants as well as by stewards and landlords.

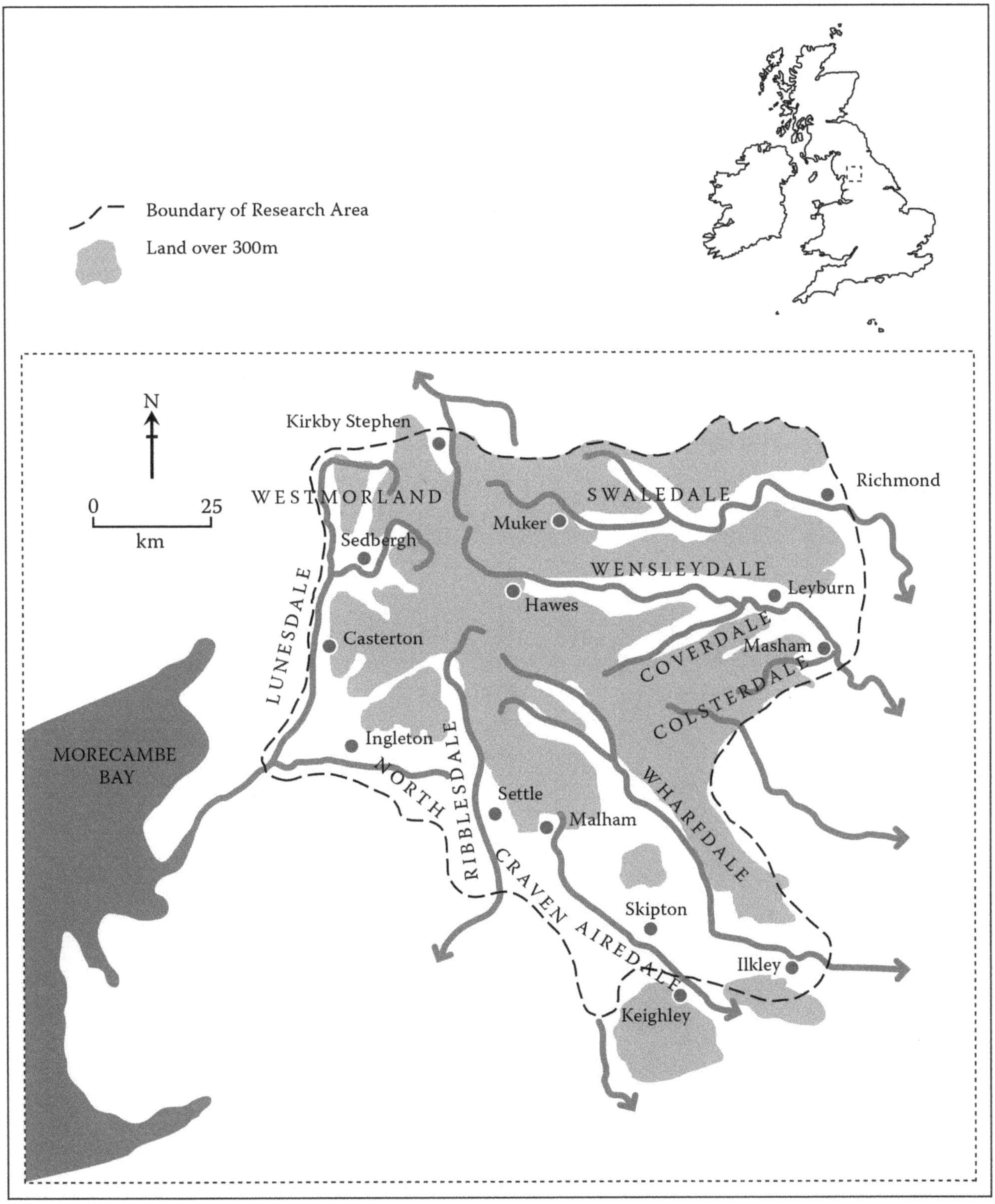

Frontispiece: Extent of the research area

Chapter 1
Introduction

1. Research aims

Justification for the research

The focus of this research is the historic use of lime as a soil improver within the Central Pennines and contiguous areas (Frontispiece). As discussed below, much farmland within this area is dominated by low-potential acidic grassland based on a range of soil types that are themselves of an acidic nature. They have either weathered from impermeable sandstones, shales, or glacial material or from limestones whose alkalinity is reduced by prevailing climatic conditions and low evapo-transpiration rates. All soils within the area require intervention to reduce acidity levels and boost their productive status. Nowadays this is achieved by spreading chemical inputs and/or animal slurry and 'muck'. Historically various methods were employed to this end, such as draining, dunging and marling, with the application of burnt lime (quicklime) being of exceptional importance across large parts of the area over a long period of time.

Many studies have been undertaken on the production of lime in various parts of the country but – as will be shown in Chapter 2 – the vast majority have been primarily concerned with lime kilns rather than the uses made of their end product. They have focussed on kilns as structures – their design, how they functioned, what remains in today's landscape – with occasional attempts to date them or to put them into an industrial context. Some studies have attempted to put kilns into a wider context, looking at why lime was being produced in rural areas, and others have made limited use of archival sources to set the process into a local context. However, the geographical and historical focus of many such studies has been limited.[1] Though of undoubted value at the micro-scale, they do not permit firm conclusions to be drawn about many aspects of agricultural lime burning, nor can they provide an overview of how the whole process of soil improvement and land management evolved through the centuries.

Added to this is a tendency, in some 'popular' publications, to perpetuate aspects of received knowledge that do not stand up to close scrutiny. It has often been stated, for example, that the use of lime on the land was essentially a phenomenon of the process of parliamentary enclosure, roughly between 1750 and 1850 (see, for example, Isham 2000, 13; Muir 2004, 145; Raistrick 1973, 71; White 1997, 91). Equally, it has been written that most lime kilns were sited on or close to a limestone outcrop, and that the lime was primarily used locally (see Buchanan 1974, 185; Fieldhouse 1980, 193; Davies-Shiel 1973, 97; Raistrick 1973, 69; Stanier 2000, 110). None of these assertions has much validity in the Central Pennines, as will be shown in the following chapters.

This study adopts a more holistic approach to land management and improvement. By taking the long view (the *longue durée* so beloved of archaeologists and prehistorians) it enables an examination of why attempts have been made to improve land productivity from medieval times onwards, how one generation built on the achievements of its predecessors, and what practical methods were employed to bring about improvement. By adopting a regional approach encompassing the Central Pennines and adjacent areas, in other words by researching what could perhaps be termed a *pays* rather than just an artificial construct such as a modern county or a set of contiguous civil parishes, it has been possible to see land improvement in the context of land ownership and husbandry. Whether it was monastic demesne, early modern landed estate or a community of yeomen farming valley intakes and open fells, management practices spread beyond – indeed took little heed of – administrative or parochial boundaries. Finally, by drawing upon a broad range of sources and methods (see Chapter 3) such as archaeological excavation, detailed field surveying across the entire research area and comprehensive archival research, it has been possible to reach conclusions inaccessible to previous writers in the field.

In short, this work makes a fresh contribution to the understanding of the historic use of lime on the land by considerably extending it in both temporal and spatial dimensions.

Aims

The overarching aim of the book is to detail the use of lime in land improvement within the Central Pennines and contiguous areas, from before 1300 to the end of the nineteenth century. To achieve this aim several strands of research were followed targeting differing elements of the whole. They can be summarised as follows:

1. to identify the time frame within which lime was applied on the land, either for arable or pastoral farming, with the aim of significantly extending the length of time hitherto generally accepted;
2. to examine whether there is evidence to confirm that limestone was burned to produce agricultural lime in areas not subjected to parliamentary enclosure, particularly in dale bottoms and the lowland fringes of the Pennine uplands;
3. to consider whether agricultural lime was produced and applied in areas where limestone does not occur as bedrock, such as in the gritstone Pennines of north-east Lancashire, Lower Airedale and Lower Wharfedale,

[1] See Chapter 2.5 for full referencing.

including the use of rather arcane methods of sourcing the raw material;
4. to examine the extent to which the benefits of liming were appreciated by landowners, managers and tenants, and the part that all tiers of the agricultural hierarchy played in its adoption.

Time frame

The scope of the book spans six centuries with the late thirteenth century having been selected as the starting point. This is not because it is associated with any specific event. There is a dearth of documentary sources for the area prior to that time, but the latter part of that century coincided with an era when monastic land improvement was underway, involving lime and/or other strategies. It was also a time when new farmland was being created by encroachment onto the waste and by woodland clearance (see Chapter 4.3).

The end date has been set at the close of the nineteenth century. This, too, is an arbitrary date that falls within a lengthy period of agricultural depression that began in the 1870s and lasted until the 1930s. However, it was partly due to the depressed state of farming that the use of lime on the land slipped into a temporary but potentially terminal decline. This did not happen overnight in the 1870s but was a gradual and increasingly apparent process through the closing decades of that century. Liming, however, was not abandoned. The year 1900 is adopted for convenience.

2. Epistemological framework

Archaeology, and to a lesser extent prehistory, have been polarised into two mutually opposing paradigms; the more recently developed disciplines of landscape history and landscape archaeology have similarly seen their proponents adopting mutually exclusive stances. There are close parallels between the two sets of views which will be explored more fully in Chapter 2: discussion here will focus tightly on where this writer sits.

Archaeologists tend to fall into the processual or post-processual camp. The former attempts to provide robust explanations for perceived changes and sees them as the result of physical or anthropogenic forces (or processes) which can be explained in a theoretical and objective way. This approach basically decries mere description. The post-processualist, on the other hand, denies the validity of objective systems-based models to explain change and sees 'man' as a knowledgeable being who throughout time possessed the ability to come up with new ways of doing things, and who has always been able to formulate ideas based on circumstance. The post-processualist prefers to interpret the past by trying to make sense of what we can see of the past, offering alternative explanations, rather than being bound by models of what might have been going on. In short, the individual is more important than the system.

The research framework behind this study leans to the post-processual rather than the processual. Throughout the period under review those concerned with the land, whether owner, steward or husbandman, made decisions based on their knowledge of local conditions, opportunities and constraints, and not on what modern theoretical models might suggest. Soil characteristics, micro-climate, details of topography, access to fuel and raw materials, were the considerations uppermost in their minds, influencing the decisions they made concerning whether to improve their land or not and, if so, how and when to do it. At the same time their actions were also influenced by knowledge they had gleaned from elsewhere (for example, from seeing what others were doing, from journals and treatises, by listening to others) while still taking heed of what instinct dictated. They were not slavishly following what someone else said was best.

It is also almost inevitable that decisions made by farmers in the past may well have been influenced by external factors, such as climatic change or market forces or advances in available technology. Elements of these externalities can be seen within the Central Pennines as will be demonstrated throughout the book.

Over the last four decades geography as a discipline has largely eschewed the regional approach in favour of a thematic or processual treatment, but one school of landscape history maintains that taking the regional view is still perfectly valid because, as we have seen, individual landscape elements combine to make up spatially wider landscape components. Regionality could be seen as a logical extension of this way of thinking and should therefore be encouraged (Taylor 2000, 161).

W G Hoskins, a pioneer landscape historian – if perhaps not the first (Chandler 2000) – is widely acknowledged for having composed a powerful evocation of English regional identity in his *The making of the English landscape* (see M Johnson 2007a, 69), a work that led to the publication of a series of regional, landscape-based, studies with the edition on Lancashire being one that perfectly enshrined the Hoskins approach. It has a mix of theoretical and ground-based description, as well as analysis of how landscapes evolved, from the perspective of the historical geographer (Millward 1955). In Hoskins' mind the chief role of such researchers should be to 'show how man has clothed the geological skeleton' over the past fifteen centuries (Hoskins 1955b, xi), and to explain how the countryside of our time has been moulded by our collective predecessors.

We shall return to the arguments for and against the Hoskins ideal in the following chapter. Suffice it to say here that he has been criticised for being too descriptive and bedded within the Romantic paradigm (see, for example, M Johnson 2007a, 40 ff). The rise of the processualists, especially in the 1970s, seeking a more rigorous and theory-based approach, led to a temporary abandonment of the Hoskins approach. More recently, however, there has been a re-awakening of interest in and appreciation of regional syntheses with a strong element of empirically-derived descriptive content (Taylor 2000, 161).

In landscape history the conceptual divide separates those who favour the empirical route from those who prefer the theoretical. Some of the latter disparagingly see the former as

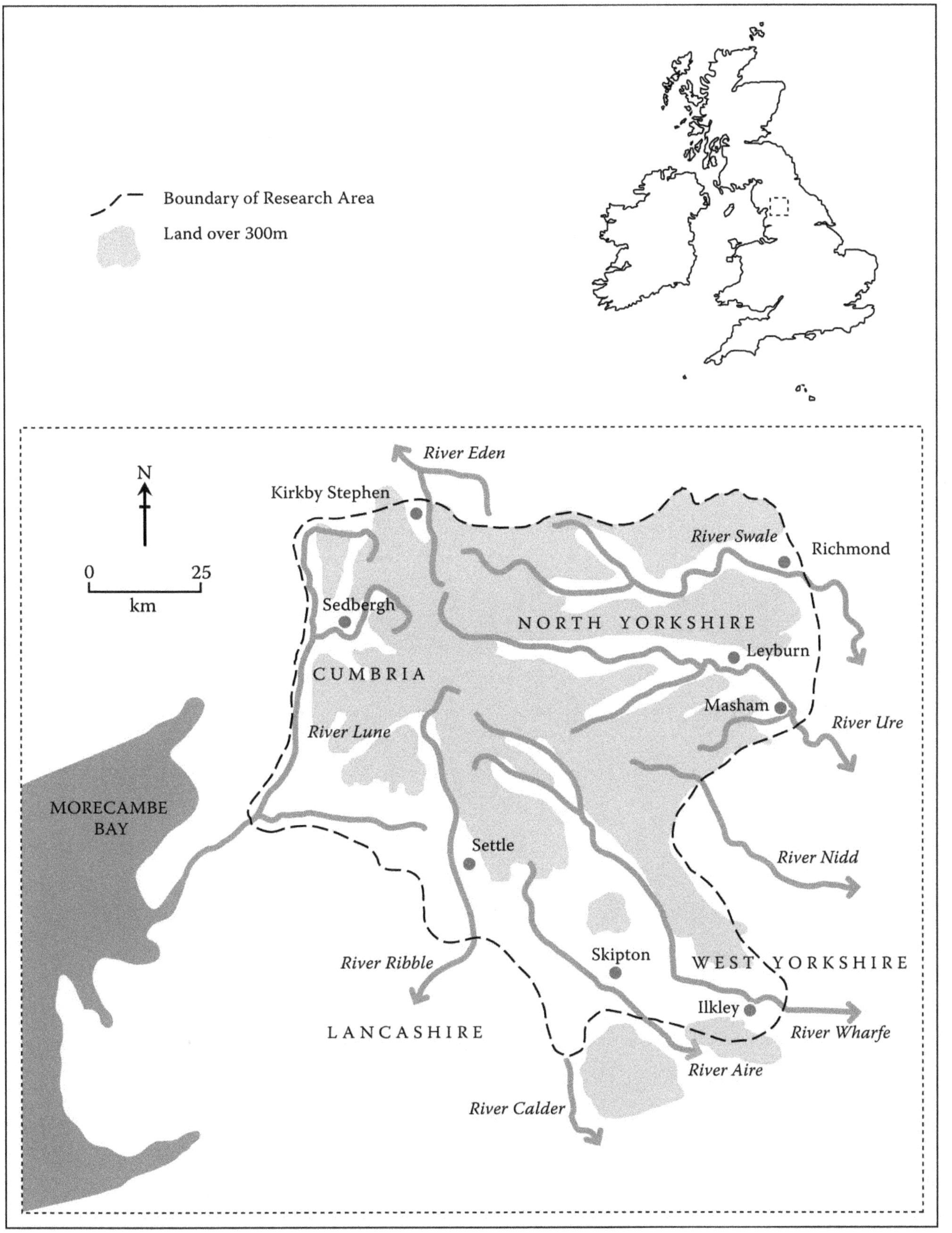

Figure 1.1 Location and extent of the research area

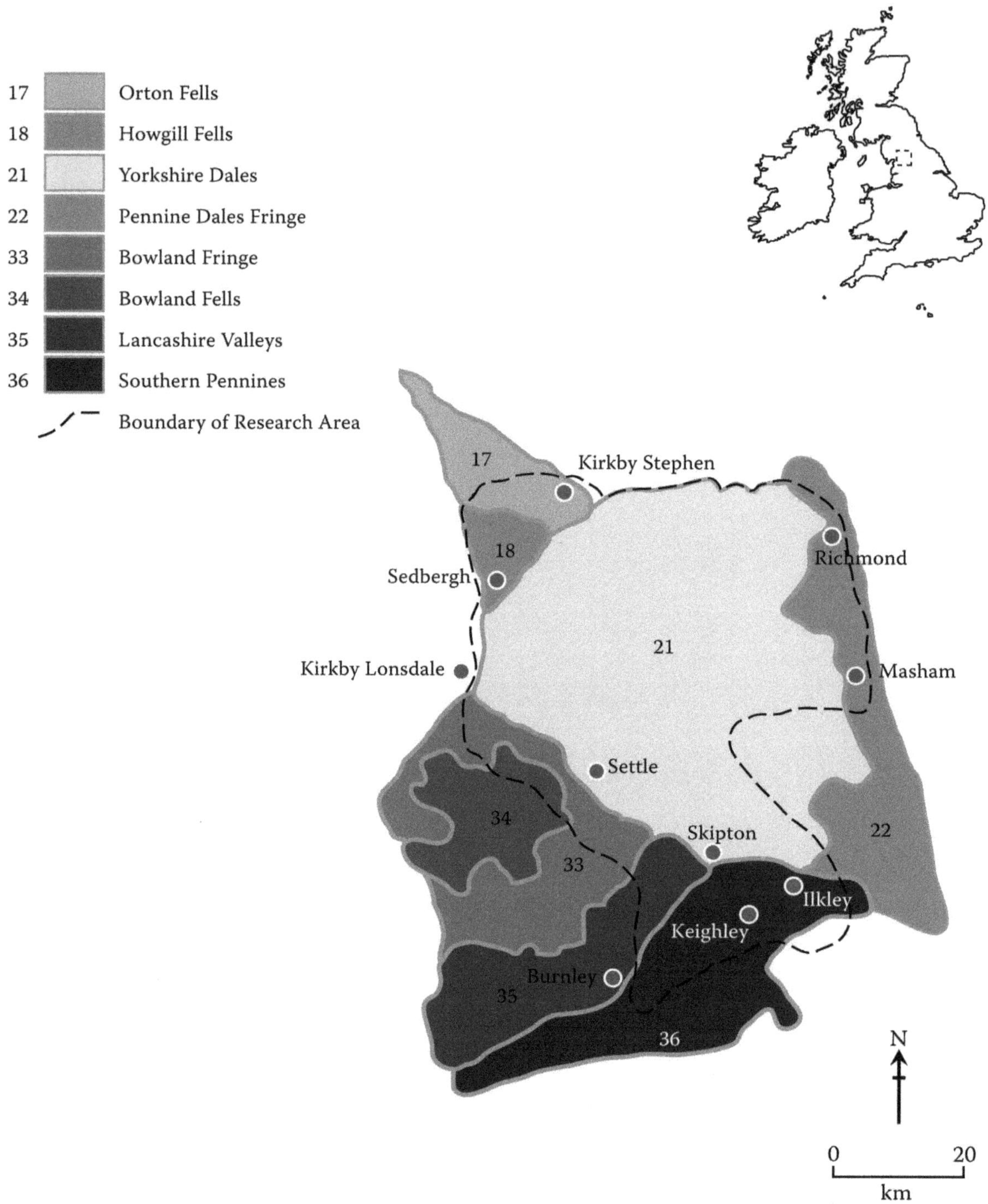

Figure 1.2 'Joint Character Areas' within the research area
Source: English Nature (1996).

'theoretically challenged', but the empirical researcher observes and records and then comes to conclusions not in a vague subjective manner but rigorously and objectively. Statistical analysis is important in making sense of recorded data but the empiricist does not need to resort to hypothetical models to explain what has gone on. 'Ground-truthing' is a vital tool in the empiricist's collection, namely gathering data from carefully thought out fieldwork and objectively interrogating them to derive conclusions. This study is firmly based in the empirical camp.

3. Spatial extent of the research area

The book takes as its geographical focus the Yorkshire Dales and contiguous areas. The research area extends across the western part of North Yorkshire within and bordering the Dales, the north-western extremity of West Yorkshire in Lower Airedale and Lower Wharfedale, the moorland edge within north-east Lancashire adjacent to North and West Yorkshire, the south-eastern part of the former county of Westmorland east of Lunesdale, that part of Lancashire which is contiguous to Lunesdale, and the part of Cumbria that was formerly within the West Riding (Fig. 1.1).

Estate	Family	Location	Date range	Theme
Swinton Park	Danby	Masham	1690-1850s	lime-coal, general improvement, lime
Croft Hall	Chaytor	Stainton, Spennithorne	late 17th c.	general improvement, lime, farm leases
Towneley	Towneley	Burnley	1620s-1804	hushing, general improvement, lime, Charles Towneley
Danby Hall	Scrope	Leyburn	1730-91	lime, farm leases, general improvement
Crackpot	Garth	Swaledale	1795-1911	lime-coal, general improvement
Ingleborough	Farrer	Clapham	1727-1884	drainage, lime, general improvement
Kiplin Hall	Crowe	Richmond	early 18th c.	general improvement, lime, leases, Chris. Crowe
Bolton Hall	Bolton	Wensley	1802-1900	lime-coal, lime, leases, general improvement

Table 1.1 Summary of case studies

In 1996 the former English Nature and Countryside Commission, in conjunction with English Heritage, published in cartographic format 159 Joint Character Areas (JCA) covering the entire country (English Nature, 1996). The research area is centred on JCA 21 (Yorkshire Dales) but also includes all or parts of seven other JCAs, as shown on Figure 1.2.

The core of the research area lies within what Winchester (2000, 1 and Fig. 1.1) described as the Central Pennines, stretching through the Dales and Craven towards Lower Airedale and Lower Wharfedale.[2] As Figure 1.1 indicates, much of the research area lies above the 300m contour and consists of rolling moorlands or open fells interspersed with deep-cut valleys of glacial origin. The valleys, or dales, are generally steep-sided, particularly so in the more confined dales of the Swale and Upper Wharfe though dales such as that of the Ribble or Ure (Wensleydale) have a more open aspect and less dramatic topography. Where the main rivers leave the Pennine massif the dales quickly open out to form broad lowland plains, as along the Swale-Ure Washlands downstream of Richmond and Leyburn respectively, in the Wenning valley to the south of Ingleborough in North Craven, and through the Aire gap between Settle and Skipton.

[2] The term Central Pennines has been applied by other writers to the hills between Craven and the Dark Peak (see, for example, Barnes 1982, 9; Edwards and Trotter 1978, 2).

The study does not confine itself to the upland zone as landed estates centred in those lower-lying parts of Swaledale and Wensleydale, east of the Pennines, extended westwards into the uplands; the same phenomenon applied in lowland Craven. Estate management and land improvement were practiced across the board rather than being restricted to either upland or lowland, and the two cannot logically be separated. The research area also sits within the 'fell or moorland' designation as proffered as part of an eight-fold division of England based on broad differences in farming which in turn reflect natural phenomena (Everitt 1985, 16-20).

Land improvement and liming were not processes unique to the Central Pennines. Indeed, in the days before railways made long-distance trade possible, lime was utilised wherever limestone outcropped. What singles the Central Pennines out, and forms the justification for defining the current research area, is the visibility of the evidence.

4. Case studies

While the process of land improvement has been recognised across the entire research area, particular attention will be devoted to a selection of examples for which evidence of the use of lime as a soil improver is compelling. The case studies have been chosen to achieve a relatively even geographical coverage across the area as well as to span the centuries over which lime

1 Danby of Swinton Park
2 Chaytor of Croft
3 Towneley of Towneley Hall
4 Scrope of Danby Hall
5 Garth of Crackpot
6 Farrer of Ingleborough
7 Crowe of Kiplin Hall
8 Bolton of Bolton Hall

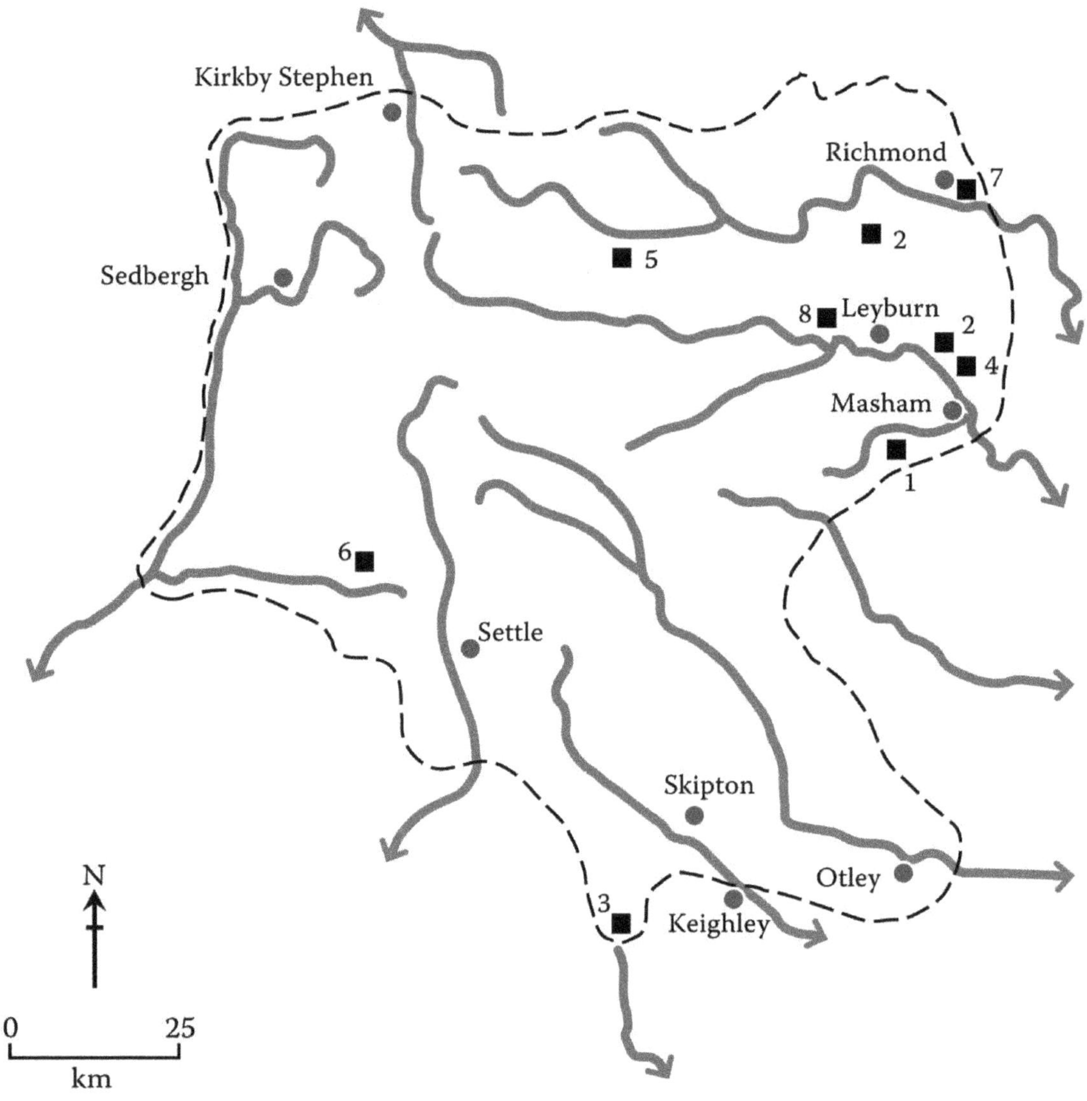

Figure 1.3 Location of examples

is known to have been used on the land. Inevitably, though, the survival of documentary evidence is heavily weighted to the eighteenth and nineteenth centuries. Table 1.1 summarises the selected case studies and Figure 1.3 locates them within the study area.

In addition to these specific examples there will also be a focus on practices involving agricultural lime outside landed estates which display characteristics suggestive of sustained usage. In north-east Lancashire, for example, hushing which involved channelling water down hillsides to flush out limestone cobbles to burn into lime is known from at least the sixteenth century, in the manors of Colne and Trawden. In Lower Airedale and Lower Wharfedale the raw material was sourced either from within river channels or from pits dug into glacial deposits on the moors. None of these activities is widely known but all illustrate the lengths people in these localities went to in order to obtain agricultural (and building) lime. Finally, archaeological excavations have provided evidence of lime burning from the sixteenth century (and possibly earlier) in North Craven. Discussion of these practices will compensate for the scarcity of early documentation.

Habitat	Typical soil	pH	% of surface area
Calcareous grasses	Rendzina or wind-blown loess	6.5	4.6
Neutral grasses	Brown earth	5.6 – 6	29.1
Heath	Acidic brown earth	5 – 5.5	5.9
Acid grassland	Acidic brown earth	5 – 5.5	24.2
Blanket bog	Podsol	<5	24.5
Others	-	-	11.7
Total			100

Table 1.2 Soil types and habitats in the Yorkshire Dales

Detailed evidence from the Westmorland part of the research area is more fragmentary: apart from the earls of Lowther lands and the earl of Thanet's properties there were no large landed estates, and much farmland remained in the hands of small yeoman farmers for whom records are sparse.

5. Geological overview

With the exception of several small pre-Carboniferous Basement inliers along the North Craven Fault, almost the entire area is underlain by Carboniferous strata (Fig. 1.4).[3]

Within the Yorkshire Dales these primarily consist of Great Scar Limestone and the Yoredale (Wensleydale) Group which is made up of alternating beds of limestone, sandstone and shale.[4] Except in the west and south, around the fringes of the Dales, these beds are overlain by the slightly younger Millstone Grit Group with Coal Measures outcropping south of Ingleton and sporadically in north-east Lancashire. South of Richmond, and running southwards down the Ure-Swale Washlands, there is a relatively narrow strip of Magnesian, or dolomitic, Limestone of Permian age. Dolomitic limestone also occurs in Ravenstonedale in south-east Westmorland. Across much of the area, though, solid geology is masked by a surface veneer of glacial till or fluvio-glacial drift deposits.

This geological framework has been critical both in bringing about the need for the application of agricultural lime as well as in enabling the development of lime burning on such a large and spatially extensive scale across the research area. Impermeable rocks, such as shales, sandstones and grits, and glacial material derived from them, weather into soils that are essentially acidic in nature and contain high proportions of clay particles.

Lowland meadows and pastures (the neutral grasslands) in the Dales typically have a pH between 5.3 and 6.3 (Smith 1988). The current writer has carried out selective pH tests in North Craven finding that brown earths for example, which tend to develop on sandstone or till, have a pH of 5.5 to 6.0 (Table 1.2). This can only sustain neutral grassland and, in turn, only support low stocking densities. In poorly drained (upland) areas these are replaced by acidic brown earths (pH 5.0 to 5.5), supporting dry acidic grassland and heath, or podsols (pH <5.0) supporting wet acidic grassland and blanket bog. Neither of these latter vegetation types has more than marginal economic stocking potential. Such soils and vegetation types dominate the non-limestone sections of the research area and thus represent a significant proportion of actual or potential farmland.

Those parts underlain by limestone fare better: here rendzina or deposits of wind-blown loess are the typical soil types supporting flower- and herb-rich calcareous grassland. However, because of the prevailing damp, cloudy and cool climate in the uplands, added to high rates of leaching and low evapo-transpiration rates, rendzinas are often still slightly acidic (pH typically 6.5). In short, virtually all soil and vegetation types within the research area, in their natural unaltered state, can support neither arable nor pastoral farming on a sustained and economically viable basis. Virtually all were – and in many instances still are – in need of periodic treatment to maintain soil productivity. Liming was one such treatment. Figure 1.5 illustrates this graphically. All the land shown in this view is based on limestone but only that enclosed within the walled intakes consists of high quality pasture because this was improved in the past whereas the rest was untreated commons. The remains of lime kilns nearby suggest it was probably limed despite being on limestone.

In terms of enabling large-scale lime burning, about half of the total surface area is underlain by broad expanses of limestone of one kind or another (Kidd *et al.* n.d., 17), most of it being readily accessible to surface working and of a high level of purity. In addition, where Yoredale strata predominate, small outcrops of limestone in many parts of the Dales have long been exploited for lime burning. Purity in terms of the proportion of calcium carbonate (as opposed to impurities) should ideally be between eighty and ninety per cent, while purity below that makes the lime of questionable quality for

[3] For detailed memoirs on solid and drift geology across the research area see Arthurton *et al.* (1988) and Dunham and Wilson (1985).

[4] A maximum of eleven discrete limestone beds have been recognised within the Yoredale sequence.

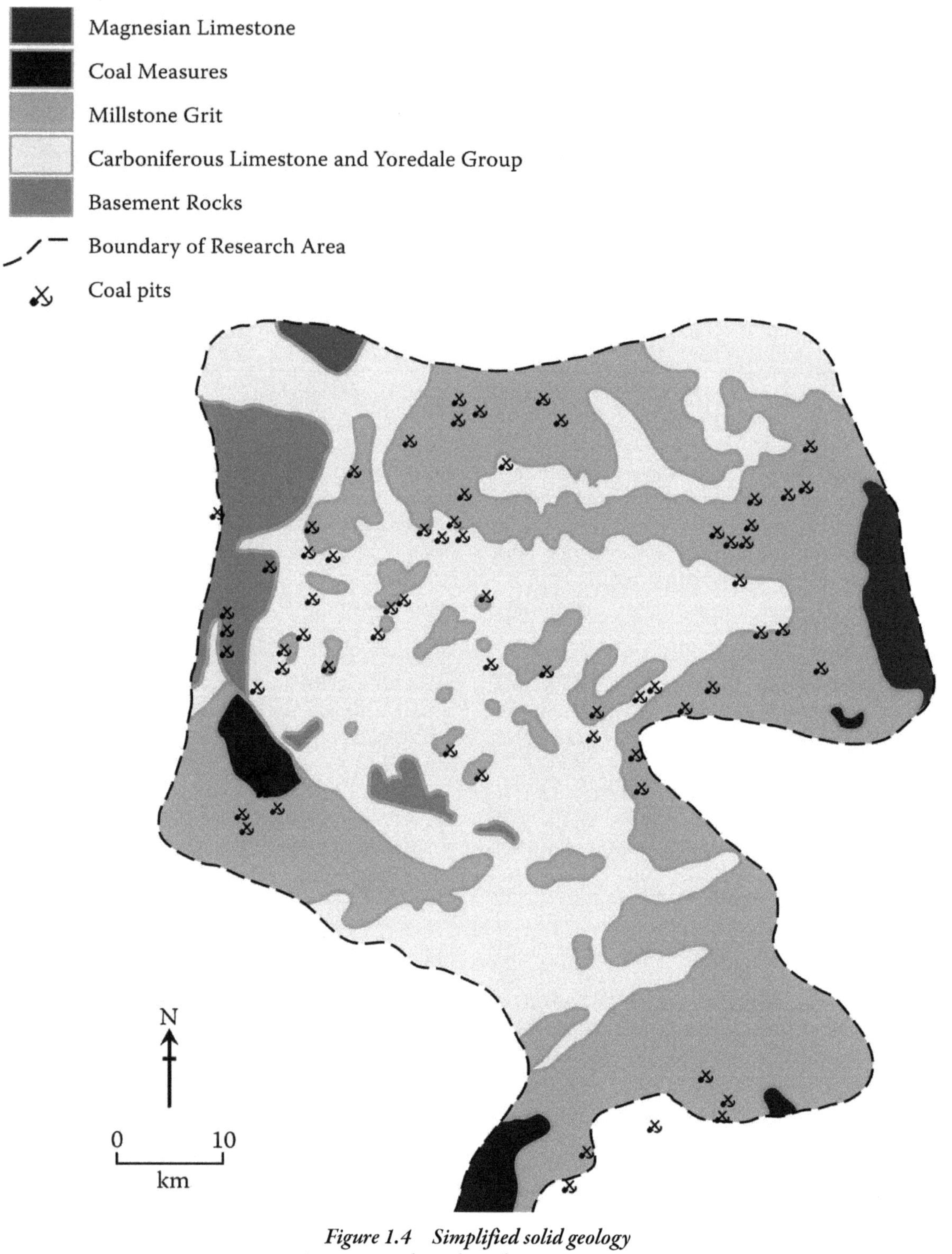

Figure 1.4 Simplified solid geology
Source: British Geological Survey mapping

liming. Robinson (1943, 137) considered limestone less than eighty per cent pure to be of minimal value.

The mineral composition of the limestone was also a significant factor in determining whether or not given reserves were successfully exploited for lime, though in the past end-users would not have been aware of the science behind such differences. Calcite-rich Carboniferous Limestone was in general preferred to dolomite-rich Magnesian Limestone as the latter reacts and breaks down rather too slowly (Pearson and Adams 1967, 177), and has lower calcium carbonate content (Robinson 1943, 140-41). Having said this, though, soils with depleted levels of magnesium will not support high quality grass or root crops and there is anecdotal evidence that some farmers in North Yorkshire where only calcitic limestone occurs brought in dolomitic lime from a commercial quarry in Wharfedale because they believed the latter to be better for their particular pastures where a lack of magnesium made hypomagnesaemia (staggers) a common disease in sheep (Eddie Webster, pers. comm., February 2008). In this case the

Figure 1.5 Potts Valley, Crosby Garrett, showing the contrast between limed (walled in) and unlimed limestone pastures. NGR NY705 080.

farmers concerned were well aware of the chemistry. On the other hand the same quarry despatched calcite-based lime to other farmers where dolomite dominated: farmers were prone to make empirically-derived decisions based on what they perceived to be in their best interests.

Contrasts between different types of limestone, in terms of their efficacy as soil additives, have long been recognised. Thomas Hitt (1761, 88) noted that lime from 'hard' and 'grey' (*ie* Carboniferous) limestone produced lime of far better effectiveness than from chalk or 'soft' (*ie* Magnesian or dolomitic) limestone. His reasoning rested on the observation that twice as much of the latter was required to achieve the same results as the former.

The solid geology in the southern part of the research area, in JCAs 35 and 36 (see Figure 1.2), is masked by varying thicknesses of glacial till which was either deposited as patches of hummocky moraine on sandstone or gritstone moorlands, for example between Lower Airedale and Lower Wharfedale in JCA 36, or piled up by southward-moving glaciers against the northern edge of the Southern Pennines in both JCAs 35 and 36. As the glaciers had flowed across limestone areas to the north, the till contains large quantities of limestone boulders and cobbles and these were picked out for burning into lime. The moraine was worked in a series of shallow quarries, locally known as delfs or limestone boulder pits; in addition to cobbles, the lime-rich glacial clay itself was dug out to be spread on the land as what Camden called 'yellow marl' (see Chapter 4.4 for a discussion of marling). The Pennine edge was also extensively hushed to expose and loosen limestone cobbles:[5] both of these practices were fundamental in the early use of agricultural lime, as will be discussed in Chapter 5.5 and 5.6.

Elsewhere within JCAs 35 and 36 upland fast-flowing streams eroded limestone cobbles from till deposits and transported them down to the Aire and Wharfe where reduced channel velocity and competence led to their being laid down as stone spreads on the inner banks of meander loops. These, too, were exploited as an early source of stone for burning into lime in riverside kilns: this topic will also be revisited in Chapter 5.5.

The geological make-up of the research area contributed to large-scale lime burning in quite a different way. At least twelve of the Carboniferous beds have coal seams, often very narrow and of poor quality but sufficient for agricultural lime

[5] Hushing involved artificially channelling water downslope to flush out and loosen limestone cobbles usually from a matrix of glacial till.

Location	NGR	Context	Date	Reference
Housesteads	NY790 690	Fort	early 3rd c.	Crow 1995
Northants.	SP885 818	Villa	mid 2nd c.	Jackson 1973
Peterborough	Helpston Quarry	not stated	4th c.	Challands 1976
Bedford	TL081 490	Mill site	2nd c.	White 1977
Womersley	SE523 202	Farmstead	3rd or 4th c.	Tomson 1992-93

Table 1.3 Excavated Roman lime kilns in Britain

burning (Ironbridge RS 1026, RS 1057; Kelly 2008, 15). Seams associated with the Middle Limestone, for example, were worked specifically for lime kilns in Mallerstang, Upper Wensleydale and on Great Whernside in Upper Wharfedale; while other Yoredale seams were worked above Garsdale, Dentdale and Deepdale (Hudson 1998; Kelly 2008). The Coal Measures have been extensively worked from medieval times in the Burnley area; in Lower Airedale from at least 1310 (Gill, 2004, 24); and around Ingleton since the seventeenth century (Bentley *et al.* 2005). Figure 6.2 shows the relationship between solid geology and exploited coal reserves.

6. The historical context: the origins of lime usage

There is broad agreement among archaeologists and prehistorians that lime has been known to mankind since the Neolithic but there is less certainty as to exactly where and when it was first used, or for what purpose. The earliest archaeological evidence comes from the Middle and Near East where Neolithic city-based peoples had recognised the value of lime as wall plaster even before they had evolved pottery-making techniques (Gourdin and Kingery 1975, 133-34). One site in particular, the settlement of Çatalhöyük in Anatolia, has surviving rooms where 'the walls still rise above your head, their white lime-plaster finish as bright now as when it was fresh' (Taylor 2002, 227). This long-lived city dates from the ninth millennium BC and the amount and quality of the lime plaster testify to a high level of technical sophistication.

The earliest confirmed date for a lime kiln is 2450 BC, in Mesopotamia, where lime was extensively used for mortar (Oates 1998, 4), but the fact that evidence of lime mortar has been found across ancient Babylon and Assyria (pre- and post-2450 BC respectively) means that kiln technology was widely known in the Middle Eastern Neolithic (Forbes 1964, 58-59).

Lime plaster is also known from Ancient Egypt between 1400 and 1200 BC (Gourdin and Kingery 1975, 135; though see Dancaster 1916, 2, for a contrary but now outdated view), and possibly from as early as 2000 BC (Cowper 1927, 3). Lime plaster can be clearly seen, for example, in the Great Pyramid of Khufu at Giza as well as in many Egyptian tombs where walls were coated with lime plaster to produce a firm base on which to paint murals (Forbes 1965, 243). It was similarly applied as wall plaster by the Minoans in Knossos from the mid-second millennium BC (Cowper 1927, 3).

Lime was widely known across the Ancient Greek empire and Xenophon recorded the sinking of a ship in 350 BC whose main cargo had been linen along with lime 'for its bleaching' (Boynton 1980, 3). By the time of the Roman civilisation lime had a multitude of uses, industrial and constructional, which now also included application on the land: whether or not agricultural lime pre-dates the Romans is not known. It has been said that the Egyptian Fourth Dynasty pyramid of the pharaoh Snofru (2575-2551 BC) was stripped of much of its limestone casing to 'produce agricultural lime in Roman and Arab times', though this comment was unreferenced (Manley 1996, 13). While no Roman writer seems to have recognised the benefits of liming,[6] later writings do point to the advantages of using 'soft chalk and marl' on pasture land in Britain (MAFF 1981, 5). Chemically, chalk is a limestone and burnt chalk performs in the same ways as burnt lime, and the term marl was commonly misapplied (see Chapter 4.4).

According to one source lime was known to the Britons in what is now north-east Wales (Brown 2004, 66-67). Brown quoted Strabo by saying that the local tribesmen coated their hair with limewash, drawing it back and thickening it with the lime so that it differed 'in no way from a horse's mane'. However, given that Strabo died in AD 23, long before the Roman invasion, one has to doubt the reliability of this source.

There is a growing corpus of excavation accounts of lime kilns across Roman-occupied Britain (Table 1.3 and see Dix 1982). Some were associated with villa sites rather than towns and, though this does not necessarily mean the lime was destined for the land, it could well have been.

After the Roman withdrawal from Britain, at the beginning of the fifth century, the use of lime as mortar and on the land may have gone into abeyance as urban and commercial systems slowly decayed, and there is a long gap in the historic and archaeological record. If we assume that lime burning did disappear from the scene, it has to have reappeared during the Anglo-Saxon period as every known stone building (church or monastic foundation, for example) incorporated lime mortar. In fact a derelict church was restored by St Wilfrid in about AD 690 and finished off 'whitewashed ... with shining lime' (Salzman 1952, 356). There is an earlier, unprovenanced,

[6] Cato (234-149 BC), Varro (116-27 BC) and Columella (early first century AD) wrote extensively on soil improvement and agricultural management.

possible use of lime at Whithorn, Galloway, where St.Ninian's late fourth-century chapel (Scotland's first) was known as the *Candida Casa* (glittering white house) (www.whithorn.com). It may have been finished in limewash – or merely constructed with stone that glistened in the sunlight.

A further piece of documentary evidence refers to an Anglo-Saxon kiln in south Devon in AD 846 when a charter described a boundary with reference to *odencolc* (Born 1991, 213). 'Oden' derived from a Celtic word for kiln and 'colc' from the Latin *calc* meaning lime. According to Born, this kiln was most probably used for agricultural lime, though this must remain pure speculation. As yet, no Anglo-Saxon lime kiln sites have been confirmed anywhere in Britain.

7. The historical context: vernacular uses of lime

The focus of this study is the historic application of lime on the land, as a means of improving soil for arable and pastoral farming, but lime was also utilised in a wide variety of industrial processes from iron and steel production to chemicals, textiles and food processing: these are beyond the scope of this study. Within rural areas there was an equally broad range of end uses: domestic, craft and agricultural.

Perhaps the most obvious and common non-agricultural rural use was in construction, either for new-build or repairs. No significant construction project could go ahead without lime mortar, and there is evidence of lime having been transported considerable distances for construction (Tyson 1998, 68). Excavation of a site in London, in 2007, revealed rare evidence of the use of lime plaster, in a house dated to the second century AD (Catling 2010 25).

In the post-medieval era small-scale building work tended to source lime from commercial or estate kilns but larger schemes invariably involved the creation of a lime kiln on site to ensure a constant supply of lime of the required quality and consistency. Rental and account ledgers of larger landed estates itemised purchases of lime for making mortar on-site where estate properties, such as tenants' houses, farm buildings or halls, were being repaired (see, for example, DL. Min. Acc. 18-19, 1440-43; NYCRO. ZAL 3/5, 1855-67; NYCRO. ZAL 3, 1876-89; NYCRO. TD 125, 1835-54; YAS. DD 146/13/9, 1707-39).[7]

Evidence of the construction of kilns for specific new-build projects within the study area has been derived from archaeological excavation and documentary sources. This writer oversaw excavation of a clamp kiln in Wharfedale whose last firing was dated to between 1660 and 1700, though this kiln bowl had been built inside a much older undated bowl (Johnson 2006, 35-40). It was sited within a building complex consisting of a post-medieval corn mill, a dated sixteenth-century building and seventeenth-century barn, at Cracoe south of Grassington. The two phases of this kiln were clearly linked with building work within this complex. Further up Wharfedale another clamp kiln was excavated, dated to between 1620 to 1670, sited close to Kilnsey Old Hall which is known to have been rebuilt from a derelict monastic complex in 1648 (Johnson 2008b).

In 1378 Sir Richard le Scrope entered into an indenture with John Lewyn, master mason, for the building of Bolton Castle in Wensleydale (NYCRO ZBO MD 1). Lewyn was obliged to supply all the necessary materials except for 'timber for the fuel of the kilns, when they are lit' which Sir Richard was to provide ('... le dit Monsr Richard luy trovera meresme pur le brandet pur les torailles qant ils serront ardz ...'). The wording of the agreement suggests a project of this size required more than one kiln.

Catterick Bridge across the Swale was rebuilt in stone in 1421-22 and seven local barons engaged the services of three masons to undertake the work, giving them the right to 'get lymstone and burne itte, and care itte, and make yair lymkilns of your own cost' (NYCRO ZRL 1/23). The masons were to cart all the necessary materials to the kiln that they were required to build on-site.

A similar situation prevailed when a new chapel was erected at Barden Tower in Wharfedale in 1515 (Dev. Coll., Chatsworth, Clifford). Very detailed building accounts were kept by the steward with payments for 'leading lymeston' to the kiln, to 'staynbrekers at lyme keilne' and to 'iij chowchers at lyme keilne'. Later that century the Shuttleworths were building their new Gawthorpe Hall near Burnley and maintaining equally detailed accounts (Harland 1856). They were carting in 'unslacked' quicklime from Clitheroe for the use of 'John the limer of the house' whose job was to slake the lime, mix it with horse hair (to strengthen it) and prepare mortar and plaster. In this case burnt lime was imported but there are also entries referring to a kiln on the site of the new hall.

Thus, lime was needed for making mortar and plaster, for internal wall covering and external render or harl. In addition, slaked lime had been used in vernacular houses since the Middle Ages for coating wattle and daub walls which were finished off with limewash. Stone-built internal walls were given an annual coat of whitewash, which brightened up a smoke-filled interior and acted as a disinfectant to rid the house of lingering bacteria.

Until the eighteenth century it had been common practice to make stone roofing slates watertight by packing moss into the interstices between individual slates but this was superseded by a process known as tiering whereby a mixture of slaked lime, sand and hair was applied to the joints and under-surface of the slates. This not only kept rain out but also anchored slates to rafters while allowing air circulation within the roof space. Tiering was not a one-off task: the Duke of Bolton's steward, William Law, wrote to his master, in 1760, pointing out that the 'nature of our slate and lime requires the roofs of houses to be pointed with lime once in three years' (NYCRO ZBO IV. 8). Later specifications for building a new barn and cowhouse at Melmerby, in 1878, included the requirement to 'point the

[7] The present writer is pursuing parallel research into the historic use of lime mortar in Yorkshire.

underside of the roof with lime and hair mortar', the mortar having been mixed from 'well burned stone lime' (NYCRO ZTS [M]).

The mixing of lime mortar, plaster or limewash needed a stone trough and contemporary maps or documents occasionally hint at the location of such troughs, or mortar pits. A survey of properties belonging to Trinity College, Cambridge, in Sedbergh, in 1568, listed all fields including 'one other close called Morter Pittes' (Sedbergh, TCC, Box 17), which name survived as 'Mortarpits' on the 1920 edition of Ordnance Survey mapping. Whether such pits were in private or communal use has not been determined.

Lime is known to have been used in the north of England during the eighteenth century as an aid to quarrying slate before gunpowder was introduced (Stanier 2000, 132). Quicklime was packed into cleavage cracks and, when slaked by rain, it would expand prising individual slates apart and making it easier for the quarrymen to separate them.

Lime mortar was also used in the construction of dew ponds and evidence from the repair in the North York Moors of a thirteenth-century dew pond showed it had been made in a virtually identical way to an undated pond excavated on Malham Moor in the 1990s (John McIlwaine, pers. comm., August 2001). Recommendations for constructing enclosure-period dew ponds described the laying of lime 12.5mm thick (Marshall 1788, 147-48 and 154-55). Puddled clay formed the base layer with a layer of quicklime or lime mortar and straw sandwiched between that and a top layer of small stones or stone slabs. The lime acted as a waterproof membrane thereby protecting the base against cattle trampling while discouraging earthworms from burrowing through the clay and making the pond porous.

Quicklime was an essential ingredient in the tanning industry which was an age-old rural craft found across the cattle-rearing parts of the Central Pennines. Hides were soaked for a minimum of eighteen hours in a bath of hydrated lime to soften the hairs before they were scraped off the skin. A common craft in Ribblesdale, north of Settle,[8] was the making of paper with lime being used to reduce the acidity and compaction of the raw material (rags) by soaking them in a liquid lime solution.

Both tanning and paper making required tanks, commonly known as lime pits,[9] and these were considered of sufficient value to be included in wills and inventories, such as that of William Browne of Settle whose will of 1576 included 'all my stone troughs which I have within my lime pitt house' (Borthwick 21, 5).

In areas where bracken thrived, soap was produced in potash kilns for use in local linen processing, fulling mills and dye houses (all of which required lime), and lime once again acted as a vital ingredient in the manufacture of this soap (Davies-Shiel 1972).

Rather more arcane uses included using lime to disorientate fish. John Brearley, a wool frizzer writing around 1760, described how river eels could be caught if you 'put some quicklime in that is fallen of itt self and all the fish will swim att top' making them easy to scoop out (Smail 2001, 26). Slaked lime was also used until within living memory for preserving eggs stored in it (Kathleen Slinger, pers. comm., August 2008), and corn seed was traditionally soaked in a solution of lime and urine to protect it against rot (Hartley and Ingilby 1975, 52).

Quicklime had long been used as a medium for speeding up the decomposition of dead farm animals and this use has also been recorded elsewhere in medieval human burials (Gilchrist and Sloane 2005, 123-24, 142). It has also been used in this way nefariously. An infamous court case, tried at Lancaster in 1778, tried to prove that the accused had murdered a cattle trader near Clitheroe five years previously and had buried the body 'among lime, with which it was so mixed' but the body had failed to decompose sufficiently and was still identifiable when found in 1776 (Williamson n.d.).

As far as stock farmers were concerned lime had a multitude of direct uses, other than as top dressing to enrich pastures, as custom maintained that it could prevent or cure a number of ailments (Table 1.4).

[8] Stainforth in Ribblesdale, for example, had fourteen paper makers listed in the 1841 census.

[9] The term lime pit was also used for pits used to hold slaked lime prior to mixing mortar or limewash.

Disease	Stock	Traditional treatment
Lactation tetany	Cattle, sheep	Solution of calcium salts, or topdressing with magnesitic limestone
Cochexia ossifraga (cripple disease)	Sheep	Solution of salts with lime
Red water	Cattle	Top dressing with lime
Hypomagnesaemia	Cattle	Solution of magnesitic lime

Table 1.4 Stock diseases and lime

8. Structure of the book

Chapter 2 offers a review of work done by others in the subject area of land improvement and lime usage in Britain within the context of landscape history and landscape archaeology as well as of agricultural and industrial history. Chapter 3 discusses the various sources and methods used in the research process. It identifies problems encountered and outlines mitigation measures taken to minimise their impact. Finally, it sets out the research schedule.

The following five chapters form the substantive sections. Chapter 4 considers the beginnings of land improvement in the pre-modern era, how land was managed and improved on monastic demesne and manorial estates. It identifies evidence of possibly the earliest post-Roman use of lime on the land within the research area. The following chapter takes the use of lime as a soil improver into the early modern period, a time when lime was firmly established as a major element within land improvement and when external influences were having a greater impact than hitherto. Chapter 6 analyses a wealth of primary data concerning the sale of lime-coal on an estate on the eastern side of the Central Pennines and draws conclusions concerning patterns of use of agricultural lime, seasonal, annual and decadal.

Chapter 7 returns to the chronological theme by examining the massive increase in the use of lime on the land during the period of parliamentary enclosure. It examines the reasons for this expansion, whether resulting from external influences or details of estate management. Chapter 8 takes the narrative through the nineteenth century, from the impact of peace in 1815 to the late-century agricultural depression. It discusses why lime usage fluctuated and then declined towards the end of that century.

Chapter 9 synthesises the various strands of research and summarises the results achieved in terms of the initial aims, then draws together the conclusions reached. It also suggests the contribution that this research exercise has made to the corpus of knowledge, while recognising any inherent drawbacks. It goes on to discuss avenues for future research, and finishes with a consideration of the state of farming in the Central Pennines at the present time.

A full bibliography of archival, primary written and secondary written sources is provided to support the narrative and conclusions; as well as a glossary of specialist terms relevant to the burning and application of lime.

9. Conclusion

This chapter has set out the research framework and summarised the writer's epistemological position; and has put the research into geological and historical contexts. It has also set the scene for:

- extending backwards the time frame of lime usage on the land,
- extending the spatial area over which land was improved by the use of lime beyond parliamentary enclosures,
- extending its application away from areas of limestone bedrock, and
- showing that the benefits of liming were appreciated by elements of all sectors of the farming economy within the research area.

Chapter 2 is composed of a survey of work carried out by other researchers on the subject of land improvement and agricultural management in Britain, in the context of a landscape history and landscape archaeology approach. The chapter considers how earlier writers have perceived historical constraints to land improvement, as well as catalysts to improvement, within the north of England, and offers an appreciation of the limitations of previously published work.

Chapter 2
On the Literature of Agricultural Improvement

1. Introduction

This chapter presents a critical review of prior work carried out in the broad field of study within which this work sits. The nature of the research into the historic role of lime in agricultural improvement has been empirical and cross-disciplinary, drawing upon the concerns and priorities of historical geographers, landscape historians, landscape archaeologists, industrial archaeologists, and agricultural historians. Relevant literature from each of these disciplines will be reviewed as landscape studies simply cannot be confined to one discipline. In the words of Bond (2000, 71):

> The framework of traditional academic disciplines continues to be a barrier rather than an aid to progress ... academic boundaries should never be allowed to become a straightjacket ... If we are ever to understand ... the landscape, multi-disciplinary synthesis will be essential.

A century ago Reid (1913) expressed a similar view, that knowledge and research should not be constrained within academic straightjackets – or 'compartments'; and much more recently Dalglish (2009, 250) stressed the need for those involved in studying landscape to maintain (or develop?) a meaningful dialogue, particularly between empiricists and non-empiricists.

Historical geographers and landscape historians would take the broadest view of a geographical area where farmland has been improved in the past.[10] They would identify the main elements of that area, including topography, drainage systems, patterns of soil and vegetation, and details of human intervention. The traditional historical geographer and the landscape historian would be keen to identify a historical dimension within a given landscape. The area in question might be seen as a palimpsest or a kaleidoscope with several identifiable layers of past activity which reveal a coherent picture based on pertinent questions, such as how and when the area was improved; whether improvement was a gradual and halting process or the result of a concentrated drive; whether it was directly improved by those who worked the land or at the behest of the landowner. The picture cannot be completed by field observation and analysis alone: recourse must be had to cartographic, photographic (especially aerial) and documentary sources which all furnish additional material to inform meaningful synthesis and analysis. Without these, erroneous interpretations could result.

[10] It has to be said that currently historical geography and landscape history have little in common: to the former, landscape history is bedevilled by traditional empirical approaches; to the latter, historical geography is drowning in obscure theory.

This whole process can be taken a stage further if the techniques of landscape archaeology are applied, though the distinction between landscape history and landscape archaeology is somewhat blurred (Newman 2001, 100-01; Williamson 2002, 21; Palmer 2007, 2). It is also important to distinguish this branch of archaeology from its parent discipline. The latter focuses on specific sites whereas the former uses archaeological techniques to identify features and patterns over a much wider area. For example, traditional practitioners might excavate a number of trenches within a medieval field system to look for evidence of dating, of how lynchets or ridge and furrow were created, and to see if any evidence of the use of lime can be isolated. The landscape archaeologist would put that site into its broader landscape context, and would see its having a (symbiotic) relationship with surrounding areas, with the ultimate aim being to deconstruct that landscape, to see how one element relates to another now and in the past. The landscape archaeologist is more likely to be concerned with an entire landed estate, rather than a specific field or lime kiln site. Furthermore, the 'pure' archaeologist, stereotypically, fails to look beyond the excavation trench or site, does not seek relationships between the site or feature and its broader historical and geographical contexts, and can thus 'appear as a mole' (Tilley 1994, 3).

Industrial archaeologists open up yet another way of understanding a given landscape. They have the skills that the generalist archaeologist may not possess. They can interpret particular structures (for example early lime kilns or coal bell-pits) so as to slot them into whatever unfolding picture is being pieced together. Agricultural history, too, plays an important part. Soil improvement resulted from a decision made by someone but it was not a decision made in isolation or from mere whim. The decision-maker was responding to external influences, possibly locally within the community or from afar. The historian can point up broader economic, political, social or demographic trends that were brought to bear on the dissemination of information and on local decision-making processes.

Each of these approaches has been integral to this research, as will be shown, so each element will be considered in this chapter.

2. On landscape theory

There is no doubt that landscapes such as those found across the Central Pennines contain an enormous wealth of detail about the past. These landscapes can be read and broken down, all the more easily because earlier elements have been fossilised within the present landscape, unsullied by later developments. Such elements may well have changed from their original form but they are still recognisable among later elements. Within the Central Pennines prehistoric settlements and droveways, medieval field systems and routeways, and signs of early modern industrial development, have not been expunged by urban or large-scale modern growth or by intensive cultivation. This applies equally well to the production of quicklime and its application on the land. Enough clues have survived the passage of time for a meaningful picture to be painted.

While this notion of landscape as an open book, of its being a 'historical text' (Muir 1999, 37) or the nation's 'richest historical record' (Hoskins 1955a, 14), is generally accepted by students of landscape, there have inevitably been differences in emphasis and approach. Few, though, would quibble with Bond's rejection of subject-based 'boxes', and many would concur with the philosophy that landscape is most effectively viewed from an inter-disciplinary perspective so that the individual elements within a given landscape can be seen in the whole, as integral subsets of a broader scene (Muir 1999, 49; Taylor 2000, 160-61).

This core idea of whether it is possible to see the whole story set out in a given landscape has in itself engendered a good deal of discussion. Some adhere to the now rather hackneyed use of the term palimpsest (a term promoted by the archaeologist O G S Crawford over half a century ago [Crawford 1953, 51]) to describe historic landscapes, where one layer of occupation or activity overlies and (accidentally) does not quite wipe out those that preceded it (Dodgshon 2000, Preface). The countryside is beyond doubt a 'complex tapestry' (Lake 2007, 28), very often carefully planned to achieve a desired objective, within which elements from successive periods are visible to a degree, but there is an alternative way of approaching this concept. Hoskins' ideas, at the time of his writing *The making of the English landscape* new and exciting, are now perceived by many landscape practitioners to be out of date and superseded by alternative approaches (Taylor, 2005, 81).

Strictly speaking a palimpsest is a document that has been wiped clean and reused.[11] This would imply that the latest layer of activity has (almost?) completely removed all signs of what came before. This may indeed be so for land subjected to modern large-scale urban development but not necessarily so for a rural upland landscape such as within the area under review, though even here earlier landscapes cannot have survived in their entirety. Some change is inevitable, either humanly or physically induced. Whether Lake had in mind his tapestry as a palimpsest is not clear but it has been argued that historic landscapes are kaleidoscopic phenomena (see, for example, Williamson 1998, 6; and Oosthuizen 2006, 27-31). This view quite correctly sees elements of successive layers preserved within the landscape but in a dynamic way with glimpses of each being visible here and there 'in a constantly moving pattern'. The term 'sheer veil' has been applied within landscape archaeology in the sense that shadows of past landscapes can still be discerned through later layers (Finch 2007, 39). Winchester (2000, 5; 2006, 75) preferred the term 'tide mark' suggesting, for example, that signs (marks) of medieval plough cultivation may be preserved beneath those of Napoleonic-period cultivation: each cultivation episode represents a 'high tide' of past activity.

A series of essays arising from a conference held to celebrate the fiftieth anniversary of Hoskins' seminal work made the point strongly that use of the term palimpsest should be discarded precisely because it implies previous layers being expunged which, as suggested above, is emphatically not the case in the Central Pennines (Gardiner and Rippon 2007, 1). To some, it may seem like semantics to draw such a fine distinction between palimpsest and kaleidoscope but this is arguably fundamental to understanding the multi-varied facets within a given historic landscape. Nevertheless, it is surely the recognition of these very facets that is important in landscape studies rather than the precise terminology employed.

The acceptance from the 1970s that archaeology can make a valid contribution to landscape studies has arguably weakened the palimpsest approach. Assuming that users of the term have not misunderstood its precise meaning, and even if one layer has seemingly been obliterated on the surface to the point where it is invisible to the eye, getting under the turf can lay it bare again. This involves either (or both) invasive excavation and study from an aerial perspective. This writer's field surveying and archaeological work illustrate the point well, as will be examined in Chapter 5.4. Cursory examination of the lowland fringe around the Central Pennines, such as in the Wenning valley between Burton in Lonsdale and Austwick, will reveal a neat landscape of irregular hedged fields and scattered farmsteads exuding a general air of relative productivity and prosperity based on pastoral farming. The land here has been improved, long before the advent of chemical inputs, but there is no reliable evidence on the ground that liming ever played a part in this. Yet, lime *was* often employed, and from an early age. Superficial signs have indeed gone, so the palimpsest idea could seemingly be invoked, but excavation has revealed the true picture: the lime kilns indeed are still there under the turf layer or as imperceptible earthworks. Muir's full 'historical text' is still there (1999, 37).

This brings us neatly to another aspect of landscape, which could be approached at different levels within a hierarchy of scale (Rippon 2004, 19-21). If we take the present line of research as an example, each individual lime kiln is a single *element* within its historic landscape but it does not sit there in isolation, any more than it was once worked in isolation. It slots into a bigger *parcel*, in our case a tract of moorland or an enclosure, or the quarry within which it was built. In turn, the parcel forms part of a wider *component* such as a whole group of fields that formed units of a particular enclosure award. The component could also be the system of tracks used to bring fuel to the kiln or to distribute the finished product to end users; or the coppiced woodland that may have supplied the kiln with chopwood kindling. The components build up to form a landscape *type*.

Each of the elements, or *facets* (Muir 1999, 50), can justifiably be studied in its own right, as when an individual kiln is

[11] The word derives from the Greek and literally means rubbed smooth again.

compared against others with the aim of recognising regional styles or variations in form (Johnson 2010b). Equally, one element can be looked at in its landscape context to identify how the role it performed helped bring about today's landscape. How, for example, did one particular kiln contribute to the improvement of a given piece of farmland, assuming they were contemporary?

It is also valid, arguably necessary, to factor in elements that apparently have left no direct and tangible imprint on the ground. Prominent amongst these is the concept of 'human agency' (Rippon 2004, 52), that is the extent to which people had a hand in creating a given landscape, and perceiving it in a particular way (Palmer 2007, 3). Lime kilns did not build themselves and people did not build or site them without reason. A conscious decision was made by landowner, steward or tenant to locate them at given points and this decision-making process involved weighing up the logistics and costs of transporting in the fuel or the stone, or both. Some are sited close to a reliable fuel source but far from limestone outcrops, others close to the source of stone, and yet others where the lime was intended for use with both fuel and stone having been brought in. Such decisions have left recognisable traces on the ground and the student of landscape should be able to recognise them.

In contrast, however, certain aspects of human agency have left no ground evidence, so these require consideration of externalities such as the influence of agricultural Improvers or the Board of Agriculture; political processes such as the expansion of monastic estates or the Napoleonic wars; economic forces like changes in market demand, introduction of cheaper alternatives, or price fluctuations; or stochastic events such as climatic deterioration or an outbreak of cattle murrain. This serves to emphasise that the study of landscape draws upon a range of academic disciplines (Taylor 2000, 160) and is truly multi-disciplinary.

Rippon (2004, 17-19) has highlighted a dichotomy within recent landscape studies, with some researchers being focussed on specific elements of the historic landscape, whilst others take a broader and more holistic view that once again invokes the concept of the *pays* associated with the thinking of Paul Vidal de la Blache, founder of the French School of regional geographical studies. Alan Everitt (1977; 1985, 13-15 and 20), Hoskins' immediate academic successor, stressed the longevity of specific areas as *pays* which managed to survive broadly intact despite profound changes brought about by urbanisation and industrialisation. In his view the 'bare fellsides' of the Yorkshire Pennines and the fertile lowland fringes are distinctive enough to be considered as discrete *pays*.

However, the *pays* concept sits uneasily with the ethos of the series of regional studies that followed Hoskins' seminal study, even though the idea of the *pays* can be teased out of his own work.

The essence of a *pays* is that, at the local scale, it has characteristics which give it an identity of its own and which distinguish it from the surrounding areas: *pays* do not respect manmade political and administrative boundaries. The first studies in the Hoskins mould were entitled *Cornwall*, *Gloucestershire*, *Lancashire* and so on: no one can convincingly argue that Lancashire has a single regional identity. A later series of books (*Landscapes of Britain*), that advertised a regional approach to landscape history, also seemed unsure of the *pays* concept: titles such as *The south-west peninsula* or *The Welsh Marches*, both of which could be described as *pays*, sit awkwardly with the volume on Cumbria that treats in one volume such totally contrasting regions as the high central fells and the west Cumberland coalfield (Millward and Robinson 1972). Counties are arbitrary constructs, regardless of how old each may be, and they do not respect facets of landscape.

Tilley (1994, 18) preferred to think in terms of *locales*, places 'created and known through common experiences, symbols and meanings'. He is an archaeologist with a focus on prehistory but his reasoning still has resonance in post-medieval landscape studies. A landscape of improvement, whether monastic, Tudor or eighteenth-century, may well have been perceived by those who were part of creating it as something special to them.

Similar to the *pays* concept is the identification of discrete landscape units called *regios* or *regiones* which have as their fundamental basis a unit of land contained within a watershed. Everitt (1977) convincingly applied this concept to Kent, and suggested it is also relevant elsewhere; within the current research area Fleming (1998) conceived the idea of Swaledale as one such unit (though he did not use the term *regio* or *regione*). It certainly has an identity different from those of Wensleydale to the south or Stainmore to the north, it is patently contained within an upland massif, and historically it formed one political and administrative unit. Dentdale and Garsdale are arguably also sufficiently distinct to warrant the nomen *regio*.

As in any healthy academic discipline, landscape historians have adopted contrasting ways of attaining the same ultimate goal of understanding how landscapes evolved. At times, though, academic debate has had a (healthy?) tendency to fracture into mutually exclusive camps. Muir was perhaps the first prominent landscape historian to draw attention to that subject's propensity for marked disagreement (1999, 299). Perhaps overstating his case, he accused certain of its practitioners of 'faddism', of indecision in attaching themselves to any passing bandwagon and of having either been 'infatuated' with quantification or lost in the realms of obscurity. He was, however, unhelpfully vague by not developing these particular criticisms.

Muir may have been the first landscape historian to express concerns about the means of studying past landscapes but historical geography has endured criticism for far longer, not least for its having at times displayed deterministic tendencies. This much-maligned school of thought saw Man at the mercy of physical conditions, such as climate and prevailing soil types, and unable to make choices, and it attracted such vociferous criticism, including from Vidal de la Blache and his possibilist school, that many geographers rejected any hint of deterministic thinking. In the context of this writer's current research theme and geographical area, those involved in farming had recognised as far back as the Middle Ages that more could be teased out of the soil and they made use of whatever was available locally – as long as it seemed to work for them. Experimentation with marl and lime, for example, enabled them to overcome the strictures of nature and to bring into enhanced use land that had formerly been seen as waste. They found ways of obtaining the raw material where seemingly none was to be found. They were indeed able to make reasoned choices based on the possibilities available to them.

Yet, to an extent, determinism is alive and well, and the finger has been pointed at historians who underrate the importance of natural processes in influencing past human responses and thereby reject determinism (Williamson 2003, 23).

The French historian Fernand Braudel asked a very pertinent question of his homeland: was France invented by its geography (1986, 265-66)? In other words were its peoples shaped by landscape opportunities or did they have the wherewithal to adapt it to their needs and aspirations? To answer his own question, he wrote:

> Pour eux, l'élément decisive, ce n'est pas la terre, la nature ou le *milieu*, c'est l'histoire, c'est l'homme ... car héritier, continuateur des actes, faits et gestes, techniques et traditions ... ont modelé le paysage, l'engageant, à l'avance, dans une série de determinisms retrospectives.

This can be translated as 'for them, the decisive element is not the earth, nature or the environment, it is history, it is man as the inheritor and continuer of the acts and actions, the techniques and traditions, that shape its landscape, and trap him in a series of retrospective determinisms' (this writer's translation). In essence, he is professing that the way a society develops (its history) is governed by geography and not by people's actions. They are locked into a process beyond their immediate control. The answer to Braudel's question is probably a bit of both: the French were shaped and they did have the wherewithal.

Landscape archaeology can have deterministic inclinations, too. It really became established in its own right as a sub-discipline in the 1990s, though the term was first used in 1974 (Aston and Rowley, 1974; Fleming 2006)[12] and it is difficult to meaningfully distinguish it from landscape history (Everson *et al.* 1998, 1; Williamson 1998, 1 and 2002, 21). Whereas the parent discipline, mainstream archaeology, was tending to move down the objective and scientific processual route, landscape archaeology largely adopted a post-processual mode of thinking, wishing to go beyond the 'hard' ground evidence of earthworks and artefacts to more subjective interpretation of the evidence to tease out what might have been going on in the past, while placing people firmly in the picture.

Post-processualist landscape archaeologists (and historians) fall into one of two schools of thought. On the one hand there are the phenomenologists who push their horizons well beyond what the hard evidence indicates. Their approach is ethnographic and requires a vivid imagination with frequent diversions into metaphor: indeed, Fleming (2006) has rather dismissed it as a 'magic garden' approach, and has suggested it is somewhat rooted in environmental determinism. The alternative, hyper-interpretative, school avoids determinism altogether. It still goes beyond the evidence and is equally imaginative in its thinking and analysis and, suggests Fleming, it treads a fine line between reality-based interpretation and conjecture.

The divide within landscape studies can be seen in another way: there are the empirical and the theoretical schools. The first owes much to Hoskins, and its followers are concerned with amassing quantities of data in the field (and from maps and documents), all of which can be used to build up a picture of what past landscapes looked like or, put another way, what the past geographies were. The opposing school relies less on the hard evidence and is content to use samples from it, adding them to possible parallels from modern anthropological observations elsewhere in the world.

Empiricists have been specifically criticised for suggesting that modern landscapes can be easily read and thus easily understood and categorised, thereby giving the impression that there is nothing more to be learnt from them (Finch 2007, 39). Whether this standpoint has any validity is questionable though, as Finch seems to be suggesting, the empirical approach is said to be superficial and lacking in rigour. Others would disagree: Williamson (2002, 1) has argued that landscape history is more firmly rooted in the real environment, and in the realities of rural life, than some 'rarified models' that purport to explain landscape from a theoretical perspective.

Fleming represents the empirical school, as does Muir; the theorists include archaeologist Matthew Johnson. In very recent times there has been a most lively public debate between Fleming and Johnson. In a compelling book, Johnson (2007a) argued persuasively that empiricists need to mend their errant ways because they are too bogged down by facts and too reluctant to theorise, too focussed on subjective narrative at the expense of objective analysis. In his view, they are too obsessed with their 'muddy boots',[13] and also steeped in determinism (M Johnson 2007a, 2 and 144).

[12] Crawford (1953, 36) noted that the term 'field archaeology' was invented in 1915 by Dr Williams-Freeman to describe archaeological work that excluded excavation. In essence this form of archaeology has parallels with modern landscape archaeology except that the latter can involve excavation.

[13] It was Richard Muir who reintroduced the concept of the muddy boot in the dedication to his *The NEW reading the landscape*, published in 2000: 'For the ordinary fieldworkers with the muddy boots and the aching joints who do most of the work, yet see little of the credit' - a cutting comment indeed. However, Beresford (1957, 249) first included 'a good pair of boots' in the arsenal of the student of landscape, though Hoskins (1955,14) had earlier talked of 'scrambling on foot'.

Needless to say, Fleming (2007), swiftly hit back in a neatly argued defence of empiricism (and of muddy boots) though he accepted Johnson's point that an empirical methodology would have more validity if tied in with greater investigative and critical rigour. Not to be silenced, however, the latter came back with a rejoinder and a plea not to 'bin your brain' (M Johnson 2007b), while Fleming (thus far) has had the final word by reiterating his earlier rejection of Johnson's 'fashionable disparagement of "traditional" landscape archaeology' (Fleming 2008).

Meaningful study of landscape cannot happen without directly observing the components of a landscape from within it, but it cannot be fully understood without a certain degree of postulation and critical thinking.

The 'brain' is needed to put people back into a past landscape. We can look at and ponder a landscape that bears all the hallmarks of improvement, namely the walled enclosures, the green and lush nature of the pasture within them, the drains, the small limestone quarry with perhaps a nearby lime kiln, but these alone do not give us the complete picture. We can turn to archival and cartographic sources to supply more details of when it was improved, and perhaps how long it took and what it all cost, but we still need to know more.

This is where the anti-empiricists are right. In many cases the ground and written records tell us nothing about who was involved in the process. True, there may be written accounts of who paid for the lime and even names of those who carried out the major work, but such records were necessarily penned by and on behalf of the landowner. They tell his story. The man on the ground, the tenant or peasant farmer, whose sweat and time actually made the landscape, is all too often invisible to us. In most cases the person who built a given lime kiln is unknown; we can have no knowledge of why he (it usually was he) decided to site his kiln at that precise spot especially if it makes no sense to our way of thinking. We cannot presume to understand his thought processes and nor should we assume that he was even thinking in what we would call a rational manner. He may not have been 'rational man' at all. The study of landscape needs to embrace all these aspects and has to rely on a degree of theory and conjecture if we feel it is necessary to repopulate the past.

3. On Historic Landscape Characterisation

Historic Landscape Characterisation (HLC) offers a radically different approach to the study and understanding of landscapes. It is an on-going process, under the aegis of English Heritage, with the aim of mapping historic landscapes across the entire country based on Geographical Information Systems (GIS) technology. In the words of one of its current practitioners it 'creates a basic understanding, *lacking until now*, of the historic dimension of the whole landscape' (this writer's italics) (Fairclough 2007). It would be fair to say that HLC has been widely condemned and there is a gulf between its supporters and 'traditional' landscape students. Part of the problem lies in the phrase italicised above as it implies that the last four decades of landscape history, not to mention even more decades of historical geography, have achieved nothing by way of understanding landscape.

Beyond that, there are basic flaws in how HLC works, and these derive from its underlying premise. It was established as a planning tool to manage rural change and as an aid in carrying out feasibility studies for prospective agri-environment schemes. All it does is map – using polygons, which in itself is a crude mechanism (Williamson 2007a, 67-68) – the dominant historic signs in today's landscape, and it is entirely a desk-based exercise. It does not seek to get below the surface to peel off layers of past activity, so it is in essence one-dimensional. It claims to map the 'time depth resonant within the modern landscape' (in other words what elements of the past have survived) but deliberately makes no attempt to reconstruct past landscapes (www.lancashire 2008), and this cannot be right.

The time-depth aspect of studying landscape is unarguably important as is fieldwork: this is essential if misinterpretation is to be avoided. An example from within the current research area will illustrate this point. A short distance to the west of the village of Dent, in Cumbria, is an enclosure called Stone Rigg (centred on NGR SD683 866). This is a tract of broken ground with outcropping limestone clearly visible and it is generally very pockmarked, and appears thus on aerial photographs. It has not yet been mapped in the HLC programme for the Yorkshire Dales National Park but, no doubt, it might be classified as parliamentary enclosure (which it is not) or enclosed grazing (which it is). No documentary sources have been located for Stone Rigg, so a desk-based search would add nothing to what aerial photos indicate, apart from the depiction on the Ordnance Survey First Edition six inch and 1859 enclosure award maps of one lime kiln on its northern edge. The only way to understand and deconstruct Stone Rigg, and therefore to correctly classify it, is to walk it with an eye for detail. It has been heavily quarried, mostly on a small-scale, and has ground evidence in the form of structural or earthwork remains of lime kilns that this writer has ascribed to four discrete phases of operation, stretching back at least to the immediate post-medieval period. The time-depth here is enormous and it can only be appreciated on the ground, from empirical yet analytical observation. Stone Rigg is a proto-industrial landscape in a totally rural upland setting.

To quote one critic of HLC, it provides us with an 'outline caricature' of landscape, providing a distorted parody of reality rather than a meaningful three-dimensional representation, it ignores cultural identity within landscapes, and decontextualises the multi-faceted elements that make up a given landscape (Austin 2007, 103). Nevertheless, HLC is an ongoing process across the country and thus one that cannot be ignored or dismissed.

4. On industrial archaeology

Industrial archaeology as a discrete discipline is of roughly the same vintage as landscape history though the term was first used in the 1950s in a plea to academia to show recognition of the role that industrialisation had played in the country by taking positive action to preserve the nation's industrial heritage (Rix 1955, 228). Since then it has experienced a paradigmatic shift in emphasis from being focussed on 'its surviving monuments and artefacts' (Buchanan 1972, 19), in other words the mines, machinery and transport infrastructure, to taking a more holistic view of our industrial past. It was not an easy transition and the shift is still current and hesitant.

Buchanan led the way in the subject's early years but, probably owing to its early parochial and amateur nature, acceptance within academia was slow to be achieved. Indeed, its profile within university archaeology departments is still very low (Palmer 2010, 16). In a review of Rix's seminal *Industrial archaeology,* Crossley (1967, 137-38) had questioned the validity of the use of the very term 'industrial archaeology' as few of its practitioners had any background or interest in excavation techniques and doubted whether or not Rix had made a convincing argument for the discipline, despite Rix's evocation of and justification for the discipline (Rix 1967).

Another of its early practitioners, Neil Cossons, lamented the lack of academic rigour and the absence of 'respectable academic origins' (1978, 19). For many years no thought was given to the wider aspects of industry such as how given industrial sites fitted into a broader local and regional context. It was deemed sufficient to study the site and its physical remains *per se.* Even less attention was given to populating the industrial past – people were excluded from the picture except, perhaps, for the entrepreneurs and magnates. Social ramifications were largely ignored.

The publishing house David & Charles produced an impressive list of regional volumes on industrial archaeology (see, for example, Ashmore 1969 on Lancashire; Harris 1971 on the Peak District; and Nixon 1969 on Derbyshire) and the journal *Industrial Archaeology* promoted itself as the journal of the history and technology of industry. All provided a valuable service in describing what then remained by trying to recreate the industrial past, but they were essentially gazetteers and 'roles of (industrial) honour'.

Pressure for change only came to a head during the first decade of this century though the process of transformation had long been gaining momentum. In 1966, for example, the Society for Post-Medieval Archaeology was founded to promote the 'archaeology of late medieval to industrial society' (SPMA 2008) which put it in direct opposition to the informal bodies engaged in industrial archaeology. However, this did not prevent the formation of the Association for Industrial Archaeology in 1973, whose mission remains the promotion of the study and appreciation of industrial archaeology. Initially at odds with each other, the two came together in 1999 to organise a joint conference on the archaeology of industrialisation (Barker and Cranstone 2004).

Further pressure came from within. Cossons became a prominent convert, at first suggesting that industrial archaeology had had its day (Cossons 2000, 13), but more recently declaring the subject to have become 'one of the most vibrant and progressive areas of research and practice' (2005, ix). This conversion, his personal paradigmatic shift, has come about owing to a recognition that more emphasis should be put on investigating the aspirations and motivations of those involved in industry, be they magnates, managers or shop-floor workers (Palmer 2004, 4; 2005, 13).

This argument has been taken further. The subject's supporters are now exhorted to look beyond the physical and consider in the broadest way the social aspects of industrialisation as expressed through 'control and ownership ... of monuments and landscapes' (Nevell 2006, 12). Equally important in Nevell's view is the study of rural landscape change brought about by past industrial growth. It is pertinent to note that the pioneering industrial archaeologist Arthur Raistrick (he would not have dreamt up that term), who worked in the Dales from the 1930s to the early 1970s, firmly believed in linking industry to its impact on the landscape, particularly in areas where lead mining had left such a lasting impression on the ground (Palmer 2000, 121).

Matthew Johnson's conclusion (1996, 12), that industrial archaeology was aloof and fragmented and thus unsuccessful, may well have been valid then but its transformation has been dramatic. People and landscape are now central (Gwyn 2009, 20; Nevell 2009, 39); monuments are but one element of a broader picture; and 'authorship' of past industrial sites is paramount (Cranstone 2004, 318). There is still, however, a noticeable mismatch between the academic discipline of industrial archaeology, with its enduring tendency to self-doubt, and tourism-oriented industrial heritage which has been witnessing a resurgence in the popular imagination over the current decade (Orange 2008).

This dual way of thinking can be applied to land improvement and the application of lime. The old school of industrial archaeologists would have sought out a lime kiln, would have carried out a full site survey of the structure and identified the source of the raw material and roadways used to transfer limestone to kiln. They would probably have stopped there. Even the type and source of fuel may not really have figured in their investigations. However, that is only one small part of the picture. 'Authorship' (Cranstone 2004, 318) is ignored: who decided to build the kiln there and not somewhere else; what was the prime locational factor, and was the siting decision based on rational thought processes or purely personal desires? We can look at a given monument and wonder why it was built in that style. Did it conform to a regional template, was it built as cheaply as possible and designed as a functional object or with a degree of aesthetic embellishment? This is all part of authorship. Beyond these considerations, we should ask why this person built the kiln at all, why they perceived agricultural lime to be worthy of the trouble and expense involved. Then we should want to know *who* had it built – the landowner, the estate steward or the tenant farmer. Each of these had an opinion, a motivation and aspirations. Finally, perhaps, there is the complex issue of the actual application, and many questions are raised here. Was the lime burnt on the estate/farm or bought in? At what time of the year was it applied, and how frequently? What was it applied on? What methods were used to apply it?

All of these are inherently valid and valuable lines of enquiry, but are they industrial archaeology? This question can be examined in three ways. Firstly, the answer would initially have been negative as industry was perceived only on an 'industrial' scale and rural crafts and trades were excluded. Nowadays the answer would be positive. Gwyn (2006, 232; 2007, 11) encapsulated one strand of current thinking: 'we need to consider the relationship between industrial archaeology and the archaeology of agriculture', and cannot ignore, for example, the enormous strides taken in agriculture during the eighteenth century (2006, 96) or the 'industrialization of agriculture' begun in the sixteenth (Bayley and Crossley 2004, 17).[14]

Secondly, this last point supports the view that industrial archaeology should not just be about the so-called Industrial Revolution. It should not just concern itself with developments from the late eighteenth century onwards but should cast its horizons further back. As will be shown in the following chapters, the production of agricultural lime can be traced back over many centuries and it is perfectly valid to consider it as both industrial history and industrial archaeology, albeit on a small-scale and localised basis.

Finally, the very title 'industrial archaeology' is now being questioned (see, for example, Newman 2001, 184; Gwyn 2006, 236-37 and Nevell 2006, 13). Maybe we should think instead in terms of the archaeology of the industrial period or the archaeology of industrial society (though both could exclude pre-1700 developments). On the other hand, maybe it would be more productive to study the process in a holistic way – the monuments, the people, the economic forces, the externalities, the outcomes – rather than remaining transfixed with labels. This was at the centre of the approach adopted in this research programme.

5. On the literature of lime burning

For decades writers have been drawing attention to the dearth of research into lime burning in this country: many years ago Buchanan (1972, 185) noted the large number of lime kilns that pepper every limestone landscape and expressed the view that they were 'worth the attention of industrial archaeologists'. Exactly thirty years later the present writer wrote that this now lost aspect of our industrial and agricultural past was still the 'cinderella of industrial archaeology' (Johnson 2002, 6). Very similar comments had been made by Cossons (1978, 221) who emphasised the contrast between lime kilns being amongst the most common landscape features yet the 'least studied'.

Since Buchanan and Cossons wrote not a lot has changed and, although there is considerable anecdotal evidence of a keen general public interest in lime burning, detailed rigorous study of it in the widest sense is still lacking (but see regional studies by Fleming 1998; Isham 2000; Leach 2002; and Barnatt and Dickson 2004).

Leach suggested that lime burning's role had been downplayed because kilns were perceived as small-scale centres of production set in isolated locations and therefore not part of the wider national economy. This very point had been picked up earlier in a very personal review of landscape history (Taylor 2000, 161), without lime kilns actually being mentioned. The thrust of Taylor's argument was that low-status sites (which must include rural kilns) far outnumbered those of high-status and thus deserved far greater attention from students of landscape.

Leach and Taylor were absolutely correct. As has been discussed earlier, a lime kiln did not sit in isolation but was inextricably linked to other aspects of the rural economy. Kilns supported coal pits and limestone quarrying (Raistrick 1968, 114; 1971, 44); they required a network of trackways to bring in raw materials and send away lime, and these served other rural enterprises. Setting aside the broad range of industrial, domestic and constructional uses (see Chapter 1.7), the role lime played in agriculture from monastic times through to the 1960s should not be underestimated – an issue that provides the basic premise for this book. The application of lime was fundamental to the improvement of existing farmland and to the successful winning of 'new' land from the waste.

Various conclusions can be drawn from the summary in Table 2.1 and by comparing these data with Figure 2.1, which shows those parts of mainland Britain with extensive outcrops of limestone (including chalk). It will be immediately clear that there has been a concentration of research into lime burning in relatively few areas, with Derbyshire and central Scotland having had perhaps the greatest attention. Large areas where Magnesian limestone and chalk outcrop, and where lime was widely produced, have had no targeted research. Areas such as Pembrokeshire and the south-west either exploited localised outcrops or shipped raw stone in from South Wales.

Almost half of the published items adopted a regional approach, either on a county or part-county basis, and seven were confined to a specific lime burning site. Two-thirds included some historical discussion. This may seem encouraging but a close analysis of content and emphasis shows otherwise. The majority of historical discussions were brief and confined to the more accessible archival sources; related to specific sites; or, in some instances, based on generalised accounts of the history of lime burning that could apply to more or less anywhere. Skinner (1969), Davies (1997) and Leach (2000 and 2002), however, do provide more meaningful historical analyses.

Half of the items concentrated on technological aspects, conforming to the 'old' style of industrial archaeology focussing on the monuments themselves. A third are dominated by gazetteer-type entries derived from field surveying but, again, the monuments were the prime object of the research. Similarly, around one third included a high proportion of material on

[14] Landscape archaeology has been criticised for not giving due prominence to agricultural history (Williamson 1998, 6).

Source[2]	Area exemplified	Scale[3]	Arch.[4]	Gaz.[5]	Hist.[6]	Tech.[7]	Uses[8]
Aldsworth 1979	West Sussex	S			Y	Y	
Barker 2008	S Cumbria, N Lancs.	L		Y	Y		Y
Barnatt 2004	Derbyshire	S	Y		Y	Y	
Bhanji 1999	S Devon	L		Y			
Bick 1984	Gloucs./ Herefords.	R		Y	Y		
Born 1991	N Devon	L	Y	Y	Y		
Brock 1987	Richmondshire	L	Y	Y			
Brown 1997	Pembrokeshire	R	Y	Y	Y	Y	
Clarke 1987	Dumfriesshire	S			Y	Y	Y
Cleasby 1994-5	Sedbergh area	R		Y	Y		Y
Daniel 2001	Somerset	L		Y		Y	
Davies 1997	Pembrokeshire	R					Y
Donnachie 1971	SW Scotland	R	Y		Y	Y	
Evans 1987	Devon	S				Y	
Gibbons 1993	C Scotland	R			Y	Y	
Green 1977	Walsall	L	Y			Y	
Groom 1993	Buxton	L			Y	Y	
Holt 1971	C Sussex	L			Y		
Isham 2000	Cornwall	R		Y	Y		
Leach 1995	Derbyshire	R	Y		Y	Y	
Leach 2000	Devon	L			Y		
Leach 2002	Devon	R			Y		
Martins 2008	Argyll	L	Y		Y		
Minchinton 1984	S and SW Wales	R	Y	Y			
Pounds 1971	Cornwall	S		Y	Y		
Robertson 1999	N Pennines	R	Y	Y		Y	Y
ScLCT 2006	Firth of Forth	S		Y	Y		
Skinner 1969	Lothians	R				Y	
Skinner 1975	Scotland	R			Y	Y	
Taylor 2009	S Yorkshire	L	Y				Y
Toft 1988	Gower	L	Y				
Unwin 1968	E Cheshire	S			Y		
Wright 1967	Yorkshire Dales	R			y		

Table 2.1 Summary of literature on lime burning and lime kilns, 1967-2009[1]

Notes
[1] Publications by this writer are excluded, and accounts of specific excavations.
[2] See the Appendix to the Bibliography for full bibliographic references.
[3] L - local scale, R - regional or county scale, S - site specific.
[4] Arch. - deals with archaeological evidence
[5] Gaz. - includes a gazetteer of lime kiln sites
[6] Hist. - provides a brief historical introduction
[7] Tech. - concentrates on kiln technology
[8] Uses - considers local uses of lime

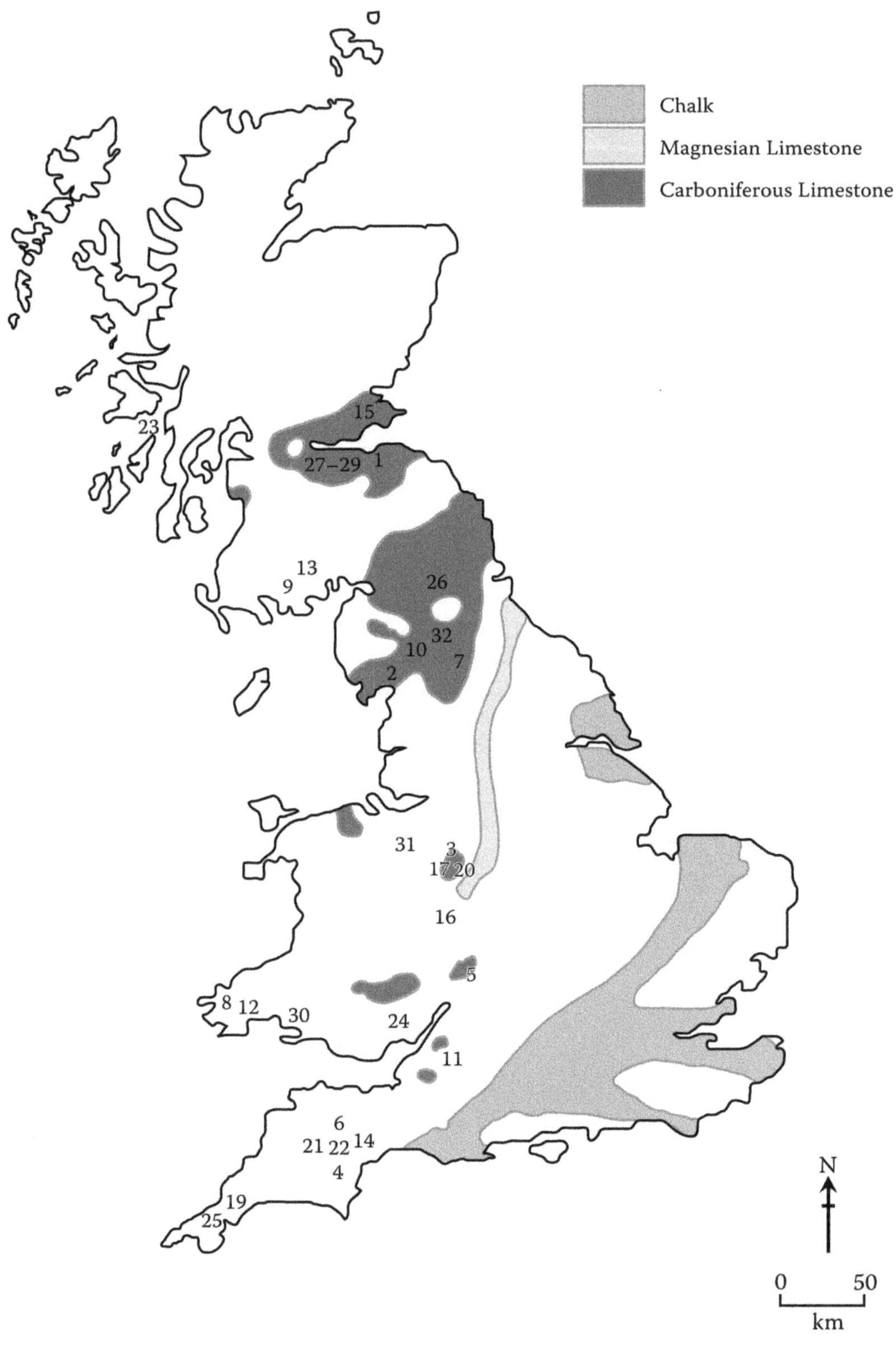

Figure 2.1 Relationship between outcropping limestones and published material on lime burning.
Source: British Geological Survey mapping.

the archaeology of lime kilns. Only six authors had attempted any discussion of the uses to which the lime was put in their respective research areas.

It will be apparent from the preceding discussion that, although there is a corpus of published material on the subject, most authors have tended to focus on specific sites for in-depth study or on regional field surveys of lime kilns. This is understandable and not to be criticised in itself. Indeed, this writer's initial research was aimed at recording the status and detail of field kiln sites across the Yorkshire Dales National Park rather than on any wider aspects. It could be argued that field surveying is a necessary prerequisite for detailed archival research. It was only on the basis of field observation that the full importance of lime in the Dales landscape slowly began to reveal itself to its full extent.

Of the thirty-two items reviewed, only three authors undertook work within the present research area. Wright (1967) provided a three-page item which was inevitably superficial and introductory; Brock (1987) concentrated on a field survey of kilns within an arbitrary 5.6km-radius of Richmond on the north-eastern edge of the Dales; and Cleasby (1994; 1995a; 1995b) summarised the field survey work of a local volunteer group, with a useful document-based discussion of lime production and usage within three parishes on the north-west corner of the Dales.

Virtually all the items reviewed here have credence in their own right as local accounts and, taken as a whole, they help to build up a picture of lime burning and the use of lime across the country. However, none attempted an in-depth analysis. This study will build up a detailed and rounded account of the application of lime in the Central Pennines in a way that probably no other published study has done for any part of the country.

6. On the process of land improvement in England

The late eighteenth century and the early years of the nineteenth have long been associated with a body of men known as the Improvers, national figures such as Arthur Young, Thomas Coke, 1st earl of Leicester, and Francis Russell, 5th duke of Bedford, as well as those of more regional significance like Christopher Crowe of Kiplin Hall near Richmond, William Danby of Swinton Park near Masham, Charles Towneley of Towneley Hall in Burnley and Thomas Eccleston of Scarisbrick near Southport. However, improvement is a process that can be traced back to the medieval period with the desire to improve land having been something that all but the most conservative farmers would have held dear (Muir 2000, 59). Medieval peasants were probably conservative as a class and may have been initially suspicious of innovation but they did not live in isolation and were subject to external pressures (Campbell 1981, 120-21). If not under direct monastic control, they would surely have observed the often dramatic change made on monastic demesne land and been able to translate that into ways of increasing productivity on the land they worked. Every society has those with vision who are able to break out of the prevailing mould, regardless of where they stand on the societal ladder.

It is broadly accepted that the medieval centuries saw a transformation of the countryside and of the rural economy (Fryde 1996, 7 and 25) with the slow growth of economic specialisation (Higham 2004, 81) and commodification (M Johnson 1996, 1). These economic and perceptual changes went hand in hand with the political and social transformation from feudalism to incipient capitalism; and the development of materialistic attitudes, so ably demonstrated by Mathew Johnson, made itself felt in the countryside with an increased urge to expand the cultivable area (by assarting and piecemeal enclosure of forest and waste) and to increase the quality of existing crop and pasture land. Clearly, such fundamental changes did not happen everywhere at the same time. Some remote or ultra-conservative areas remained immune from change into the early modern era.

Key to much of this improvement was the introduction (or increased use: the evidence does not allow us to say with certainty which) of drainage and soil inputs such as marl and lime; the planting of legumes; and the development, at least by the sixteenth century, of convertible or ley husbandry involving the rotation of crops and pasture on the same piece of land (see, for example, Brenner 1976, 23-24; Cantor 1987, 49; and Thirsk 1997, 16).

Equally important in bringing about a desire to increase output was demographic pressure (Lewis *et al.* 1997, 169). Rising populations,[15] and the slow but steady growth of urban centres, put pressure on landowners and farmers alike to produce more and therefore to improve, while stimulating the notion of profit. The symbiotic relationship between population and economic performance was not a phenomenon unique to Britain but has been recognised in continental Europe (M Johnson 1996, 21). Growth in both was apparent up to c. 1350 and from c. 1450 to c. 1650 in Britain and on the continent with serious decline and stagnation in the intervening hundred years.

Up to c. 1350 the role of the monasteries in land improvement and profit generation cannot be overestimated (Bond 2000, 63). One quarter of England and Wales was in monastic ownership with six foundations having control of extensive tracts within the research area. They had the vision and the wherewithal to take the long view, they had absolute security of tenure, and were able to create for themselves economic empires built on wool. By any definition, they were pioneers in the field of agricultural improvement. However, a note of caution should perhaps be sounded. In the words of Joan Thirsk:

> At a distance of more than five centuries, the context and interrelationship of agricultural innovations in the later Middle Ages can only dimly be perceived, and they certainly cannot be satisfactorily measured. (Thirsk 1997, 20).

Taking monastic vision and progressive peasant attitudes on the one hand and population growth on the other, we can perhaps accept that the many lynchet systems within the Yorkshire Dales represent medieval cultivation of former marginal land during the climatic optimum of c. 1150 to c. 1300, with place-name evidence (for example thwaite, ridding or ley) indicating encroachment into the woodland estate.

After c. 1350, or possibly several decades earlier, deterministic tendencies can be recognised (Williamson 2003, 23). From 1315 to 1326 the Central Pennines were ravaged by a series of Scots raids, and climatic fluctuations led to harvest failure and outbreaks of cattle murrain. These disasters culminated

[15] The population of England is estimated to have risen from 1.25m in 1100 to 5m by 1300 (Thirsk 1997, 7).

in plague epidemics later that century (Kershaw 1973a, 7-27 passim; Stratton and Houghton Brown 1978, passim; Cantor 1987, 14-15). Stock numbers were decimated, the human population was weakened and reduced,[16] and the political system broke down. Landlords were deprived of their labour pool and it is hard to see how they could have felt any incentive to invest in land or to carry out any form of land management (see Brenner 1976, 48). Improvement can hardly have figured highly in anybody's mind.

Undoubtedly marginal land was abandoned and crop land, including lynchet systems, reverted to pasture, but Dyer (1993, 143-44) offered a different view. He believed the setbacks led to increased intensification of farmland and consolidation of peasant holdings. This idea of there having been positive outcomes from years of trouble, as forward-looking peasant farmers became free tenants keen on improving their holdings, has also been expressed elsewhere as has the emergence during the later fifteenth century of a class of tenants sufficiently wealthy to finance land improvement (Brenner 1976, 61; Cantor 1987, 15). The increase in pasture land eventually brought about an increase in stocking levels which gave the peasant a greater supply of manure with which he was able to improve soil fertility and productivity.

Deterministic ideas can possibly be invoked again for the fifteenth century with further epidemics between 1390 and 1439, serious crop failure from 1437-39 and famine for two years after that (Fryde 1996, 116), with depression in agriculture being felt widely from the 1430s to the 1470s.[17] It was only economic and agricultural revival after the 1470s that paved the way for renewed investment in land and farm improvement. Perhaps this catalogue of disasters is why there are no documentary or written accounts of soil improvement between the monastic period of high farming and the early sixteenth century in the research area: investment in land had not been a priority.

The sixteenth century was a time of mixed fortunes with further recorded harvest failures up to c. 1550 in southern England but signs of progress within the research area, despite repeated outbreaks of typhus (or possibly dysentery) and plague from the 1540s to the 1590s (Wrigley and Schofield 1981, 668-69). Much open field arable had already been enclosed in the West Riding by 1540 (Smith 1970, 13) with a concomitant surge in the number of individual smallholders, especially in the north. In the manor of Colne, for instance, the number of small tenants exactly doubled from fifty-one in 1527 to 102 in 1617; and similar patterns have been recognised within Swaledale and Wensleydale (Outhwaite 1986, 9); and in that part of the West Riding which now lies within Cumbria where a new class of yeomen began to emerge towards the end of the sixteenth century (Millward and Robinson 1972, 54-55). Their land holdings were limited in size but they were able to develop specialisation in cattle and sheep rearing. In Lower Airedale and the Worth valley signs of development in farming were evident as, for example, large areas of common land had already been enclosed for more intensive use by 1615 (WYAS Bradford, Ferrand C/218 1847).

A link has been made between the resurgence of the cattle droving trade from Scotland to England, after the Union of the Crowns in 1603 brought peace to the borderlands, and land improvement in the north of England (Simmons 2003, 93-94). Droving both engendered a demand for improved pastures along the regular droving routes and helped bring it about as accumulations of dung left by stock on overnight stances were spread on nearby grassland. Once farmers had accepted the benefits that soil inputs such as dung could bring about, there remained only a small step to their seeking ever better methods.

It is not beyond the bounds of possibility to postulate that, even then, the more percipient pastoralists were able to recognise that their soil and pastures were lime-deficient and in need of palliative treatment. Whether they gained the knowledge of liming from the growing output of agricultural treatises, from empirical observation of grass growing more lushly where piles of lime for mortar had been stacked, or through slow word of mouth dissemination, is arguable. The fact is that lime usage became more widespread during the late sixteenth and early seventeenth centuries, as will be shown.

The ethos of and drive for improvement gained momentum during the eighteenth century, either 'by gentle prodding, propaganda, and grants of one sort or another' (Thirsk 1997, 29) – elements of which can be identified within the Central Pennines, as again to be discussed later – or by advances in 'human ingenuity' (Marshall 1929-30, 41). The prodding came from estate owners and stewards, the propaganda from quasi-government bodies and learned societies, the grants from landlords as allowances to tenants, and the ingenuity from advances in kiln technology (permitting increased production at a higher quality) and methods of application of agricultural lime. All will be revisited in later chapters, but the matter of ingenuity will lead this discussion, in the next section, to an examination of the notion of the so-called Agricultural Revolution.

Prior to that, however, it is pertinent to address the matter of tenure as farmers' willingness to invest in their holdings was strongly influenced by security of tenure. Tenurial patterns varied throughout the research area with some manors being dominated by leasehold tenure with others subject to customary tenant right. Those who enjoyed customary rights tended to jealously guard them and strongly resist any attempt by the lord to whittle them down or to convert them to leasehold. Leasehold tenancies were imposed either for the life of the current tenant or for a set number of years: former Fountains Abbey lands, for example, followed this pattern (Jennings 1992, 105). In Westmorland it was common for leases to be set at seven, nine, fourteen or twenty-one years which was a clear disincentive for the tenant who rightly felt

[16] It took until 1500 to pass the 2.5m mark. It regained the 5m level by 1650 (Outhwaite 1986, 1), though Cantor (1987, 44) set the population in 1600 at 4m, only achieving 5m in 1700. Wrigley and Schofield (1981, 566-77) put the 5m threshold around 1670.

[17] Being largely pastoral, much of Craven probably escaped the worst of this recession (Hoyle 1986, 230-31).

that such short leases could not give him time to recoup any financial outlay on improvements (Bouch and Jones 1961, 229). To make matters worse landowners here were entitled to eighteen months 'improved value' on leasehold agreements for every seven-year period. Effectively this reduced the income-generating potential of a seven-year tenancy to five and a half years, too short by far to benefit from investments made.

Through the first part of the post-Dissolution sixteenth century many leases in Craven were vastly increased in length. It was not uncommon to find leases set for three hundred, one thousand or even five thousand years. This gave the lord an assured long-term source of income and a greater chance of having tenants working his estate with a real sense of commitment and belonging. For the tenant there was guaranteed security of tenure along with *de facto* rights to demise or sell his holdings and to have them passed down in a hereditary manner (Hoyle 1986, 45-48). This process can be illustrated by an indenture settled in 1677 by which William Foster of Stainforth in Ribblesdale demised to his son, Thomas, a messuage, tenement and various closes then in William's tenure for a period of 800 years (Stainforth, private).

Across much of the research area customary rights prevailed. In eighteenth-century Westmorland, for instance, two-thirds of all manors were dominated by customary tenancies with rights guaranteed by 'ancient' case law (Searle 1993, 128). Such tenants had the inalienable right to sell or demise their holdings in return for payment to the lord on entry and inheritance fines and heriots. Exactly when such rights had become enshrined in the manor court system is difficult to gauge but they pre-date Dissolution and in the Forest of Knaresborough, for example, they were certainly established by the 1490s and were confirmed by royal decree in 1562. In Craven and Wensleydale such rights were confirmed during the seventeenth century. In effect customary tenants could be equated with freehold tenants, and in the West Riding and Westmorland it was this form of secure tenancy backed up by long-established customary rights that fostered the development of a stable and influential yeoman class (Jennings 1992, 129; Winchester 2000, 16). As Whyte (2003a, 36-37) has demonstrated, customary tenants acting in unison were sufficiently powerful to either promote enclosure of the waste during the period of parliamentary enclosure or prevent the lord from forcing enclosure through.

7. On the Agricultural Revolution

Study of the use of lime on the land lies firmly within the domain of agricultural history and it is appropriate to consider here the matter of the much-debated Agricultural Revolution. Use of the term 'revolution' must imply recognisable and dramatic change over a limited period of time. In farming these changes encompassed the development of new crops such as turnips, legumes, sainfoin and clover; selective breeding of more productive cattle and sheep; the introduction of machinery; and increased awareness of how soils could be improved. The end result was that improvements in labour and land productivity should have brought about increased yields, increased farm/estate income and higher profit margins, thereby enabling greater investment in farm enterprises.

Controversy has surrounded the issue since the publication of R E Prothero's seminal book almost a century ago (Prothero 1912),[18] a book which has been so influential that it was reissued as recently as 1972. He can be seen as the first to draw attention to the idea of a revolution but his assertion that it spanned the years 1750 to 1850 has been both rejected as a myth (Kerridge 1969, 464-65; 1990, 463; Allen 1999) and accepted as fact (Chambers and Mingay 1966, passim; Overton 1996, 3; Williamson 2000a, 110).

Kerridge expressed the view that the revolution began around 1560 and can be traced through to the 1760s (1967, 15 and 347; 1969, 463) though it is this writer's opinion that changes spanning 200 years preclude use of the term revolution. Outhwaite (1986, 7) challenged the notion of major concentrated change between 1500 and 1650, suggesting that there was an element of economic regression during that period, in favour of even earlier dramatic changes. Support for this latter interpretation was provided by scientific analysis of animal bone size data which suggested that noticeable improvements in farming (represented by selective breeding) began in the sixteenth century leading to industrialisation of agriculture before 1600 (Bayley and Crossley 2004, 17). The suggestion has also been put forward that the development of monasticism and 'Christian corporatism' brought about an 'early medieval agricultural revolution' (Low 1992, 64-65). In Low's view use of the term 'improvement', as related to a possible revolution, can be dated to the sixteenth century when it was understood to mean to 'turn *land* to profit; to inclose and cultivate' and therefore to make waste land better (1992, 36). Evidence will be presented in this book to show that lime was being applied on the land in the area under review before1600 and, furthermore, it is suggested that this practice supports the idea of marked change in farming practice coinciding with the start of the early modern period.

This ties in with Kerridge's assertion that the use of liming increased in scale after 1560 'and still more after 1590' (1967, 248), though the choice of those two very specific dates would not necessarily stand up to scrutiny in the Central Pennines. Whether or not he was correct in writing 'in early modern England private enterprise found half a dozen ways to break this circle (of poverty and want in winter) and with characteristic vigour adopted them all in an agricultural revolution of unparalleled achievement' (1967, 348) is a matter for debate. The changes were indeed profound but it may be premature to describe them as a revolution. After all, change of one kind or another can be recognised right through from the sixteenth to the 'High Farming' era of the mid-nineteenth century.

This writer tends to the view that discussion of a specific Agricultural Revolution verges on the specious and prefers to think in terms of a gradual evolution of ideas and techniques punctuated by episodic technological advances and occasional

[18] Though he first set forth his ideas in an article dated 1885 (see Virgoe 2005, 93).

regression during times of economic downturn.[19] This process can be identified in the sixteenth as much as in the nineteenth century.

8. On improvement and the use of lime beyond England

Agricultural improvement was not a phenomenon unique to this country and similar processes have been identified elsewhere within the British Isles and in mainland Europe. This writer has examined lime kilns in various parts of Europe (in Ireland, northern Cyprus, Greece, Slovakia and Spain) and in northern Morocco, though whether these were producing lime for entirely agricultural use cannot be determined. Lime was also historically important for improving acidic soils elsewhere in the world (see, for example, Day 2007 on its use in Antigua).

The use of agricultural lime has been well documented for Scotland by examination of estate papers (see Dodgshon 1978; Whyte 1979a; 1995a; Whyte and Whyte 1991; and Johnstone 2007) and from more general primary sources. Though the first mention of liming there dates from 1604, in Ayrshire, its use seems only to have been taken up widely either later on that century (Whyte 1979b, 199 and 201; Johnstone 2007, 212) or during the next (Handley 1963, 29 and 69), with knowledge of its benefits having filtered across the border from England. Other contemporary (English) commentators stressed that Scottish farming was less advanced than its southern counterpart at Union in 1707 (see Smout and Fenton 1965, 86) but such comments may have contained a proportion of anti-Scottish propaganda. Smout and Fenton were aware of early sixteenth-century references to liming but they believed it was not generally taken up until after 1750 (1965, 82), a view shared by Dodgshon (1978, 2) who saw the 1760s as the time when liming replaced marling (see Chapter 4.4) as the preferred choice for improving soils. This view is reinforced by the contribution of James Donaldson (1788) who described improvements made in the Carse of Gowrie, upstream of Dundee, 'within the last thirty years' that had involved draining, liming and the introduction of clover.

Generalisation is rarely a safe course to follow and it is probable that the take-up of lime in Scotland occurred at different times from one area to another, depending on pre-existing soil fertility and drainage levels, the degree to which farmers and landowners were progressive in their thoughts and actions, and the ease of accessing new ideas from elsewhere, not to mention access to raw materials.

If the first known reference to liming in Scotland is from 1604, it is remarkably coincident that the first in Wales was from the previous year. In 1603 George Owen published a description of the county of Pembroke in which he wrote 'now ... I will speak somewhat of the naturelle helpes which are found in the countrie to better the lande ... the cheffest thereof I recon the lyme' (quoted in Miles 1994, 67). His use of language is evidence enough that liming was already an established practice in south-west Wales by at least the late sixteenth century. This state of affairs has been confirmed with reports of liming in south-east and south-west Wales before 1600 by Moore-Colyer (1988, 54) who repeated a rhyme penned by Thomas Churchyard in his *Worthiness of Wales*, published in 1587:

> They have begun of late to lime their land
> And plowe the ground where sturdie okes did stand'.

The *Extent of Cemaes* in Pembrokeshire, set down in 1594, was an account of that lordship and it contains a number of references to agricultural lime (Howells 1977, passim). Estate tenants were required to set aside from one to four days annually to carry lime for the lord's (*ie* George Owen's) use as he strongly believed that liming the land would increase estate income and enable him to raise farm rentals.

Emery (1967, 133-34) quoted from a 1616 source from south-east Wales which backs up Churchyard's and Owen's comments, namely that liming of corn was 'first practised within the memory of the fathers and grandfathers of men yet living' thereby extending lime usage here to perhaps the third quarter of the sixteenth century.

Despite these early references to liming in Wales, farming there prior to the Stuart accession has been somewhat denigrated for having been traditional and unimproved, with very 'limited' use of soil additives and 'possibly some lime' along with marl, dung and shell-sand (Gwyn 2006, 38). In fact, Gwyn saw Welsh farming not really taking off before the eighteenth century: perhaps he was being too pessimistic.

In conversation with an archaeologist from the Office of Public Works in Ireland in 2008, this writer was told that almost every farm had a lime kiln for agricultural use in south Donegal, Leitrim and east Sligo. Many more can still be seen in all limestone regions of that country. Lime kilns are, in fact, considered to be Ireland's most common industrial monument, found across most of the island (Rynne 2006, 157). The practice of liming here dates from the mid-sixteenth century when colonisation by the 'New English' set out on a determined programme of radically transforming Ireland's 'neglected' farmlands (Smyth 2006, 3). As these Plantations gained pace and expanded out from the Pale during the earlier Stuart era and the Interregnum more land was brought into cultivation or improved pasture, with lime having played a prominent role, especially on new landed estates. Aristocratic and gentry landowners improved their own vast estates and also encouraged their tenants to do likewise. This culminated in the 1750s by which time the country was 'in the middle of a revolution in ... estate management' with massive 'landlord intervention' powering it all from a desire to commodify the land and turn it all into a profitable business enterprise (Smyth 2006, 418). Despite this, Ireland did not undergo any form of general 'revolution' in farming (Rynne 2006, 187).

Sir William Brereton undertook a series of travels abroad in 1635 and kept a detailed diary (Hawkins 1844). His writings suggest he was impressed by the benefits of liming: he noted its

[19] Beckett (1990, 9) also questioned the validity of thinking in terms of a revolution in farming, for the same reasons as this writer.

use to 'enrich their ground' while pointing out the improvement that could be brought about in other parts of Ireland.

Soon after Brereton, Gerard Boate, also journeying through Ireland, compiled an impressive and detailed natural history of the country.[20] He was impressed with the quality of the grasslands across Galway and Connacht in the west where lime 'doth so warm the ground and giveth it so much strength' (Jordan 2006, 49). As in England a range of soil additives was employed, from pigeon and animal dung to mud dug out of drainage ditches, to sea-sand, in all corners of the country. He painted a very rosy picture of the state of the 'new' farming but, like Brereton, can hardly be described as a disinterested observer, as his express mission was to encourage English settlers to Ireland's new plantations (Rynne 2006, 109-10).

Arthur Young, the notable English Improver, visited Ireland from 1776 to 1778 and was less than complimentary in traducing the state of Irish farming (Young 1780, xii). It was, he said, 'five centuries behind the best cultivated' English farms. His opinion of settler farming, however, was quite different, especially of large estates owned by English aristocrats, but it should be borne in mind that he was opinionated and, to an extent, blinkered in his views (see Chapter 7.3 for discussion of Young's role as an Improver).

As a result of the plantation system, much of the (Catholic) Irish populace was pushed westwards into marginal or unproductive areas, such as Galway and Connacht, and expected to make do. This by and large was impossible but the government did little until the French wars demanded an increase in food production in Britain and Ireland, and various harebrained schemes were put forward for draining bogs and bringing them into cultivation (Robinson 2007, 73-74). Needless to say, peace saw them swiftly abandoned.

Sir Richard Weston of Sutton Park in Surrey was another noted agricultural Improver who had spent time in the Low Countries where he closely observed methods in use there. Clearly impressed, in 1645 he wrote his *Discours* (Weston 1645) and imported from there to his English estate a novel crop rotation system and a new strain of hay grass. His ideas were widely disseminated in this country. It may be that ever-increasing pressure of population in the Low Countries had acted as the catalyst for improvement and, if one takes Weston's account at face value, agriculture across the Channel was significantly more advanced than here. In one letter sent home to his sons, Weston discussed all manner of soil additives with lime mixed with ash being promoted as a valuable means of reclaiming heathland.

This positive picture of agriculture in contemporary Belgium contrasts markedly with an anonymous letter submitted to the *Annals of Agriculture* in 1793 (Anon a 1793). The correspondent noted that Belgian farmers made 'no inquiry towards improvement – no combined system of agriculture; they are chiefly guided by common observations and old practices', which all suggests that farming there had become atrophied to a degree – or that it was so advanced anyway that no further improvement was necessary.

Four years after the end of the Napoleonic wars a comprehensive report was compiled on Flemish agriculture for the Farming Society of Ireland (Radcliff 1819). This was a follow-up to an abortive survey carried out by Sir John Sinclair, a Scot who became a prominent establishment figure in Britain, with the aim of helping to improve farming in Ireland. Radcliff itemised a variety of methods of improvement he had seen in various parts of Flanders, especially *vidange* or *merde* (night-soil), *court graisse* (he should have written *courte graisse*) which was a liquid mix of slurry and rape-cake, and lime. Interestingly, he compared the costs of different additives delivered to the field: dung cost five francs per cart load, ashes eight francs and rape-cake fifteen francs with lime topping the list at twenty-four francs. Unfortunately the value of these data is lessened by his omitting comparative cartage distances and details of crop yields achieved.

The situation in northern France was little different from that in Britain in the early modern period though in the sixteenth century it had been in a sad state owing to the depredations of the Hundred Years War (Cunliffe 2001, 525). Cunliffe quoted a contemporary observer who noted that vast expanses were devoid of people and overgrown with secondary woodland growth leaving fields largely uncultivated. Arthur Young made a similar comment for a later era pointing to the consequences of Louis XV's military campaigns on farming (Young 1769, 39-41). Predictably for an Improver, Young waxed lyrical about the large number of agricultural societies that were springing up across France and about tax concessions that had been introduced for expenditure on improvement. He was also (predictably) impressed with the major role played by the aristocracy in bringing about change in farming.

As a result of his own journeyings through France in the 1780s, he drastically changed his opinion, directing scathing criticism at the lack of investment in the land (Young 1794, 96-102). Much was, to his eye, 'miserably cultivated'. When not castigating French farmers, however, he did make time to observe methods used to improve soils: marl was dominant in many parts; burnt straw, bracken ash and dung in others; with lime widely applied wherever limestone was found.

Whereas Young was totally Anglo-centric, Baron Picot de la Peyrouse (1819, 13) was the antithesis. He wrote a treatise on the state of farming in south-west France where 'tillage was much improved these few years' with a broad range of improvements underway. Lime was not spread on the land, even though it was burned in kilns there and used for steeping cereal seed prior to planting, marl was also found but not applied, and the baron preferred to spread on his pastures what he called *platre marne*, which is basically burnt lime plaster stripped off walls before a new coat was applied. Apart from that, farmers applied only dung. However, in certain parts of the nation land was reclaimed from the waste by *brîse*, or paring as it was known in Britain (Lake 2007, 37).

[20] It was written in 1645 but published posthumously in 1652.

Improvement of French agriculture in a sustained manner has been identified within the period 1850-79 when *défrichement* (taming of the countryside) was underway (Sutton 1977, 250). Until then the ethos of improvement had largely passed France by, and extensive liming only came with the growth of the railway network and was still deemed something of an innovation even in the 1860s and early 1870s (Sutton 1977, 281).

Young's dismissal of farming on the continent did not stop at France. He was equally dismissive of what he had seen in Catalonia and less than complimentary about the state of agriculture in Lombardy, though the time he spent in either was hardly sufficient to draw any worthwhile conclusions. A more realistic impression was provided by a French aristocrat in a series of letters written from his grand tour of Italy in 1812-13; the perceived importance of his observations led to their translation into English (Rigby 1819). He travelled widely across Italy maintaining a keen focus on agriculture and methods of husbandry (which mainly involved the use of manure) but he had nothing to say about attempts at improvement other than to comment that on the Roman *campagna* the farmers 'make no attempt at change or improvement' (Rigby 1819, 150).

9. Conclusion

The purpose of this chapter was to review existing literature on the historic use of lime in agricultural improvement and the broader context within which it sits. To achieve the latter aim it was necessary to consider writings from a number of academic perspectives, in particular landscape history and archaeology, historical geography and agricultural history. As Rippon and Gardiner (2007, 234-35) have so succinctly suggested, in landscape studies there has been a palpable synergy between the first two disciplines, which must not only be upheld in the future but should be strengthened.

The burning of lime in the Central Pennines in past eras is immediately obvious to even the most casual field observer in the form of lime kilns dotted across the landscape, but signs of the application of lime on the land are less apparent. It requires a deeper understanding of landscape processes to make sense of an improved landscape, just as it needs archaeological excavation to appreciate the full geographical extent of such practices within the more intensively farmed fringes of the uplands. Much of the research area has been subjected to determined and organised improvement and lime played a crucial role in this process, in association with other forms of land management such as marling, draining, paring and burning and the development of new crop and stock strains.

Agricultural improvement has often been seen in the context of parliamentary enclosure and the pressures of the early nineteenth-century French wars but in fact it extends far back beyond then. What was happening in the Central Pennines was not unique but part of a more general trend and the use of lime here was paralleled by its use in other limestone-bearing regions. Much of the fine detail (whether field, sub-surface or archival) has hitherto remained hidden and unrecognised while the corpus of published material on liming is conspicuous by its absence, not just within the research area but across the country. Historical evidence is scant and, as will be shown in the next chapter, some written and documentary sources need to be understood as possibly biased, and certainly not disinterested, accounts. It is an old adage to state that political and military history is written by the victors: it is similarly the case that much agricultural and industrial history was compiled by those at or near the top of the hierarchy. Those lower down the scale tend to be invisible and silent, but their contribution to lime burning and land improvement can still be teased out.

It will have become clear from the foregoing review that there are many gaps in our knowledge of agricultural lime burning largely because its significance as an improver and in the basics of rural life has not yet been widely realised. Only minimal research has been carried out, especially in the Central Pennines, and the resultant absence of a sound knowledge base can lead (and has led) to errors of interpretation and judgment and to the perpetuation of misconceptions.

Few writers of agricultural or landscape history have acknowledged the key role that liming played in land improvement and most of those that did tended to tie it in with the period of parliamentary enclosure (see, for example, Havinden 1974, 130; Dent 1983, 91, who stated that lime kilns mostly date from the enclosure era; Whyte 2003a, 71). Others have placed liming in a slightly earlier period associated with piecemeal encroachment and enclosure on a temporary basis in the eighteenth century (Darby 1976, 50-51). Darby provided a set of data to support this view comprising a list of documentary references to marling and liming around 1800: Westmorland had four references for lime and none for marl, north-east Lancashire two and one respectively, Yorkshire within the current research area two and none (Darby 1976, 116). It is all too easy to criticise others, but these figures must be rejected as distortions of reality, as will be seen in ensuing chapters.

Yet others have placed the beginnings of liming in the seventeenth century (for example Dilley 1991, 526-27; Wrigley 1993, 43). The former has gone on record by repeating two misconceptions, saying that lime was the only soil input before 1700 (it was definitely not) and that it was a fertiliser in itself (strictly speaking it is not a fertiliser but acts as a catalyst for fertilisers).

At least, though, none of these writers ignored the contribution that lime has made over the years unlike one, otherwise thorough, landscape history of Swaledale whose author confessed to ignoring lime kilns (Fleming 1998, 158); and one book on medieval archaeology that ignored them altogether despite devoting an entire chapter to rural crafts and trades (Steane 1984).

Dyer (2005, 244) stressed the usefulness of recognising specific 'turning points' in making us aware of 'chronological benchmarks' that separate one era from another. Inventions of new tools and machines, developments of new crop strains and stock breeds, the introduction of tile drains, changes in landlord-tenant relationships, can all be more or less pinned down to their own decade. Breakthroughs in soil management are harder to isolate and this has perhaps contributed to the dearth of in-depth research on liming.

Some of the more obvious elements of improvement have captured the collective imagination since the heyday of the Romantics, not the mundane and imperceptible de-acidification of soil or the slow change in grass cover, but the sweeping schemes in the 'landscape gardener mould' (Darby 1976, 44-49). This is the aesthetics of improvement, the embodiment of grandiose taming and transformation of landscapes steered by the likes of Vanbrugh, Repton and Lancelot Brown and so beautifully satirised in Jane Austen's *Mansfield Park*.

This book does not concern itself with such aesthetic considerations. Rather, it sets out to fill in the many gaps in our knowledge. It will demonstrate that liming has been practised in and around the Central Pennines from before 1600 and that it has indeed had a key part to play in transforming fells and valleys into productive farmland.

Chapter 3
Research Methods

1. Introduction

Chapter 1 discussed the broad aims of the book and Chapter 2 reviewed relevant literature. A number of points were raised that are deemed germane to the various source materials used. Firstly, existing literature has tended to associate liming with the period of parliamentary enclosure, roughly from 1750 to 1850, whereas the practice can be traced with confidence much further back in time.

Secondly, following on from the notion that most liming was an integral part of modern enclosure, it has generally been assumed that lime was used to bring into productive agricultural use land that had hitherto been regarded as waste, in other words moorland and upland fell. It will be shown that liming also occurred, at a much earlier period, in the lowland fringes of the Central Pennines and within the major valleys, probably as part of a drive to bring 'new' land in such areas into production.

The third issue concerns the relationship between the location of lime burning and the proximity to limestone reserves. Again, it has generally been thought that kilns were sited where limestone outcropped on or near the surface, thereby linking stone getting and lime burning in a vertically integrated structure. The corollary of this was that the fuel probably had to have been brought in from elsewhere. Again, it will be shown that a considerable body of evidence supports the view that kilns were often sited close to the source of fuel, and that access to affordable fuel was a significant factor in determining whether (and when) lime burning could be practised. It will also be shown that, in certain parts of the research area, liming was done far from any limestone bedrock and that arcane methods were employed to source the raw material: these included hushing (normally associated with lead mining) whereby water was channelled to loosen and flush out limestone cobbles from glacial till at the foot of hillslopes; river stonary, extracting limestone cobbles from within river channels; and limestone delfs, picking limestone cobbles or lime-rich clay from shallow pits within deposits of glacial till on moorlands. The significance here is not the technology *per se* but the fact that such methods illustrate the extreme lengths people went to in order to obtain agricultural lime.

Finally, there will be a focus on the issue of how widely within the farming community the benefits of liming were understood and promoted. We will consider who the prime movers were and demonstrate that it was not just landowners and estate managers but also tenant farmers and independent yeomen and husbandmen who fully appreciated the importance of lime in soil improvement.

As explained and justified in Chapter 1.2, this research essentially adopted an empirical approach and the discussion and conclusions in the following chapters will be firmly evidence-based. The sources of the evidence will be detailed in Section 3 of this chapter after a discussion in Section 2 of the methodology employed. Section 4 is concerned with limitations and constraints.

2. Methodology

Spatial extent

In order to reach conclusions that could stand up to detailed scrutiny it was deemed necessary to define a research area on a regional rather than a local scale. Unsafe and unsubstantiated results could all too easily be teased out of areas too restricted in extent to have a substantial corpus of relevant data. Likewise, the decision was also made not to select an artificial area such as a county or district or the Yorkshire Dales National Park. County boundaries since reorganisation in 1974 do not correspond to those applicable in earlier times and present administrative districts are a relatively new phenomenon. Much of the boundary of the National Park does not conform to either administrative or topographic lines and as such the Park is a totally artificial construct.

The research area was therefore delineated to include the Central Pennines, which by and large correspond to the Park, with those parts to the east, south and west that are geomorphologically tied to the Pennines, so the major lowland river plains of the Swale, Ure, Wharfe, Aire, Ribble, Wenning and Lune were incorporated. Historically, large landed estates extended from these lowlands into the heart of the Pennines, or were dependent on them for supplies of kiln fuel or agricultural lime. This was not so much the case in Nidderdale, however, which historically tended to look east to Ripon and Harrogate rather than to the uplands of the Dales, hence its omission from this research programme.

Time frame

Similarly, this research set out to show that liming in the area was not just practised as part and parcel of the parliamentary enclosure process but had a much longer history. It therefore required as long a time frame as practically possible. Clearly, the further back one searches, the sparser are the written sources so the late thirteenth century was chosen as the symbolic starting point of recognisable land management. No relevant monastic charters or manorial documents have been located for the research area in the centuries prior to the late thirteenth century, whereas documentary evidence for the opening decade of the fourteenth century has been sourced. By the end of the nineteenth century farming had changed beyond recognition compared with the beginning of that century, and farming in the Central Pennines was mired in severe depression from the 1870s. The year 1900 is set as a symbolic and convenient time marker.

Examples

A region as geologically, topographically and climatically diverse as the Central Pennines, when viewed across 600 years, is bound to display differences in the detail of land management and agricultural improvement. Differences in geology are reflected in soil type and drainage regime; differences in topography are also reflected in soil characteristics as well as in weather. These have an impact on land capability and soil productivity. In turn, it all affects what palliative treatment a given soil requires. To take account of such variations a number of case studies from across the research area have been selected (see Chapter 1.4 for their location and justification).

Inevitably, any selection of examples will be flawed to a certain extent, and the actual choice is necessarily dependent on the availability of evidence, namely on the survival of archival material or on the visibility of field and archaeological clues. There are spatial gaps within the research area where, for instance, there were no landed estates and therefore minimal documentary evidence of what was going on in the past. After all, it is rare for yeoman records to survive, partly because few yeomen kept them. There are also temporal gaps and this applies particularly to the post-medieval era between the ending of monastic accounts and the widespread inception of organised estate records in the later seventeenth century.

Case studies by definition can only give an incomplete picture and it is important not to over-generalise about improvements being made across the whole region if the evidence is patchy in area and time. On the other hand, if conditions are very similar from one area to another (say at different localities within one dale) generalisations within limits could well be valid. This point will be revisited in Section 4.

Gaps in previous research

The point was made in the previous chapter that very little work has been published about the Central Pennines on liming or lime burning, and what has been placed in the public domain has mainly concentrated on gazetteer-type or site-specific accounts with the minimum of historical context or indications of what the lime was used for. It would therefore be more realistic to state what has been written rather than to explain what gaps exist. In terms of liming and historic methods of soil and farmland improvement in the research area, the field is wide open in both spatial and temporal dimensions.

The state of agriculture in the Central Pennines

As discussed in Chapter 1.5, one quarter of the Yorkshire Dales National Park consists of acidic grassland where soils have a pH of less than 5.5 and this can be even lower in those parts of the research area where sandstone is the dominant bedrock. The ideal pH for pasture land is above 5.5 or 6 (MAFF 1981, 25). With only minor and isolated exceptions farming in the research area is based on stock (store cattle and sheep) rearing.

No data are available for the nineteenth or earlier centuries but lime deficiency in the period 1918-39 was estimated at eighty per cent for uplands in England and Wales with around 7.3 million hectares having been classified as deficient (Robinson 1943, 139). The situation must have been very similar in the past on land where lime was not applied and during periods when its usage was generally minimal. Hard data may be lacking but the lime requirement of vast tracts of the research area was inevitably rarely met unless lime was applied. To raise pH from 5.5 to 6 over a two-year period requires an application of lime at the rate of 5.65t per hectare; to increase pH to 6.4 requires double that amount. Until the last century no farmer or scientist would have been aware of such terms as lime requirement or lime status, but they emphatically would have recognised that lime-deficient land was not producing at the optimum level.

The recognition that soil potential could be augmented, by liming or other inputs. has relevance for the earlier debate about determinism and possibilism (see Chapter 2.2). Determinists would tend to argue that acidic soils could not be improved, or at least that farmers in the past would not have been aware that they could, whereas possibilists would accept without question that farmers could intervene to boost soil potential – as indeed they did.

On average 450-560kg per hectare per annum of lime are leached out of the soil in upland areas so an average input of 2.25 tonnes per hectare per decade is needed to restore soil's optimum lime status.[21] Farming in most of the Central Pennines is beset by a host of natural adversities – high levels of cloud cover, low average temperatures, high precipitation totals, inadequately drained soil, and steep slopes – and farming at times still appears to be marginal, so it is in farmers' best interest to maintain soil fertility as much as possible. Current farming practice sees stock farmers regularly spreading slurry on pastures but this can render the soil 'sewage sick', which means it has become acidified and in need of neutralising – liming at 4.50 tonnes per hectare would rectify this.

Examples can be seen across the region of formerly improved fields which have not been maintained and are slowly reverting to their natural state as unpalatable soft rush (*Juncus effusus*) replaces grass species as the dominant growth on wet acid ground, or bracken (*Pteridium aquilinum*) on dry acid ground. Figure 3.1 illustrates both processes in Kingsdale, north-east of Ingleton. Invasive bracken has squeezed out all other species to the left (west) of the road while rushes are expanding their area bottom right and on the valley floor in the middle distance.

[21] The actual liming rate has been given as 2.24 t per ha in liming experiments begun in 1883 (Thurston *et al.* 1976, 1047).

Figure 3.1 Degrading improved pastures in Kingsdale, North Yorkshire.

The methods employed in this research, and the sources consulted, have been designed to investigate how management practices since c. 1300 have dealt with the problem of lime-deficiency, long before anyone was aware of soil chemistry or how lime modified soil characteristics.

3. Sources

Summary

The research programme incorporated a number of approaches. Comprehensive archive-based searches were undertaken of primary documentary material including manor court rolls, monastic *compoti* and cartularies, enclosure awards and maps, tithe apportionments, collections of papers from landed estates, and miscellaneous cartographic and documentary material likely to be of relevance, including early Ordnance Survey mapping. In addition, letters and contributions of contemporary writers, submissions to the Georgic Committee and to the Board of Agriculture were searched. Each will be discussed more fully along with an analysis of their relative virtues and drawbacks. In all, over thirty archive repositories were visited ranging from national collections through county record offices, learned societies and publicly accessible family muniments, to collections in private hands (Table 3.1).

A second strand of research involved consulting contemporary published material from the monastic period to the end of the nineteenth century: agricultural treatises and manuals, diaries and observations written by travellers, journals and Board of Agriculture reports.

A third strand required field survey. Over the past fifteen years this writer has undertaken comprehensive field surveying of lime kiln sites across the vast majority of parishes within the Yorkshire Dales National Park as well as in contiguous areas to the east, south and west. The initial starting point was the Ordnance Survey First Edition six inch map series which marked many kilns together with documentary and other cartographic sources. Other sites, though, were identified by field walking.

The final strand was archaeological excavation of early modern clamp (sod, or sow) kilns across Craven, supervised by this writer. In all, seven such sites were investigated directly with a further seven having been examined by him on an advisory basis for the commercial contractors charged with all archaeological work along the wayleave of a new gas pipeline, between Addingham in Wharfedale and Austwick, in 2006.

Documentary research

The court rolls or records of sixty-four manors were consulted covering the period from 1331 to 1900, though many manors have only patchy survivals. In the vast majority of cases court rolls were read in their original parchment form, often in Latin for earlier series. The rolls for fifteen manors (twenty-three per cent) contain relevant material, thereby making the process

Category	Repository	Location
National	National Archives	Kew
	British Library	St. Pancras, Boston Spa
	British Museum Library	Bloomsbury
County	Cheshire	Chester
	Cumbria	Carlisle, Kendal
	Lancashire	Preston, Wigan
	Norfolk	Norwich
	Sheffield	Sheffield
	North Yorkshire	Northallerton
	West Yorkshire	Bradford, Leeds, Wakefield,
Societies	Royal Society	The Mall
	Royal Ag. Soc. England	Stoneleigh
	Yorkshire Archaeol. Soc.	Leeds
	Northern Mines Res. Soc.	Sutton-in-Craven
Libraries	Borthwick Institute	York
	Bradford University	Bradford
	Brotherton Collection	Leeds
	Durham University	Durham
	Ironbridge Gorge Museum	Ironbridge
	Local Studies, Cliffe Castle	Keighley
	Lancaster Univ. RBA	Bailrigg
	Skipton Reference	Skipton
Private (accessible)	Devonshire Collection	Bolton Abbey, Chatsworth
	Kiplin Hall	Kiplin
	Towneley Hall	Burnley
Private (closed)	Garth Day Books	Swaledale
	Hornby Castle Muniments	Hornby Castle
	Ursula Towneley	-

Table 3.1 Archival sources accessed during the research

worthwhile: six have entries dating from the sixteenth century, six from the seventeenth, seven from the eighteenth and one from the nineteenth. In addition to direct entries concerning aspects of lime production, other court records refer to early encroachments on to the waste and to early improvements of farmland, as will be shown in the next chapter. Manor court rulings have helped to paint a fuller picture of earlier agricultural lime usage where no other written records are likely to have been kept.

Much of the research area was held as demesne or tenanted land by monastic foundations and several of these have surviving cartularies (registers of charters) and *compoti* (account books) which contain material of relevance to the work in hand, in particular the Cistercian foundations at Fountains and Furness, the Augustinian priory at Bolton and the Premonstratensian house at Cockersand. Relevant items from the monastic record are usually the only local material available for the medieval period and are thus invaluable.

Variable	% of total sample[1]	Number of awards
Provision for lime burning	14.7	22
General use of limestone	30	45
No provision, but kilns built	25.3	38
Not on limestone	16.7	25
Stinting agreements	3.3	5
No relevant data	18.7	28

Table 3.2 Summary of enclosure awards

[1] the total does not add up to 100 per cent as there is overlap between rows.

Enclosure awards and maps have been found equally invaluable for building up a detailed picture of lime usage especially after c. 1750. Every available award within the research area was consulted, both documents lodged in record offices and accessible collections and a number in private hands. In all, 150 awards were consulted along with corresponding maps where available: Table 3.2 summarises their elevance to this research.

One-fifth can be discounted immediately as they were either stinting agreements, with no provision for improvement or carving up, or were concerned with areas not on or anywhere near limestone resources (Table 3.2). One quarter of the agreements made no actual provision for getting or burning limestone but field surveying has established that kilns were, at some point, constructed in all of these enclosed areas. As will be seen in Chapter 7, it is considered valid to assume that many of these kilns were directly associated with the enclosure process.

A third of all awards did make specific general provision for householders within each township to exploit limestone reserves, in allocated public quarries, for their own use but not for sale. Some awards stated that the stone could be used for building new field walls or for repairing existing walls and buildings, or for road making; others simply allowed the exploitation of limestone as people saw fit, which could have led to burning for lime.

The final category (14.7%) stated that quarries were to be set aside for the prime purpose of getting stone to burn into lime for improvements to the newly enclosed land.

Analysis of all enclosure awards indicates that sixty-one stipulated that lime was to be burned and/or had kilns erected at some point, either directly or indirectly, with the years 1694 to 1884 being covered by these awards. Two bi-decadal periods stand out (Table 3.3, column 2), each having nearly twenty-three per cent of the sample – 1801-20 and 1841-60 – though the hundred years between 1760 and 1860 are well represented overall. Table 3.3 shows a clear positive correlation between the total number of enclosure awards within the research area (column 4) and the number with lime burning associations (column 3). In percentage terms (columns 2 and 5) the periods 1801-20 and 1841-60 are almost identical, with 22.95 and 22.8 per cent respectively for the earlier periods and 22.95 and 21.5 for the latter. The significance of these data will be explored in Chapter 7.4.

	Awards with provision/kilns built as % of total sample	**No. of awards with provision/ kilns built**	**All awards (no.)**	**% all awards**
Pre-1760	8.20	5	10	6.7
1760-80	14.75	9	21	13.4
1781-1800	13.11	8	20	13.4
1801-20	22.95	14	34	22.8
1821-40	14.75	9	20	13.4
1841-60	22.95	14	32	21.5
1861-80	3.29	2	8	5.4
1881 -	-	-	5	3.4
		n = 61	n = 150	

Table 3.3 Date of awards with provision for lime burning and/or with kilns erected

Tithe apportionment awards and accompanying maps have been found less useful in identifying the distribution of kilns or the relationship between kiln sites and improved fields. A representative sample of such awards across the research area was examined but few maps depicted kilns and the impression was gained that surveyors might have included kilns through personal choice: no evidence was found to suggest that the presence of a kiln in a given field affected the level of tithe commutation of that field. On the other hand, tithe maps have provided detail of field names with the words lime or kiln in them. One such example is the award for Wray with Botton in the Wenning Valley (NA. PRO. IR30.18/349)). A modern track, called Kiln Lane, leads north from Wray village and heads towards a meadow named in the tithe schedule as 'Lime Kiln Giffils': the remains of a kiln can still be seen (at SD5986 6824).

It has been suggested that tithe maps can be used to tentatively date kilns (Cleasby 1995a, 5) if they are compared with enclosure or Ordnance Survey First Edition maps. The justification is that if a particular kiln is shown on one map but not on an earlier one, it could have been built in the intervening years. This view is considered untenable as this writer's fieldwork has shown that Ordnance Survey surveyors occasionally omitted (presumably missed) some kilns that were marked on earlier enclosure or tithe maps.

The archival element of this research programme included a thorough search of cartographic sources other than enclosure, cadastral and Ordnance Survey mapping. The rationale for this was twofold. Firstly, some maps commissioned by landowners marked kilns, or fields with kiln names, as did a proportion of maps of lead and coal mining grounds (see Appendix 1). In both cases the location of these kilns was correlated (through fieldwork) with the level of improvement of surrounding land. In addition, many surveyors of estate maps were required to include field names. In all, 116 estate and mining maps were examined as well as eight surviving field valuation surveys. All instances where field names indicated the past existence of a lime kiln were noted and checked in the field, except for fields now under forest plantation. In many instances field surveying failed to find any ground evidence of kilns in these fields. Nor were many of them depicted on Ordnance Survey First Edition maps. Both of these observations suggest that the kilns were early forms which had disappeared by the time of Ordnance Survey field mapping.

Typical of the surveys is a field book compiled for the earl of Thanet in 1823 in the lordship of Silsden south of Skipton (YAS DD 121/Add/6). The book named every field on all the estate's tenanted farms: five separate fields were called Lime Kiln Close. Smith (1961) recorded twelve discrete forms within the West Riding part of the current research area that included the appellation 'Lime Kiln' along with such descriptors as 'Close' (twenty-five entries), 'Hill' (three), 'Field' (four) and 'Plantation' (five), the latter suggesting a recent change of land use since those particular kilns were built.

Surviving legal papers can also shed some light on the distribution of kilns, and thus lime application patterns. Kilns are rarely mentioned in agreements and it is more likely that a kiln or adjacent quarry would have been subsumed within the generic legal phrase 'and all other appurtenances'. A kiln may well have been perceived as having insufficient monetary value or significance to warrant a separate mention. What does happen, though, is that the current or former presence of a kiln is often indicated by mention of field names in estate surveys and legal agreements.

One such example from a legal agreement is seen in a conveyance in 1652, from Sir Christopher Wyvill of Constable Burton Hall in Wensleydale to Anthony Durham, a yeoman from nearby Spennithorne, of various parcels of land including one called 'Lyme Kilne Garth' (NYCRO ZQH 2/2/4). Similarly, the same estate was disposing of lands at Spennithorne in 1790, including one field known as 'Lime Kiln Close' (NYCRO ZQH 2/2/165). Prior to this, in 1774, an entire farm at Spennithorne was up for auction with specific mention of a 'lime stone quarry with a kiln' thereby suggesting, perhaps, that it was a substantial working kiln producing lime for the market rather than being a simple field kiln for on-farm use (NYCRO ZQH 2/2/124). Similarly, an indenture from 1783 concerned various tenements and parcels of land in Rathmell, south of Settle, all of which were named in the agreement (WYAS [W] 1783). One such field was named 'New Limed' providing conclusive evidence that it had recently been subjected to improvement by liming. Had it just been top-dressed, that appellation would not have applied.

Of far greater value than simple legal agreements and estate maps have been collections of estate papers. The eastern fringe of the Central Pennines was well endowed with landed estates, some spasmodically occupied by local gentry, others in the hands of absentee aristocratic landlords. The latter invariably employed an estate steward charged with running the estate on a day to day basis, ensuring that tenant farmers practised sound husbandry, collecting rents twice yearly, and promoting the landowner's interests by fostering improvement. Part of his duty was to keep the master fully informed of the local situation, a task achieved by the writing of weekly letters. For the larger estates these have survived more or less intact and they provide the researcher with a valuable insight into how the steward was thinking and how successful his endeavours had been. On the other hand, it is rare for the landowner's replies to have survived in the public domain so the picture we have is only partial. Reading between the steward's lines can hint at his master's views and wishes but it is sometimes difficult to decipher the full story.[22] The letters can, though, provide detailed information concerning liming rates and patterns, rental allowances for liming carried out, and methods employed.

The Swinton Park estate near Masham operated coal mines in Colsterdale and from 1796 to 1851 detailed lime-coal accounts were kept. Fortunately these have survived in their entirety and have provided a great deal of material on various aspects of liming. These have been used (see Chapter 6.6) to analyse and understand seasonal, annual and longer-term patterns in lime usage. They are considered to be totally reliable and without any form of bias. Though, at first sight, they may seem to be just piles of (uncatalogued) weekly and monthly production and sales statistics, as dry as the coal dust that still coats them, they have enabled analysis that would otherwise have been impossible.

Some estate paper entries can be rather terse yet can still provide useful detail. As an example, a feoffment, dated May 1623, recorded the conveyance of several closes at a farm called Deanfield Slack, in the Worth Valley (SD996 376), including two called 'Old Lymde' and 'New Lymde' (WYAS [B] HEA/A/346). A further conveyance, dated March 1739, concerned the same two closes, by then spelled as 'Old Lymed' and 'New Lymed' (WYAS [B] HEA/A/462). The field names must show that these two fields had been limed, possibly brought into production by liming, and the very fact that they were called 'Lymde' may be taken as evidence that the practice of liming here was worthy of note. The inclusion of old and new in the earlier conveyance could conceivably push back the liming of 'Old Lymde' to before the start of that century.

A valuable (and unexpected) source of data has been seventeenth-century post-mortem inventories for Ainstie and Craven in Yorkshire and for the manor of Colne and the Forest of Trawden in north-east Lancashire, with wills covering a longer span of years (see Chapter 5.5 and 5.6). Items listed and given a cash value in inventories include a 'lyme delfe', 'limes stones', '163 loads of lime' and 'in lyme'. As will be seen in Chapter 5, the early date and geographical spread of these inventories point up the importance of liming within the research area during the seventeenth century. Having said that, however, a set of 210 wills covering the period 1522 to 1600, for Swaledale, produced not a single entry relevant to liming and only one oblique entry to the application of dung (Berry 1998). In a very late will, dated 1840, for William Procter of Close House to the west of Settle, there is a list of enclosures bequeathed to his eldest son: one of them was described as 'the first Scar with the Limekiln thereon' (WYAS [W] vol. 202, f. 119). This says nothing about whether the kiln was still in use, or if it had produced agricultural lime, but it does indicate the existence of a kiln in a purely gritstone area.

In the latter years of the Elizabethan era the inception of a new way of thinking can be identified that led to profound changes in English society. First seen in poetry and then in science and religious reform, the Georgic 'revolution' fostered a mode of thinking and behaviour that stressed 'the value of intensive and persistent labor (*sic*) against hardships and difficulties' (Low 1985, 12). The movement had an impact on all aspects of (knowledgeable) society and its effects have been recognised in increased rural prosperity and agricultural innovation. After the brutalities of the Civil War came the foundation of the Royal Society, in 1660, and the creation of the long-lasting Georgical Committee charged with promoting all that was innovative and morally acceptable in farming. Members of this committee encouraged submissions from those involved at the cutting edge of farming, and responses have been preserved in their original form in the library of the Royal Society (Royal Soc. CL. P/10 iii). Coverage across England was patchy but rare insights into farming in Lower Wharfedale and around Ripon can be gleaned from two letters to the committee's successor, the Georgic Society. It could be argued that these responses are of extreme importance to the story unfolding in this work as they came from ordinary farmers and not from a representative of the gentry or aristocracy.

The Society was established around 1793 with the initial aim to 'promote and extend to a knowledge of the theory and practice of Agriculture and Husbandry, and the arts which

[22] See Chapter 1.4 for details of landed estates selected as case studies for this research project.

have a tendency to improvement thereof ...' in Britain and Ireland (Anon b 1793).

Some yeoman families maintained farm records, sometimes on a daily basis, that can be referred to as Memoranda Books (Turner *et al.* 2001, 43). Such books have been dismissed as 'repetitious jottings about the mundane and infuriating nonchalance about the significant' (Fox and Butlin 1979, 7), but an opposing viewpoint has been expressed (Turner *et al.* 1996, 27 and 29), namely that if such records have survived (and this is considered exceptional) they must represent large entrepreneurial farmers and owner-occupiers rather than lowly tenants. This certainly applies to three generations of the Garth family who farmed in upper Swaledale and kept a day book with continuous, if concise, entries for every day between 1795 and 1911 (Barker MSS). In addition to mundanely relating daily farm chores, and sometimes household chores, the day books are rich in detail of land improvements carried out on the farm and of the application and sourcing of agricultural lime, as will be discussed later (see Chapter 8.6). The Garth records are anything but 'infuriating'.

The Fox family of High Tatham in north Lancashire kept what they called a 'commonplace diary' in the later eighteenth century and, though it is somewhat chaotic and disorganised, it does provide useful snippets (Fox, unpub.).

The Board of Agriculture

As the eighteenth century drew to a close, with the French wars looming, government came under increasing pressure to formalise its efforts to promote agricultural improvement by instituting the Board of Agriculture. Formed in 1793, in the face of determined opposition from within the ranks, including from Prime Minister Pitt, the Board was never able fully to meet its obligations or fulfil its founding objectives (Virgoe 2005, 96-98). Inadequately financed, dominated by the great and the good rather than by the appropriately experienced, and with ground surveyors of variable quality, it very soon abandoned its ambitious plan to produce detailed parish by parish accounts in favour of opting for county-based surveys. Eighty such surveys were submitted to the Board with the first pilot surveys published as early as 1794. Though they are of variable quality and thoroughness (Turner *et al.* 1996, 22), they did adopt a common format and thus permit county and district comparisons to be made.

To compound its inbuilt woes the Board was bedevilled by incessant infighting between two of the most influential Improvers of the time, Arthur Young and William Marshall (see Chapter 7.3c). The Board was wound up in 1822 (Chambers and Mingay 1966, 121) having failed to live up to initial expectations. Despite this, the four county reports relevant to this research – West Riding, North Riding, Westmorland and Lancashire – rate highly and as a result have considerable importance in helping to understand the state of farming at the turn of the eighteenth century.

Contemporary published material

There is a large corpus of contemporary written works on farming covering the whole period under review and, not surprisingly, there is a direct correlation between the number of published books and journals and the scale of land improvement. Over 100 have been consulted for this research: only two are medieval while eighty-three post-date 1700. They are summarised in Table 3.4.

The earliest surviving agricultural treatise was written by Walter of Henley, probably in the mid-thirteenth century, and he emphasised the value of marling and manuring (see Chapter 4.4) (Lamond 1890, 19-21; Oschinsky 1956, 302), as did a treatise published during the reign of Edward I, known as 'Seneschaucie' (Lamond 1890, 91). Oschinsky (298-99) noted that similar treatises were produced for estate managers of the Duchy of Lancaster in the fourteenth and fifteenth centuries.

Fitzherbert[23] was the first post-medieval proponent of improvement and his two works, *The Boke of Husbandrie* (1523) and *The Boke of Surveyeing and Improvements* (1539), ran to many editions over the following centuries.[24] Thomas Tusser's pioneering work on agriculture, issued in 1557, was reprinted thirteen times before 1600 and a further nine times up to 1812. The German Conrad Heresbach's book was first published in English in 1577, and again in 1601. Each of these early works has worth by illustrating the value attached to land improvement in its time, though it is doubtful if they were read by more than a handful of the landed elite. More works appeared through the seventeenth century, especially after 1650 as emphasis on the value of labour and the influence of the Georgic movement gathered pace. Of twelve books consulted, eight placed emphasis on marling while nine discussed the role of lime in agriculture, with Camden's *Britannia,* published in 1610, being the earliest to unequivocally highlight the application and value of lime to soils.

As Table 3.4 shows, the quantity of published works increased after 1700 but rose quite dramatically after 1750 (1700-1750 n=3, 1751-1800 n=42). All but two of these books considered liming to be important in either winning new land from the waste or improving existing farmland, or indeed both. The situation in the nineteenth century is a mirror image of the eighteenth with a distinct tailing off after 1850 (1801-50 n=32, 1851-1900 n=7). The reasons for this will be discussed in Chapter 8.3 and 8.4, but suffice it to say here that the increased availability of artificial soil inputs from mid-century and the onset of severe agricultural depression in the 1870s made serious inroads into the practicality and cost-effectiveness of liming. This situation is reflected in the volume of published material on improvement in general and liming in particular.

---

[23] There is disagreement as to whether he was John or Anthony, so Master Fitzherbert is often used.

[24] Full referencing for primary works noted in this section will be provided in the following chapters.

Published	Number of books
Pre-1600	6
1600-1699	12
1701-10	0
1711-20	1
1721-30	1
1731-40	1
1741-50	0
1751-60	4
1761-70	7
1771-80	9
1781-90	6
1791-1800	16
1801-10	5
1811-20	8
1821-30	2
1831-40	4
1841-50	13
1851-60	4
1861-70	1
1871-80	1
1881-90	1
1891-1900	0

Table 3.4 Published agricultural works consulted for this research

In addition to treatises and general rural 'cyclopaedias', the landowning and farming classes were bombarded with agricultural journals: Goddard (1983) identified seventy-seven such titles in print between 1768 and 1880, some very short-lived but others that have endured to this day. Most appeared at the turn of the eighteenth century, though Goddard (118-19) concluded that the greatest number in print at any one time came after 1830.

For the purposes of this research nine journal titles were searched for articles and correspondence on liming and land improvement:

1. *Annals of Agriculture*, published from 1784 to 1815,[25]
2. *Journal of the Royal Agricultural Society of England*, first published in 1840,
3. *The Agricultural Magazine* published between 1799 and 1811,
4. The *Transactions of the Highland and Agricultural Society*, published since 1799,
5. The *Transactions of the Yorkshire Agricultural Society*, also published since 1799,
6. *The Farmer's Magazine*, published in Edinburgh from 1800 to 1837 and in London until 1881,
7. The *Monthly Magazine* and
8. The *Lonsdale Magazine*, which were both short-lived publications of the first decade of the nineteenth century,
9. *Letters and Papers* of the Bath and West of England Society, first published in journal form in 1780

As suggested earlier, it is not possible now to determine how broad a readership such journals reached, though the very fact that titles came and went with some frequency may indicate that they either failed to reach their intended audience or did not provide what farmers wanted. One surviving account book for a small estate in Swaledale does show that some at least were keen to broaden their knowledge (NYCRO OUTFAC). One (undated) item included the purchase of various books, namely *Book of the Farm* (author not given), eight volumes of the *Journal of Agriculture*, and two technical works on farming by (J F W) Johnston.

As will be shown in later chapters, many of the books and articles were written by Improvers who can hardly be described as dispassionate. They had a very firm agenda to push agricultural improvement and thereby to suggest that time-honoured methods were in dire need of replacement if farming was to achieve the heights of productivity desired. They have been described as propagandists who displayed clear bias in their writings (Horn 1982; Whyte and Whyte 1991, 55; and Whyte 1995a, 136, 143), and it is difficult to disagree with such sentiments though perhaps one should give them the credit for having been sincere in their drive to create a new countryside and a new era in farming.

One of the most prominent and prolific Improvers was Arthur Young who has been labelled 'perhaps the greatest agricultural writer of all time' (Gazley 1973, 16) and who maintained a missionary zeal until his death in 1820, despite the fact that he failed to make a success of either of his own farming projects.[26] Gazley's view has not gone unchallenged: others have cast doubt on the role he and other Improvers played and have accused Young of being simply wrong at times (Allen and Ó Gráda 1988). His creation, the *Annals of Agriculture*, which Young hoped would be ground-breaking, rarely exceeded a circulation of 350 (Mingay 1975, 18).

Contemporary treatises, in book, article or letter form, cannot be dismissed out of hand though they should be viewed with a degree of circumspection. They were issued with a genuine desire to bring about change and they do provide the researcher with an insight into farming practice and ways of thinking in the past. Whether or not they achieved the stated aim of the Gilbey Pamphlets, namely to reach the 'ordinary class of practical farmer', is a moot point and it could be argued that many small farmers had neither the disposable cash nor the education (or the inclination?) to subscribe to such

[25] The last volume (no.46) was begun in 1809 and only finished in 1815.

[26] Young had good reason to favour lime. On a tour of France in 1789 his chaise came off the road and nearly fell over a precipice but he was saved by some nearby limeburners (Gazley 1973, 239).

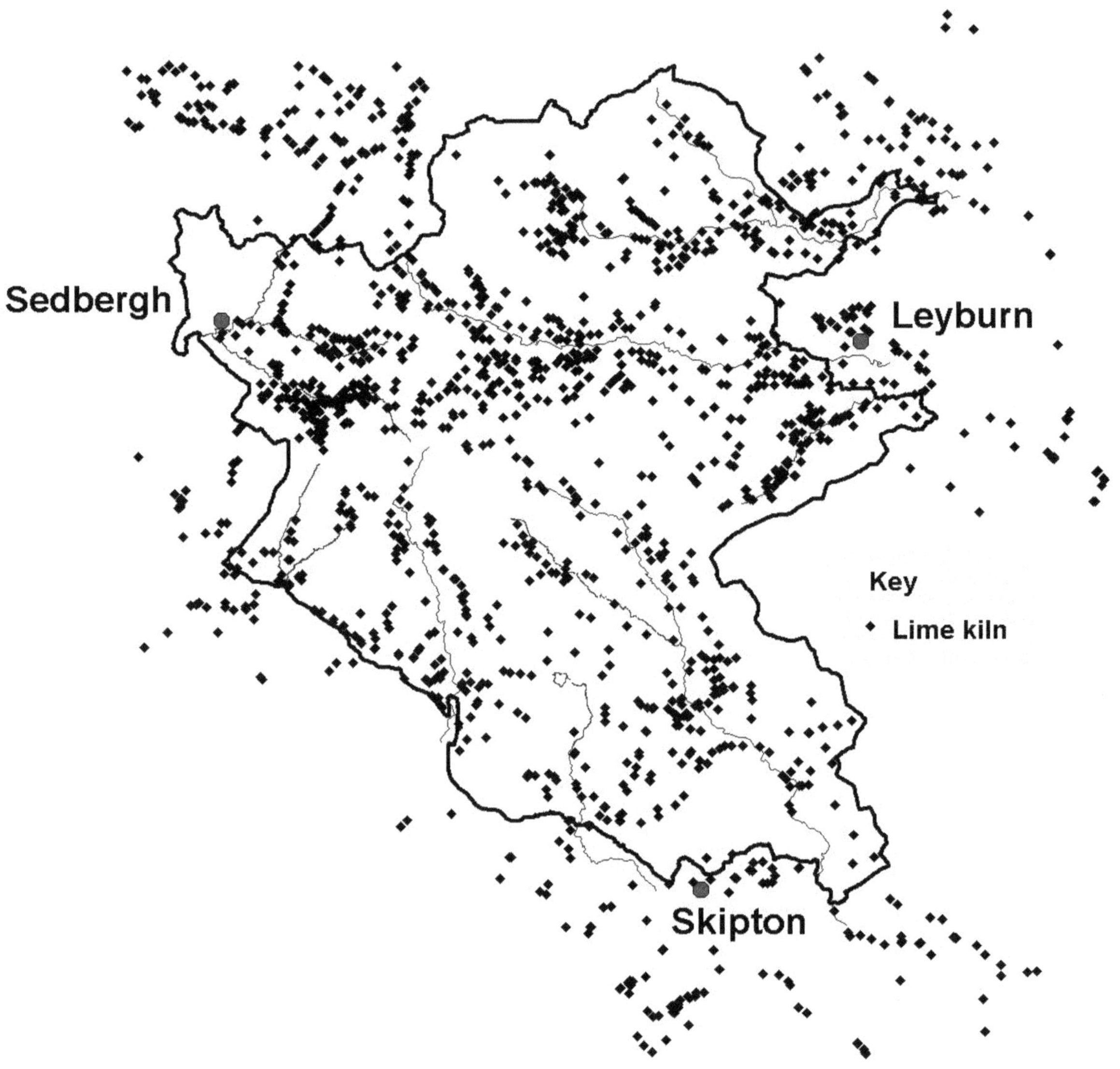

Figure 3.2 Distribution of lime kilns within and without the Yorkshire Dales National Park

material. Many of these written sources, however, are directly relevant to the current research area and will be referred to in ensuing chapters.

Field surveying

This writer began a programme of field recording of lime kilns in 1995, initially within the ancient parish of Giggleswick, but slowly extended across the entire research area. Field surveying has been completed across almost 170 parishes, and over 1500 individual kilns and sites of kilns now lost have been recorded in detail (Fig. 3.2). The aim at the beginning of this exercise was merely to record the status and details of known kiln sites on a proforma survey form for submission to the Sites and Monuments Record[27] held by the Yorkshire Dales National Park Authority and relevant county councils. As the survey progressed it became clear that it could do far more than simply record raw field data.

A number of elements of the survey are directly relevant to the research aims. The distribution of kiln sites in general sheds light on where lime was being produced while the siting of individual kilns, set in context, illustrates where lime was being produced for agricultural purposes. Identifying distributions from historic cartographic sources alone cannot provide the necessary contextual framework.

Context here refers to such variables as land use (past and present), altitude, solid geology and proximity to fuel sources, and is therefore pertinent in showing that liming was practised in dale bottoms and lowland fringes as well as on moorland and fells, and that much lime was burnt and applied far from where limestone outcropped. Though more than half the kilns surveyed have not survived, there are sufficient standing examples to enable valid conclusions to be drawn concerning kiln design and style.

These aspects are not important in isolation but can be used to inform a preliminary dating or phasing of lime kilns within the Central Pennines. In turn, this supports the contention

[27] Now the Historic Environment Record.

that lime was being burnt for a far greater period of time than hitherto widely believed.

Two examples of field kilns illustrate the point being made here. Figure 3.3 is a well-made structure with semi-dressed quoin stones, neatly coursed stonework, a carefully designed segmental arch, and it is clearly a substantial structure. This kiln is of late construction, almost certainly built during the second quarter of the nineteenth century. Figure 3.4, in contrast, is much cruder and of rubble construction. It has been 'thrown up' rather than designed, and is also much smaller: this kiln will date from the early eighteenth century.

Material from this field research will be considered in Chapters 5 to 8.

Archaeological excavation

The final element in the research programme has been excavation of early modern lime kilns, directed by this writer (Fig. 3.5).

The type of kiln in question has been referred to as a clamp kiln, sod kiln or sow kiln; primary historical sources, such as manor court records, tended to describe them as 'sod' or 'sodd' kilns. In essence they are bowl-shaped hollows, dug into a low banking, sometimes stone-lined with an internal stoke-hole to maintain airflow and to permit clearing out after each firing event, and domed above ground level with a turf or clay covering. The first in the series was investigated in 2003 as part of the excavation of a Romano-British settlement near Ingleton by the Ingleborough Archaeology Group (Johnson 2004), a dig which spawned the investigation of six clamp kilns by this writer with the Group (see Johnson 2006, 2007a, 2008a and 2008b for technical discussions).

These kilns, and others excavated within the research area by professional contractors and advised on by this writer, are of direct relevance to the discussion within Chapter 5 as radiocarbon or archaeomagnetic dates obtained from five of them spanned the period from the late fifteenth to the late seventeenth century, thereby extending the production of lime in the research area much further back in time than generally believed. The rural context of most excavated sites strongly suggests that they were producing agricultural rather than constructional lime.

4. Constraints

With any historically based research, documentary sources must be treated with caution. Some of the sources consulted can be seen as reliable, such as the coal accounts for Colsterdale or

Figure 3.3 A nineteenth-century kiln in West Witton parish, North Yorkshire

Figure 3.4 An early eighteenth-century kiln on Ickornshaw Beck, Cowling parish, North Yorkshire

Figure 3.5 Excavated early to mid seventeenth-century sod kiln at Kilnsey, Wharfedale

estate accounts that list quantities and dates of lime purchases or application rates. Equally reliable are the day books within which the Garths entered the minutiae of their daily business. Other sources, however, may represent a biased view and must be used with circumspection. Many contemporary writers and agricultural Improvers had their own distinct agenda and, in modern parlance, were promoting the 'party line' of discarding the old in favour of radically different and sometimes untested approaches. They no doubt believed everything they wrote but that does not mean they necessarily always wrote the truth: nuances of authorship of written material must be borne in mind.

Absentee landowners relied on regular correspondence from their stewards but the usefulness of surviving letters depends entirely on how comprehensive and detailed and honest the steward was. An official on the Hornby Castle estate in north Lancashire deceived his master even though the latter was not an absentee, so George Smith's memoranda books (see Chapter 8.5) may, at times, not reflect reality. This landowner was perceived at the time as being 'feeble-minded' and Smith took full advantage of this. Sir William Chaytor of Croft (see Chapter 5.3) in the old North Riding languished in a London debtor's prison for his last twenty or so years and he was totally at the mercy of his self-appointed steward who ran the estate down while advancing his own interests. Fortunately, a friend of Chaytor kept the unfortunate debtor abreast of happenings at home and his writings are certainly much more authentic than the steward's.

A further constraint is imposed by the survival, or actual creation, of estate records. As we have seen earlier, comprehensive day books such as those kept up by the Garths are rare indeed and others that may have been kept are quite likely to have remained in private hands, unknown to researchers, or have been destroyed. This writer's own enquiries have revealed on several occasions reports of records having been rescued from bonfires or post-mortem house clearances, and of sets of documents relevant to the Yorkshire Dales surfacing in salerooms in London, or online. It is only by sheer chance that these instances are noted and one is bound to wonder how many sets of papers have been 'lost' to research. Inevitably, such situations lead to gaps in the picture that is being built up, but this is so clearly the case with much historical research.

In terms of survival, it has already been seen that return correspondence from landowner to steward is rarely available, again leaving glaring gaps.

The selection of case studies has attempted to be representative both spatially and temporally but one is inevitably constrained by availability of sources. There is the danger of assuming one estate, or locality, was representative of the whole dale or area, thereby potentially leading to the drawing of unsound conclusions, and such considerations have been borne in mind throughout. Coverage through the centuries and across the entire research area is as balanced as it has been possible to make it but it is emphasised that written and documentary sources cannot always be fully relied on.

It is, partly, for this reason that the field survey and archaeology strands of the research are seen as being of such great value as sources of empirical evidence. Indeed, the research value of the targeted excavations simply cannot be underestimated.

A number of questions were posed, and issues raised, at the beginning of this chapter and the sources used and methods employed in the research were designed to enable meaningful outcomes to be achieved. The following chapters describe and analyse methods of land improvement and the application of agricultural lime through the six centuries under review. Chapter 4 begins this discussion by examining the situation from the medieval period to the Restoration.

Chapter 4

Soil Improvement Practices and the Advent of Liming, late thirteenth to mid seventeenth Century

1. Introduction

This chapter addresses the issue of when lime is first known to have been used in farming in the research area. The starting point is the late thirteenth century, when medieval agricultural practices had peaked in terms of productive capacity, and before the ravages of the following century. The end point is the middle of the seventeenth century as new-found optimism and security after the restoration of the monarchy in 1660 empowered tenants and landowners alike to invest in buildings and in land purchases and improvements on a scale as yet unseen.

Also traced here are developments in land management and soil improvement through the medieval and post-medieval eras on a broader geographical scale as a means of setting the Central Pennines in a wider context. The focus will be on a selection of examples within the research area, using them to identify trends and processes. The case studies are selected from the northern dales of Swaledale and Wensleydale, Craven, Lower Wharfedale and Airedale, and north-east Lancashire. While there are recognisable similarities between them, there are also clear differences of emphasis, approach and evidence.

Long-term climatic patterns will be considered along with demographic, socio-political and pestilential trends as their impact undoubtedly influenced attitudes to land improvement as well as the ability of those involved with the land to respond positively to new opportunities, or to retrench in the face of adversity. When tenants were struggling rentals would have been affected, thus reducing the flow of income to landowners. Each party's priorities would inevitably have changed in response to prevailing circumstances.

Before lime became widely accepted for use on the land a range of options had been tried in attempts to tease more out of the soil,[28] to boost output and increase productivity and, from the landowner's perspective, to give him the justification for increasing rents. After all, if tenants were selling more at a higher price, why should he not also benefit? Some of these early inputs may seem rather eccentric to modern minds but they must be seen in the context of an unscientific age: even though lime was increasingly used from the Middle Ages, it was not until 1814 that the science behind its efficacy was adequately explained (see Chapter 7.3).

Far less eccentric than some early inputs was the use of marl, and its significance in soil improvement and land management cannot be overestimated. Though eventually superseded by liming, as will be shown, marling endured for centuries as a means of top-dressing pasture land and improving soil quality. Chemically it shares essential constituents with lime.

For the medieval centuries written records are sparse and the researcher is reliant on a restricted range of sources, such as monastic cartularies and account books. Hard facts are rare thus necessitating an element of hypothesis and postulation in the unfolding argument. A nuanced approach is also required with fragments of hard data contributing to the final, but conjectured, picture of this period rather than to indisputable conclusions.

As the centuries progressed, surviving documentary sources become more frequent with manor court rolls being of particular importance. These recorded encroachments on to the waste, and pains laid against unlawful activities within manors, all of which provide useful insights into what was going on at a local level. Occasionally personal documents have proved of value, sixteenth- and seventeenth-century wills and post-mortem inventories being of especial interest. Estate records are few and far between but the Lowthers in Westmorland and the Towneleys in north-east Lancashire are notable exceptions from the middle of the seventeenth century. However, as Jones and Page (2006, 242) have written, those engaged in trying to deconstruct past landscapes are 'only too aware of the fragmentary nature of the evidence on which they must base their conclusions'. This thought applies equally to the process of understanding how past landscapes were managed and improved.

The validity of conclusions reached in this chapter owes much to fieldwork. This writer has undertaken extensive field surveying across Craven over a number of years identifying the location of early lime kilns, the context of which helps inform each site's usage for agriculture or construction. Furthermore, this writer's archaeological excavations of six such kilns with proven scientific dating, and his advisory role on similar (some dated) sites unearthed in 2006-07 along the easement of a major gas pipeline through Craven, have provided convincing evidence of the use of lime in this area, definitely during the sixteenth and seventeenth centuries and, in several cases, prior to that. However, as many of these early sites extended beyond the time frame of this chapter, they will be discussed in detail in Chapter 5.

2. Historical context

For most of our history, and certainly for the period under review now, farmers in the north of England have largely been at the mercy of the weather and one fundamental feature has been climatic fluctuation. From around AD 550 to the

[28] These included human urine, torn rags, horn and bone shavings, dregs from ale, and blood. Night-soil (*stercus humanum*) was also in use from medieval times, though often controversially (Woodward 1994, 106).

late twelfth century average summer temperatures were 1°C above normal though winters were colder than the norm and summers more liable to storms. Warming continued through the thirteenth century, but the years 1270 to 1310 were slightly colder than average, and climate deteriorated markedly through the remainder of that latter century, and in the early years of the fifteenth, not warming again until around 1500 (Lamb 1965; but see Nielsen 2005 for a contrary view).

Farmers then had no way of combating extreme cold or damp and the success of their harvest was frequently jeopardised by the vagaries of the weather. In the late Middle Ages, though, famine was a rare event in England, except in the 1430s in the aftermath of two decades of cold conditions (Dyer 2005, 14), because climatic conditions were generally favourable to ripening and harvesting.

The period 1310 to 1375, on the other hand, witnessed a spate of harvest failures as 'years with no summer' took their toll. Prolonged rainfall and low summer temperatures led to crops either not ripening or rotting in the fields with consequent food shortages and famine. A given population can survive one such disastrous season but not repeated ones. Weakened bodies readily succumb to disease, and epidemics – of plague or other contagious diseases – spread through the land. If the situation was serious in the Central Pennines, it was more so in Scotland and the end result was a series of well-documented raids south of the border into the Central Pennines starting in May 1318. These raids brought devastation to an area already mired in famine and civil unrest, caught up in political and military turmoil between the two nations.

Much of Wharfedale, the area around Skipton, Ribblesdale from Horton in Ribblesdale south to Sawley Abbey, the Wenning valley, and the Sedbergh area, all suffered from these raids (Kershaw 1981, 231-38). The greatest havoc was created in Lower Wharfedale and through the Aire Gap with incursions into the upper valleys relatively rare. The impact of the raids was documented at the time for those monastic houses with demesne land in affected areas, especially so for Bolton Priory. Indeed, in 1320 most of the canons there were granted permission to temporarily go elsewhere in case of further incursions from Scotland (Legg 2009, xix). The *compotus* shows that rental levels in 1473 were little different from levels when (surviving) accounts were first kept in 1340 (Kershaw 1970, xiv). Cattle stocks were severely reduced, crops ruined, settlements laid waste, populations sliced by famine and disease, with drastically reduced income for the priory and a pronounced shortage of labour to work demesne land (Kershaw 1973b, 14-16). The direct result of this latter reality was the conversion of in-house granges into vills with land being leased out to tenant farmers. The possibility that monastic houses were exaggerating their degree of financial peril cannot be discounted, however, as they were surely sufficiently astute to recognise the opportunity of securing reductions in their overall tax burden.

Many tenements remained vacant through the fourteenth century, there was an acute shortage of labour to work the land, and rents remained depressed until the later fifteenth. Set against this dismal picture, however, was an increase in the remuneration that surviving farm workers could demand in this time of 'demographic catastrophe' (Kershaw and Smith 2000, 20). Those with foresight and drive were able to capitalise on the situation and many saw the opportunity to increase the size of their holdings and to raise their own standards of living. Out of decades of want came a (renewed?) desire to improve land quality and raise levels of husbandry.

One major result of the fourteenth century's problems was a conversion of land from arable to pastoral. This has been demonstrated for the Midlands in particular (Dyer 2005, 66-67) but there is no reason to doubt that similar processes were at work further north. A smaller population required less food to be grown and provided fewer people to till the land, village desertions or shrinkages saw arable open fields going out of cultivation, and the market demand for meat and wool products began to grow. This latter opportunity was seized upon by monastic houses with lands in the research area – dominantly the abbeys of Fountains, Jervaulx, Sawley and Furness, and Bolton Priory. Though wool prices dipped in the middle of the fifteenth century, the general trend was the gradual emergence of a market-oriented economy as larger and more systematically managed estates provided more marketable produce (Dyer 2005, 196-97).

In Dyer's view the fifteenth century also witnessed the rise of the progressive farm tenant and of agricultural capitalism, and the growth of industrial trades, phenomena that became more apparent during the following century when the idea of improvement (by landowners and tenants alike) gained ground in terms of investment in new buildings, enclosure of waste land and open fields, and a concomitant growing appreciation of new forms of husbandry (Dyer 1997, 61-76). Dyer (2005, 200-01) also argued that tenants were often able to negotiate rental allowances for improvements they had undertaken. By act of parliament small encroachments (up to 1.3ha with a cottage) on to the waste were permitted as long as they 'doth no hurte' to those tenants who enjoyed rights on the waste, thereby formalising and giving impetus to the process of enclosure (Statutes of the Realm 3 & 4 Edward VI).

There are signs nationally that lime burning was also blossoming during Tudor times. An act passed in 1548 included the trade of 'lyme-burner' in a list of building and allied operatives who were granted the right to operate as such in any town regardless of where their home was, suggesting that government was reacting to a surge in demand for lime for mortar in a time of urban growth or that the demand for lime in general, including for farming, was rapidly increasing (Statutes of the Realm 2 & 3 Edward VI). This trend was clearly not short-lived as further legislation in 1562-63 included 'lymeburner' in a list of artificers permitted to take on landless young men as apprentices, again presumably to cope with rising demand (Statutes of the Realm 5 Elizabeth). Forty years previously Fitzherbert (1974, cap. xxxiiii) had vividly described how lime was burned and how farmers in different parts of the country did 'set it upon their landes as they do their dong and do sprede it in lyke maner, the whiche they call moche better than dong'. Liming was clearly an established and widespread agricultural practice by then, even if only in the South.

Evidence suggests that the last thirteen years of the sixteenth century saw further collapse of wool prices in England along with high food prices and high mortality rates, each of these factors interplaying to bring about high levels of stress among the poorer sections of rural sheep-based communities, with clear implications for land management (Drake 1961, 431-35). These processes have been illustrated for southern Westmorland and Cumberland between 1587-88 and 1597-98 (Appleby 1973, 419-20; 1978, 95-133 passim), where food shortages resulting from poor harvests and another spate of Scots raids, and vastly reduced income from a further collapse in wool prices rendering many tenants unable to pay farm rents, brought about epidemics of plague and/or typhus. The year 1597, in particular, was one of an acute '*crise de subsistances*' (Appleby 1978, 183).

Demographic crises, food shortages and epidemics continued into the seventeenth century as for example in 1613-14, 1617 and 1621-23 (Drake 1961, 432-35; Appleby 1973, 429-30). The West Riding bore the brunt of the repercussions of depressed wool prices in the 1620s, and Appleby (1973, 407) has shown that 'poor and remote' Westmorland was at the bottom of all English counties in terms of tax assessments in the sixteenth and seventeenth centuries. However, the reliability and impartiality of land tax returns is questionable (Whyte 2003a, 90). The area around Ripon suffered an outbreak of plague in 1629;[29] the West Riding endured a severe outbreak of plague in 1645 (Drake 1961, 431); the Ribble valley likewise in 1630-31 (Howson 1961, 34); while Westmorland was hit four years after that with an estimated 30,000 families lacking food, money and seed to plant (Thirsk and Cooper 1972, 51-52).

Many of these demographic and agricultural setbacks may be explained, at least in part, by climatic downturn across northern Europe (Whyte 1995b, 47-50). This has been identified, in the absence of reliable climatic data, from a range of proxy indicators such as variations in the size of cod stocks landed in Baltic fishing ports (Mackenzie *et al.* 2007, 106-19). The years 1554-76 and 1593-1630 recorded exceptionally low landings, the cause undoubtedly having been unusually high sea ice cover resulting from significant atmospheric cooling.

Meanwhile, sixteen discrete episodes of high mortality have been suggested for Lancashire between 1540 and 1654 (Sharpe France 1939, passim) showing, if true, that the array of setbacks affected many parts (if not all) of the research area.

Thus, late medieval and immediate post-medieval farmers and landowners in the Central Pennines and adjacent areas faced a series of challenges, some with minor and relatively short-lived effects, others with a devastating impact. The husbandman's travails and struggles against adversity were even eulogised in an anonymous ballad, written about 1500, called 'God spede the plough' (quoted in McRae 1996, 1). Repeatedly, those living off the land had had their previous efforts and successes thwarted with mere survival coming to the fore in their thinking. Viewed from this perspective it is perhaps remarkable that any achievements in land management and improvement were recorded during this long period. It is to these processes that the following sections turn.

3. Evidence from the monastic period

By the middle of the thirteenth century monastic estates were renowned for the level of land improvement being carried out, a process which can be likened to an agricultural revolution with massive programmes of woodland clearance, assarting, consolidation of scattered holdings and enclosure of waste land (Platt 1969, 13; Kershaw 1973b, 117). Marsh land was drained, new land brought under the plough, and acidic moorlands made good for stock rearing (Cottam 1928). All this was associated with the beginnings of a scientific and methodical approach to the maximising of agricultural productivity. Waites (1967, 34-35), writing of the Vale of Pickering, expressed the view that the thirteenth century saw the culmination of a process of improvement that had its origins two centuries earlier. Meanwhile the later fifteenth century witnessed a resurgence of interest in the works of Classical agricultural writers, such as Cato, Varro and Columella, whose books were duly reprinted many times and whose ideas formed the basis of much late medieval land management (Thirsk 1992, 18-19).

The process of improvement had included bringing ever more marginal land into use in a process eloquently described as 'the journey to the margin' (Beresford and St. Joseph 1958, 91). Many landowners had brought leased-out land back into direct control at the start of the fifteenth century in the (then correct) belief that more profit could be made by directly working their holdings on a market-oriented basis than from collecting rents (Jones and Page 2006, 132).

While the generality of these improvements has been well documented, there is a dearth of evidence concerning actual methods employed to achieve the desired ends within the research area, if not elsewhere. Bond (2004, 52) has noted the use of liming to reduce soil acidity by Canterbury Abbey on Romney Marsh, the use of 'beat-burning' by Tavistock Abbey to provide ash for mixing in with soil on reclaimed wetlands, and the use of lime and marl at Tintern, Abbey Dore, Beaulieu and Fountains abbeys. For the area under review in this book a degree of conjecture must be employed to try and build up a coherent picture from isolated facts in the few surviving sources.

Charters are one such source. For example, the cartulary for Cockersand Abbey recorded the gift of lands in Caton in Lunesdale by Vivian of Heysham, between 1221 and 1246. The parcel of land so gifted ran '... usque ad Acstub juxta limkilne' ('thence to the oak stump next to the lime kiln') (Farrer 1898, 832 and 857). It cannot be concluded with certainty that this particular kiln was producing agricultural lime but its isolated

[29] Information supplied by Ripon Heritage Centre in 2008.

location suggests it may well have been. Similarly, the grant of land to the abbey between 1241 and 1251 by Robert Gernet of Caton of parcels of land '...et altera pars inter Stanikilne et terram dictorum canonicorum ('...and the other part between Stony kiln and the land of the said canon') may not even refer to a lime kiln – it could have been a corn drying kiln – but the latter would hardly have been built within a, presumably open, area whereas a lime kiln would. If this were the case, this too could have been burning agricultural lime.

Occasional land charters for Fountains Abbey describe the bounds of gifted land by reference to kilns. One such grant, made in Stainburn near Otley during the reign of Edward II (1307-27), located the land as 'one acre between the land of the same monks and the land of Adam son of William, beside Le Lymkylne in the same territory' (Ironbridge RS1039). The wording here, as with Vivian's gift, suggests the kiln was isolated and thus producing lime for use on the land.[30] A later grant to Fountains concerned lands in Threshfield, Upper Wharfedale, in 1517 which included 'half a rodde lyinge north and sowth upon the lyme kylne in the said bank betwixt the rayn and the watter; ...' (Michelmore 1981, 63). A rayn was a boundary strip adjoining an open field, meaning this kiln presumably stood in open ground between arable (or formerly arable) selions and a watercourse so, again, it may well have been designed for agricultural usage.

Entries in the bursar's account book for Fountains covering the period 1446 to 1458 refer to purchases of lime but, as is the case with many such accounts, the entries are terse and leave questions unanswered (BL MS Add. 40011A). One entry (folio 71) can be tied down to Kilnsey in Upper Wharfedale: 'debetur eidem pro labore suo ad Kylnsay cum calce, iijd.' ('to the same person for his labour at Kilnsey with lime, 3d.'), but there is no way of knowing what end use this lime had. A further entry (folio 89) is in a sense even vaguer, though it does name the payee: 'Th^{ae} Schepard pro labore ad lymkyln, viijs.' ('to Thomas Schepard for his labour at the lime kiln, 8s.'). The difference in amounts paid between these two entries is marked and the latter was clearly being paid for many days', if not weeks', work burning lime yet there is no hint as to why the lime was required. As will be seen in Section 4, contiguous folios noted payments for marl, which did have agricultural relevance. However, no firm conclusion can be drawn about the lime.

Carte blanche permission to take in land from the waste for improvement had been granted as early as the 1230s when the Provisions of Merton recognised barons' need to 'make their profit' (Statutes of the Realm 20 Henry III). Progressive landlords took up the invitation and there soon developed a new class of men who acted as mediators between landlord and tenant or labourer, namely the farm steward, many of whom would in time elevate their families to the status of landed gentry (Hainsworth 1992, 6). Among his roles was ensuring tenancies were filled, with good tenants being encouraged to stay especially during times of economic recession.

Two medieval treatises have survived, aimed at stewards rather than landlords, and certainly not at ordinary farmers, with advice on many aspects of husbandry and land management. Walter of Henley's *Le dite de hosebandrie* exhorted stewards to mix manure with soil, especially on sandy ground, and to apply marl which he considered far superior to dung as a long-lasting soil treatment; while the *Seneschaucie*, written anonymously in the reign of Edward I (1272-1307), urged farm stewards to ensure lands in their care were 'marled, folded, manured, improved, and amended ... for the good and bettering of the manor' (Lamond 1890). It was accepted practice to manure crop lands using dung very often mixed with domestic midden deposits (Jones and Page 2006, 92). Fieldwalking undertaken in Craven over the years by this writer can confirm the frequency with which one finds scatters of shattered ceramics wherever medieval ridge and furrow or lynchet systems have been recently ploughed for reseeding.

Clearly, works of this kind were aimed at the steward class. It is unlikely that many of the medieval seigneurial class would have wanted to be involved in mundane estate matters, and tenants and labourers would hardly have been in a position to read them let alone put their ideas into practice. Having said this, though, it has been postulated that farm tenants, and earlier even peasant farmers, were driven by the survival instinct to do whatever they could to ameliorate the condition of the soil in their care (Muir 2000, xiii). What the treatises do, however, is point up a nascent realisation that soil and land could be improved to gainful effect, and it is quite feasible that new ideas did filter down from steward to husbandmen or up to the landowners.

The Scropes of Bolton Castle and Masham, without doubt the most powerful magnates of the fourteenth-century Dales, are perhaps exceptional in that they took an active interest in the betterment of their vast estates under the guiding hand of their steward, which enabled the estate to withstand the mid-fourteenth-century ravages (Vale 1987, 53-54). They were actively involved in enclosing ever more land and improving the quality of new and existing farmland at this time though this was in fact a continuation of long-established practices on the estate. By a deed sealed in 1299, for example, Henry Lescrop had been granted various parcels of land with the right to freely enclose and improve them (ChRO DCH/X/15/1).

4. Marling

To a geologist marl is calcareous mudstone. It is commonly associated with post-glacial lake deposits such as in the Ribble valley between Settle and Long Preston, and near Otterburn where deposits 400mm thick are exposed in stream sections (Arthurton *et al.* 1988, 93). Current active deposition of marl occurs, for example, within Malham Tarn. More loosely

[30] There is documentary evidence from monastic foundations in the Dales of lime kilns sited within abbey or grange precincts specifically producing lime for mortar or use in a tannery. See, for instance, an indenture of 1532 between the abbot of Fountains and Richard Paver, tanner, who was allowed 'to have lyme of theym, when thei have ony to spare bysydes there necessaries for 12d. one quarter' (Michelmore 1981, 241-43). See also Kershaw and Smith (2000) for the accounts of Bolton Priory (1286-1325) which refer to payments for lime burning for mortar at Kildwick on the Aire.

it is lime-rich mud or mudstone with unequal but variable proportions of clay and calcium carbonate compounds. It can vary in colour from khaki through red to blue. Historically, the term marl has been applied to a broad range of carbonate-bearing clays, calcareous shales and friable limestone (Eckel 1922, 305). The term was even applied to chalk in Norfolk (Trimmer 1847), the Downs (North 1930, 338-39; Williamson 2002, 67) and the Yorkshire Wolds. There is also circumstantial evidence that glacial till, dug in boulder pits either side of Lower Wharfedale and in Craven, was perceived to be a form of marl but was in fact lime-rich clay: a conveyance, dated 1681, concerned the 'messuage and tenement of Marlecloughbecke with liberty to get limestone in the delfs next unto the said messuage – situate in Ragill' (Wilson 1972, 58). Raygill, in Lothersdale, later developed into a commercial quarry but early exploitation here was of limestone dug from within the stream for burning into agricultural lime. It was not true marl, despite the term used in the conveyance. An assignment from 1300 in Glusburn included *marleriis* (marl pits) in the list of appurtenances but this, too, is unlikely to have been true marl (WYAS [B], Glusburn). Given that the swathe of glacial till that extends from Glusburn and Cowling to the Trawden area has a bluish tinge in its unweathered state, and turns yellowish when exposed to the atmosphere, and contains a high proportion of limestone cobbles, the marl was probably just lime-rich glacial clay.

'Rich' marl beds were also reported to have been worked along the Swale below Whitcliffe Wood, west of Richmond, and near Bedale but both deposits were probably also glacial clays (Cochrane 1795, 2-4).

The use of marl is known from Roman Britain (Gardner and Garner 1953, 12-13; Fussell 1955; Rackham 1986, 370-71); its later widespread use in Norfolk and the Lancashire-Cheshire plain has also been demonstrated.[31] Rackham (1986, 37-71) has drawn attention to marl pits in Anglo-Saxon charters, referred to either as marl pits or 'lampytts'. Lampitts is a term noted elsewhere and it translates literally as loam pits, so this may again hint at past confusion of terminology between marl and clay loam. According to Donkin (1973, 92), the earliest written reference to marling occurs in the chartulary of St Pancras Priory at Lewes in 1095. Its perceived value was formally recognised during the reign of Henry III by a statute requiring 'Every freeman from henceforth (to) make in his own wood, or in his land ... marlpits' (*marleram*) (Statutes at Large 9 Henry III). In 1345 the general right of chiminage (free passage) was granted to farm tenants on royal estates in Lancashire 'across the lord's meadow for marling their lands' (Cunliffe Shaw 1956, 185).

The value of marling has also come down to us in verse form. One age-old and anonymous rhyme, quoted in Davies (1960, 110), emphasises in a memorable manner that marl should be applied to greatest advantage on light sandy soils to which it adds body, but would be wasted if applied to heavy clays:

He who marls sand
May buy the land;
He that marls moss
Suffers no loss;
He that marls clay
Throws all away.

Bowden (1967, 659) recalled advice given by James Bankes to his son in 1600 at Winstanley in Lancashire: 'the charg of the marlying of an acare wyll stand you in fyfe marks so that therby you shall incres your rentes in good sort'. In other words, if his son were to apply marl each acre of treated land would increase in value by £3 6s. 8d. at contemporary rates, thus enabling him to raise his tenants' rents. Perhaps this helps to explain why Fitzherbert had earlier suggested that the practice of marling had largely died out in early Tudor times, as tenants feared landlords would use improvement as an excuse for increasing rents (quoted in Fussell 1955, 99). It had come back into common use, though, by around 1600.

A further, and also anonymous, rhyme ranked marl as the most efficacious soil additive: 'a man doth sande for himself, lyme for his sonne and marle for his graunde childe' (quoted by George Owen in 1603 – see Miles 1994, xxxviii – and by Mathew 1993, 109). In other words, calcium-rich sea sand had short-term effects, lime had medium-term value but the effects of marl lasted for several decades. Modern chemical experiments would dispute these conclusions, and whoever first penned the verse may well have had an ulterior motive, such as relative costs of application, or simply may not have appreciated that it was the wrongful application of lime that did more harm than good, not the lime itself. Anecdotal evidence has suggested that marling pastures may actually have had deleterious effects on sheep causing them to abort (quoted in Williamson 2002, 81-82). Nevertheless, in some areas and at some times marl was highly regarded: Holt, for example, who authored the Board of Agriculture's survey for Lancashire, quoted one 'intelligent' farmer who said marl was a 'never-failing friend' to most parts of that county (Holt 1795, 115).

Early maps and legal documents within the research area include place-names associated with marling, such as 'Marlehole' in Hornby in 1544, 'Marl Field' in Thornton in Craven in the early seventeenth century, and a field known as 'Marled' or 'Merled Earth' on the Towneley estate in the early eighteenth. A grant of lands to William le Scrop (of Bolton Castle) in Thoresby in Bishopdale, sealed in March 1298, concerned a parcel of arable land known as 'Le marle Pottes' (ChRO DCH/X/15/1). Another deed, undated but of approximately the same period, confirmed the grant of lands including common pasture, meadow and woodland by Wymer of Thornton Steward in Wensleydale to his uncle, John de Thornton (NYCRO Bolton 437 MK/2). The deed stated that, should a turbary, marl-pit or quarry be found on the commons, Wymer could make use of it as he saw fit.

[31] See, for example, Prince (1964) and Williamson (2003, 33-34) on Norfolk; and Davies (1960, 110-12), Fussell (1955), Grantham (1864), Hewitt (1923), Houghton (1727, 178), Long (1968, 89-90), Phillips and Smith (1994, 79-80), Shannon (2009, passim) and Winchester (2006, 94-95), on Lancashire and Cheshire. Leadbetter (1739, 174) described ceremonial activities on Lancashire farms once a particular marling operation had been completed.

In the Forests of Trawden and Pendle tenants living within the manor of Colne were granted the right to collect a range of materials from the common from the fourteenth century onwards, such as 'turf, slate, limestone and other stones, sand, gravel, marl, clay, sods, heath ... and gorse' on condition they did not sell any elsewhere (quoted in Swain 1986, 37).

Exactly how long marl remained in use is unknown but there is archival evidence that some farmers were still applying it during the eighteenth century (Overman 1842), even though some contemporary commentators believed it had long since gone out of use. For example, an open letter was written by James Cochrane, of Houghton Farm, just north-east of the Yorkshire Dales, in response to the Board of Agriculture's requests for information from around the country (J Cochrane 1795). He refuted the observation recorded in the report for the North Riding that marl was no longer applied in the county saying he knew it to be still in use below Whitcliffe Wood west of Richmond, at Hipswell just south of that town, and near Bedale in Lower Wensleydale. He clearly had his own agenda as he (incorrectly, according to the science) repeated a claim that marl would yield excellent crops or pasture for twenty years whereas 'lime cannot bear without injury above three or four crops'. As will be seen, farmers increasingly disagreed with this perspective.

An entry in the diary of George Smith, an official of the Hornby Castle estate in Lunesdale, for July 1844, records that one Brockbank was 'killed in a marl pit' (Hornby Castle Mun., Smith). Whether or not the pit was still being worked or was full of water is immaterial in the sense that Smith knew it to have been a marl pit.

The point has been made elsewhere that, nationally, liming 'grew so greatly in extent, frequency and volume, that it became effectually revolutionary' after 1560, and more particularly after 1590, though the practice was not new (Kerridge 1967, 248). It has been shown in Section 3 above that there is a certain body of evidence that the putative use of agricultural lime within the research area is known from the monastic period. The following sections will examine the situation in more detail by recourse to a number of case studies, namely the northern dales; North Craven; and Lower Wharfedale and Airedale.

5. The northern dales

A useful starting point for identifying early agricultural lime usage is made by examining the evidence for land improvement in general. Though lack of evidence makes it impossible in many cases to state with certainty that a given area was improved by liming, one can safely conclude that it was improved in some way, whether by dunging, marling, liming or a combination. Given the prevailing climatic conditions in the Central Pennines, and the resultant acidic nature of most soils, there is very little farmland that could have been productive on a sustainable basis without some form of ameliorating treatment. Documentary evidence of pre-modern enclosure and improvement is extant for much of the research area, and there is also evidence for early lime kilns.

In 1573 a case was heard at Richmond concerning a dispute between landlord Robert Bowes of Aske and one of his tenant farmers (NYCRO ZQH 7/53/3). The disputed land at the centre of the case, called Pykepurse or Easegill, had been enclosed 'three years last past' specifically for growing corn. It has not been possible to precisely locate the field, but it is known that considerable amounts of lime were being burnt in that vicinity during the sixteenth century. An agreement between Sir George Bowes and the burgesses of Richmond, in March 1562, delimited the boundary of the manor of Aske with reference to lime kilns as boundary markers: '...and so stretchinge west harde by the northe side of the nethermost lymekilnes, and at the westermoste lymekilne' (NYCRO ZQH 7/53/1). The phraseology employed in this agreement clearly points to there having been a cluster of kilns and it has been possible to locate this part of the boundary in what is now known as Mouldron and Coalsgarth north of Richmond. Within this area there were three discrete lime kiln place-names surviving at least into the nineteenth century, all in isolated situations, namely Limekiln Field (occurring twice) and Limekiln Plantation. When these place-names were first applied, though, is unknown.

A degree of caution is required here and it *may* be premature to link the Aske kilns directly to agricultural usage. The historical record is very clear in showing that the area north of Richmond, in Richmond and Aske parishes, was a significant producer of lime as early as the sixteenth century, but the records say nothing of what the lime was used for. The borough council maintained a Coucher Book which is extant from 1590 and it contains many late-Elizabethan entries concerning the burgesses' repeated attempts to control the lime trade (NYCRO CRONT 1840). Directives required all limeburners to have a licence, to sell lime at prices set by the council and to have it weighed using official measure sacks each containing the 'Towle Booth bushell', to pay a toll on each sack sold, and to fill in and level off any kiln or quarry hole that was no longer used. These directives were reissued in 1605. The fact that kilns, which would have been sod kilns, had to be levelled off explains why this writer's field walking here has failed to locate any early kiln sites.

Lime production here was tightly controlled, partly because the kilns and quarry holes were owned and leased out on short-term contracts by the borough and partly, no doubt, as the burgesses saw it as a lucrative source of income for the town. This practice was unique within the current research area. The earliest extant example of an actual lease dates from 1660: Matthew Buckton, Thomas Cape, Cuthbert Frier and George (surname illegible) paid 2d. per 'lime pit' and kiln, with Frier having leased three and the others two each (NYCRO DC/RMB 2/1/1).

Work further up Swaledale has shown that land was being taken in for more productive agricultural use from the mid-fifteenth century, either by agreement or by illegal encroachment, and by the end of the Tudor era most high quality, valley bottom grazing land had been parcelled up with considerable intaking on the lower fell sides (Fieldhouse and Jennings 1978, 62-145 passim). This had occurred in the townships of Ivelet, Walburn,

Satron, Gunnerside, Melbecks, Healaugh, Harkerside and Reeth, with intakes reaching the 300m contour in places. Fieldhouse and Jennings (1978, 156) also expressed the belief that lime was 'almost certainly' being burnt in Swaledale from the late sixteenth century.

This view of enclosure and improvement in Swaledale is confirmed by the contents of a survey carried out in 1561 which indicated that most valley bottom land had already been carved up and that by then virtually all arable land had been converted to pasture (McDonnell 1990, 29-30) with each tenant having grazing rights on valley sides, enclosed stinted pastures (named as 'Pasture' or 'Side' on modern maps) and on unenclosed moorland commons. Farmers in the dale were so progressive that they had recognised the (financial) benefits of improving their enclosed land.

A contemporary account also adds credence to this picture of a valley with most of its land in productive pastoral use. John Leland journeyed through the area between 1535 and 1543 on what he described as 'my laborious yourney and costely enterprise' (Chandler 1993, xvii) and, while he noted in his diaries 'sum good corn, and much more' between Richmond and Ravensworth to the north (Toulmin Smith 1964, 79), in Swaledale proper he observed that the land 'norisith many bestes' (Thwaite 1972, 7) confirming its prosperous pastoral emphasis.

Immediately to the west of the hamlet of Stainton, between Swaledale and Wensleydale, is an area now known as Stainton Moor, formerly Stainton Common. The eastern flanks of the moor are divided up into a series of small, irregularly shaped enclosures extending up to the 270m contour. These are shown on an estate map, dated 1694, but the dates of enclosure are not known (NYCRO ZAZ [M] 1). They are all underlain by sandstone and the unenclosed sections of the moor adjacent to the intakes are still dominated by unproductive acidic grassland and heather. In fact they were long since given over to military firing ranges with very low stocking densities of sheep. High up on the moor a site was labelled on the 1850s Ordnance Survey First Edition six inch map as 'Sowkiln Gill'. A sow kiln is one term for an early clamp lime kiln and this very imperceptible valley would surely not have been so named if it had not held a kiln. This writer had located one ruined lime kiln close by, shown on the 1850s map as disused, lying on a small outcrop of limestone, and excavation of an earthwork lying at 350m OD, resembling the typical sow kiln form, was carried out in 2006 (Johnson 2006, 44-46). However, this feature proved to have been a prehistoric pit. Nevertheless, the lower enclosures were initially improved by liming and the lime can only have come from this outcrop on Stainton Moor, perhaps from the ruined kiln.

Prior to selling the entire manor of Abbotside in 1613-14 the duke of Lennox commissioned a full survey which reported that forty-two per cent of the total area was already enclosed, much of it improved as were large parts elsewhere in Wensleydale. Leland had noted that the Middleham-Jervaulx Abbey region consisted of 'enclosed pastures' (Toulmin Smith 1964, 80), and Lennox's survey had noted that widespread enclosure had been a feature of the area north and west of Hawes even before the dissolution of the monastic estate (Willan and Crossley 1941, xxi-xxii).

The lordship of Middleham, along with that of Richmond, laid down numerous conditions by which tenants were bound, both during its possession by the crown and after it was sold to the citizens of London in 1628 (Willan and Crossley 1941, 150). Of direct relevance to this discourse was one that stated 'they likewise covenant eyther by themselves or other sufficient men to inhabite, occupie and manure theyre severall tenements'. There is no mention of lime here but it should be borne in mind that use of the term manure was frequently vague, and even lime was wrongly perceived as a fertiliser – using the word in its technical sense – even as late as the nineteenth century.[32] The key point here is that improvement was not an optional matter for farm tenants: it was mandatory.

Fieldhouse (1980, 172, 178; 1982, 114) has also demonstrated that enclosure within Wensleydale had been ongoing since medieval times, and accelerated during the Tudor years, and that by about 1600 roughly ninety per cent of dale bottom land had been fully enclosed and parcelled up, with only 8.5 per cent still under open field ploughland. Set against this, however, is his assertion that there was no evidence of widespread improvement before 1800 (1980, 192-93). These two statements do seem rather contradictory.

Direct evidence of lime burning from documentary sources is somewhat sparse for Wensleydale but, of course, the dearth of evidence does not necessarily mean it was not going on. The earliest located reference to lime burning dates from 1464 and took the form of a sixteen-year lease to Richard Pygot and Thomas Gaile from their landlord Miles Bowes, of several enclosures and meadows bounded partly by 'the Pelegarth end Trendill well and Lyme potte', the latter referring to a lime kiln (NYCRO ZFW 4/1). These place-names have long since been lost so it is not now possible to locate the land in question, or to determine if the location was rural or otherwise.

Two centuries later a further conveyance used a lime kiln in the description, this time from Sir Christopher Wyvill of Constable Burton to Anthony Durham of Spennithorne, yeoman, dated 1652 (NYCRO ZQH 2.2/4). The conveyed land, with tenement, was called 'Lyme Kilne Garth'. As with the earlier agreement, this cannot now be located on the ground, so this item is of limited value in determining if the lime was destined for the land.

A further hint of lime burning was contained in a verdict of Middleham manor court at its sitting in October 1677 (NYCRO Z 775). This forbade anyone from graving 'any sodd upon Middleham Moor unless it be for common butts or Lyme Kilns 16s. 8d.'. This ruling, and the monetary value set, are of more importance than might appear at first sight, as

[32] One account of lime burning in Cornwall, published in 2000, includes a section heading 'Burnt lime as a fertiliser' (Isham 2000, 6), so the misconception that lime is a fertiliser still persists.

the practice of digging sods for lime kilns must long have been commonplace for this degree of control to have been imposed, whether to discourage it or as a means of raising income. Similar manor court rulings elsewhere never indicated the intended usage of the lime, but the area where these particular kilns were burning lime (for centuries, in fact) is known: it was several kilometres from Middleham town, suggesting at the very least partial agricultural use.

6. Craven

As in the northern dales the evidence for Craven is patchy and to an extent circumstantial. Estate papers are lacking – in large part owing to the fact that large landed estates were not a feature of Craven in the post-medieval era – so recourse must be made to archival evidence in the form of legal documents such as wills, conveyances and court rulings. Nevertheless, it is possible to build up a preliminary picture of land improvement and the use of agricultural lime from the later sixteenth century. As a starting point for this discussion, a comprehensive analysis of the immediate post-monastic scene in Craven provides invaluable background detail (Hoyle 1986).

Much of Craven had been managed as monastic demesne and almost all was sold off between 1537 and 1547 to largely absentee landlords who in most instances soon sold their new estates either to what Hoyle (1986, 40-42) has described as the new gentry or to established tenants. This process accelerated after 1560. The vendors had made a sizeable profit on their earlier investment while the purchasers were able to consolidate their previous small and possibly scattered holdings by engrossment, thereby increasing their economic potential and providing the funds needed for improvement. If lands had not been sold on, they were leased out on new long leases (of 300, 1000 or even 5000 years) which gave the landowner a guaranteed source of income and the tenant new-found security of tenure, which enabled the latter to increase his scale of operations and capacity to improve. In Hoyle's words, 'a tenant liable to sudden, uncompensated, eviction was unlikely to invest surplus income in agricultural improvements' (1986, 57). Many of these new tenants established lineages that lasted through the centuries, with progressive improvements throughout, and the reality that most arable land was now converted to permanent pasture enabled Craven to withstand the worst of the late-Tudor recession and the harvest failures of the 1590s.

Collections of assignments and conveyances have survived for Craven, from at least the 1590s to the 1650s, confirming how common was the process of new intaking, engrossment and rationalisation of holdings (Skipton MSS 601).

The sixteenth century also witnessed the inception of stinting in Craven whereby open commonly-held pastures were enclosed by walls with a carefully regulated allocation of beast and sheep gaits (Winchester 2000, 68-69). It was in the interests of those tenants with an allocation of grazing rights within a given stinted pasture not to abuse their rights, which may lead one to assume that they would have been keen to improve such pastures in the interests of improving their own stock, though in reality a lack of communal effort was often a major factor behind the later carving up of stinted pastures into smaller, privately-owned allotments. This vested interest could also explain why the process of individuals buying up gaits to consolidate their 'ownership' of such pastures gained impetus during the sixteenth and seventeenth centuries; it also explains why the ancient regulatory practice of levancy and couchancy was gradually abandoned in the Settle area in favour of gaits being rented out by owners of stinted pastures.[33]

A similar set of processes is apparent from adjacent areas within the Forest of Bowland (Porter 1977, 4-15 passim). Wastes were actively enclosed and divided up from the mid-sixteenth century and Porter identified cases where lime kilns were specifically constructed on these new enclosures for land improvement in the second half of the century. If this was happening in Bowland, where limestone resources are not readily available, it is reasonable to suggest that it may also have occurred in limestone-rich Craven.

Occasionally, legal documents provide tentative evidence of the application of lime. In 1572 John Kydd of Kilnsey, yeoman, demised to two brothers, Richard and Robert Tombson of Malham, lands including 'one little close "lymekylnebancke"' (YAS MD 247/6). This close has been identified on the ground, near Chapel House in Upper Wharfedale, as the field name still existed at the time of the tithe apportionment in 1845. It has a remote situation and was adjacent to a field which is known to have been arable well into the nineteenth century. There are signs in the adjacent wood that limestone was quarried here, suggesting there was indeed a kiln. There is, however, no means of determining if the lime was used on the enclosures here.

Evidence from Hornby manor, adjacent to and west of Craven, is also suggestive of early lime production, with no hint provided as to its end use. The court roll for May 1579 contains a plea: 'Jacobus Kempt de Claughton queritus versus Thomam glooes junior de robrondaile in placito debiti vs. viz for one stone of lyme to him sold ut declaret' (*sic*) (Hornby Castle Muniments vol. 2). Kempt had complained that Glooes, presumably a farmer in Roeburndale, owed him 5s. for lime not yet paid for.

In 1617 William Foster of Arncliffe in Littondale, yeoman, demised to William Bincton and family of Litton various properties in Arncliffe which included one parcel 'of meadowe or pasture grounde in the myires called Lymekilne Close' (WYAS [W], Hammond). This field name also persisted and it has been possible to identify its location and the fact that a kiln still survived within it in the 1840s when the first Ordnance Survey mapping was undertaken here – but was it an agricultural kiln?

[33] This ensured that the number of livestock that farmers were allowed to graze on the commons in summer was in direct proportion to the number they could over-winter at the farmstead.

An analagous, if vaguer, conveyance recorded the transfer of a tenement at Burnsall in 1622 from Stephen Tempest of Broughton Hall to a group of nineteen yeomen. Among the parcels of land were one called 'Lymgill' and another called 'Lymgarth' (Ironbridge RS 1057.25). Place-names such as these are more likely to refer to lime rather than to limestone rock, as documentary sources tended to be explicit when writing of limestone. This interpretation is supported by the geology in the locality, but whether the fields were so-called because they contained a lime kiln or because they had been limed remains speculative.

A note of caution must be sounded with place-names as false assumptions can so easily be made. A case in point is an indenture whereby William Caterall and colleagues demised to Ralph Backhouse a tenement called New Hall, between Giggleswick and Rathmell, in 1628 (LRO RCHY 2/2/128). Among the many appurtenances listed is 'Kilne thwaite'. It would be premature to conclude that the kiln was a lime kiln: it may have been but it could equally well have been a corn drying kiln. Two decades later William Hill of Settle, yeoman, drew up his will and among his bequests was one to his son Hugh of 'all my pte in the New Kilne' (Borthwick 1647-48). The farm is not identified in the will and there is no possibility now of saying what kind of kiln this was.

However, wills have proved to be a useful source of data. Two examples from across the research area will illustrate the point. Thomas Bullock of Tosside bequeathed to his son, William, *inter alia* 'his limes stones down the near dale in the haies' (Borthwick 1570). The last word means enclosures and here dale probably meant parcel so one can tentatively suggest either that the limestone in question had been brought in and stacked in the field ready to be burned (certainly in a clamp kiln at that date) or that it referred to lime already burned but not yet applied. It is difficult to explain this brief entry in any other way as there is no limestone geology here.

In 1621 Henry Holmes of Hebden, east of Grassington, bequeathed 'one lyme kilne and turves' valued in his post-mortem inventory at 12s. (Ironbridge RS 1039.1). There is no doubt here that this was a lime kiln, but one cannot assume it was just producing agricultural lime. The inclusion of turves, however, does strongly indicate this was a clamp kiln as they were sealed with turf in Craven at that time.

Finally, manor court rulings can help inform prevailing practices. A copy of a royal decree was included in the manor court sitting for Newby, west of Clapham, held in April 1573 (NYCRO ZUC). The former tenants of Furness Abbey had complained because they had been sorely 'troubled and impoverished by Inclosures and improvements of their Commons of Pasture' over the preceding decades. The ruling does not exemplify what improvements had been implemented but evidence to be presented in the next chapter suggests that liming was involved, and archaeological evidence will demonstrate the widespread burning of lime in a very rural locality within the manor of Newby.

By around 1600 the manor court of Lothersdale had enshrined for all future land and farm conveyances the right to 'digg and get limestones ... needfull for the meantaining and manuering' of the land in question (Wilson 1972, 28). This is unequivocal: tenants here had been given the inalienable right to delve for limestone locally to burn into lime for application on the land. This is perhaps the clearest indication that the practice of liming was well established in Craven before the end of the sixteenth century as the ruling was giving a clear legal basis to what had been a *de facto* habit hitherto.

It is known from Skipton Castle coal accounts that lime was being burned at Carleton Park, south-west of the town, around 1605, and field examination here indicates that several fields have been improved, by draining and probably liming, at some point, though there is no documentary evidence to directly link these kilns with the improved fields (Gill 2004, 27).

At about the same time the bylawgraves of Giggleswick (this township had no formal manor court system) reiterated regulations first set forth in 1574. These have been lost since being reproduced by Brayshaw and Robinson (1932, 83-84). The 1602 issue included two bylaws of direct relevance to this discussion. Firstly, 'none of the towne of Giglesweeke ... shall build any lime kilne within the same towne nor in the gardens orchards or crofts adjoyneing thereunto in paine of xxs.' The second was very different: '... the inhabitants of Meweth before the 10th of August next shall repaire every one their parte of the kilne called stane kilne and so after uphold it in paine of iiis. iiijd.'. Both bylaws specifically referred to lime kilns, Mewith was and still is a rural, pastoral township, and Giggleswick was a major producer of lime until 1974. Whatever end uses these kilns may have had, one can safely assume that agricultural lime was one of them. Two further conclusions can be drawn from these bylaws. By 1602 lime burning in Giggleswick had reached such a scale that pollution was perceived to be a serious nuisance in need of control; and the Mewith kiln must have been provided for communal use.

7. Lower Wharfedale and Airedale

To some extent, the lower reaches of Airedale and Wharfedale, as far downstream as Otley and Keighley respectively, demonstrate similar historic land management processes to Craven and the northern dales. The paucity of firm evidence is equally apparent here, though isolated entries within the documentary record do help to build up a clear picture of what was going on, especially references to place-names, legal cases and manor court rulings. Unlike the case studies discussed earlier, there is a corpus of evidence (archival and archaeological) that stone for burning into lime in this area was obtained locally either from boulder pits or delfs, dug into boulder clay on the moors, or by a practice called river stonary whereby limestone cobbles were picked from within river channels. It is known that some of the lime so obtained was applied on the land, and the nature of these two practices illustrates the lengths farmers and/or limeburners were prepared to go to in sourcing lime. As these practices continued right through the seventeenth century, however, they will be discussed in more detail in Chapter 5.

Askwith is typical of the townships along the Lower Wharfe in terms of environmental conditions and land potential. It is also typical in terms of land management: most of the open fields within the manor had been divided up into individual enclosures by 1596, for improvement and increased productivity (Pickles and Bosworth 1991). Similar processes can be recognised across the watershed in Airedale as, for example, in Oakworth where most of the common land was carved up in 1614 and 1615 (WYAS [B], Ferrand C/218), again presumably for improvement.

Contemporary correspondence confirms the contrast between the more productive valley bottoms and lower slopes on the one hand and the moorlands on the other. Between 1587 and 1590 James Ryther, lord of the manor at Harewood, wrote a series of letters to Lord Burghley (reproduced in Craig 1984). In one letter, written at Barden on the Wharfe in September 1589, Ryther commented on the 'much fine and fertill pasture & meddowe, som in their hilltops' that he had observed on his journeyings as well as 'som grounde that will bear no kynde of graine', partly owing to the extreme climate. Little has changed here in the intervening four centuries and he would undoubtedly make similar observations today.

None of these sources made mention of lime, or any other soil additive, but Ryther's contemporaries, William Camden and John Speed, did point to lime's virtues. Camden started working on his *Britannia* in 1577, publishing it in Latin in 1586 and in English in 1610 (Camden 1610, 862-63; Taylor 1951, 374). He described the organised and large-scale production of lime around Brotherton and Knottingley, south-east of Leeds, and – most importantly for this narrative – its transportation *in great quantities* westwards (into the Aire valley) where 'they manure and improve the soil' (Camden 1610, 862). What cannot be determined from his writings is exactly when he had made these observations but it was certainly prior to first publication in 1586, thereby providing one of the earliest definitive references to the application of agricultural lime in the research area.

The year 1610 is of especial significance: John Speed had completed his *Eboracensis Comitatus Descriptio* (Description of the County of York) which was later to make up Chapter 39 of his great atlas of Britain, published in full, in Latin, in 1612 (Speed, 1610). On the reverse of the map of the county Speed added a description. Though he did not locate precisely where he was writing of, it is highly probable that part of his description, the quarrying, referred to the Magnesian Limestone belt (see Figure 1.4) because he described these rocks as *molliores*, which means soft. Carboniferous limestone could never be thus described whereas Magnesian limestone is much softer both physically and in surface colouration. The remaining part of his description, talking of where the lime was used, referred to either Craven or the moors around Lower Airedale and Wharfedale. Of the Magnesian belt he wrote 'Una ipsius pars Latomia quadam singulari nobilitatur ...' ('one part of it is ennobled by a certain quarry'); of the latter he wrote:

> Altera vero non minus calce viva illic passim effossa celebris, cuius excoctae iam sive exustae magna vis in alias partes quae montosae et frigidiores convecta ad agros laetificandos et foecundandos plurimum facit.

It can be translated thus: 'The other part, however, is no less celebrated for the quicklime (*calce viva*) dug up there from time to time, the great power of which when baked or burned, when it has been taken to the other mountainous and colder parts, has very great effect in making the fields rich and fertile'.

In the margin, adjacent to this paragraph, are the words 'Calx, laetificat agros' (Fig. 4.1). These three words speak volumes because they mean 'lime gladdens the fields'. If it was added by Speed himself, which is most likely, this is the earliest attributable, first-hand, comment on the *value* placed on lime as a soil improver – as opposed to just its usage – for the north of England: it was written by 1610 but the ground surveying for the map must have preceded it by several years. Its significance cannot be overestimated.

Lime production in Lower Wharfedale is known from as early as 1307 when an Extent, an estate inventory required when a manor changed hands, was compiled for the manor of Otley by Robert de Blunderstone, royal commissioner (Otley Museum O/O/dc/7/13)). Listed is a pasture called 'Lymkilnbanks', valued at 6d., with a note to the effect that Richard Ketill held '1 acre of meadow ... *Item* he holds freely 1 rood of meadow below Lymkillbanck and pay 3d'. This meadow has been identified and lay immediately south of Otley, close to the banks of the Wharfe. A valuation such as this would have no reason to state if a lime kiln was still operational unless it had a taxable value in its own right. The field name, however, points to there having been one at some point in time, and its location within a meadow may point to agricultural usage.

Other pertinent place-names include 'Ley Lympittes' at Reynard Ings in Addingham on the Wharfe, mentioned by Smith (1961) for 1421-22. The term lime pit on occasion meant a tanning pit where hides were soaked in lime but it was also used to refer to early clamp lime kilns. A survey of Denton Estate, undertaken in 1596, itemised all fields with their area and name: two such were called 'Lyme Stone Hill' and 'Kiln Flatte' (NYCRO ZFW, Denton). Taken in isolation they mean nothing but together they could indicate that the limestone in one field was being, or had been, burned in the kiln in the other.

Several rather later documents included relevant place-names. For example, the post-mortem inventory for William Hill of Cottingley in Airedale, finalised in 1627, listed two small limestone quarries or delfs (Ironbridge Box 14 RS 1057.20). One was a 'lyme delfe in bean banke' and the other 'another delfe in Heley belonging to John Hill'. Leases on the Horsfall estate at Burley-in-Wharfedale concerned fields such as two 'Lime Delf Closes' used as meadow in 1651; and a parcel called 'the Limestone Bed' in 1659. It is impossible, so far removed in time, to determine why lime was being produced in these places but, nevertheless, it is valid to suggest that some of it could well have been applied to the land, given their isolated

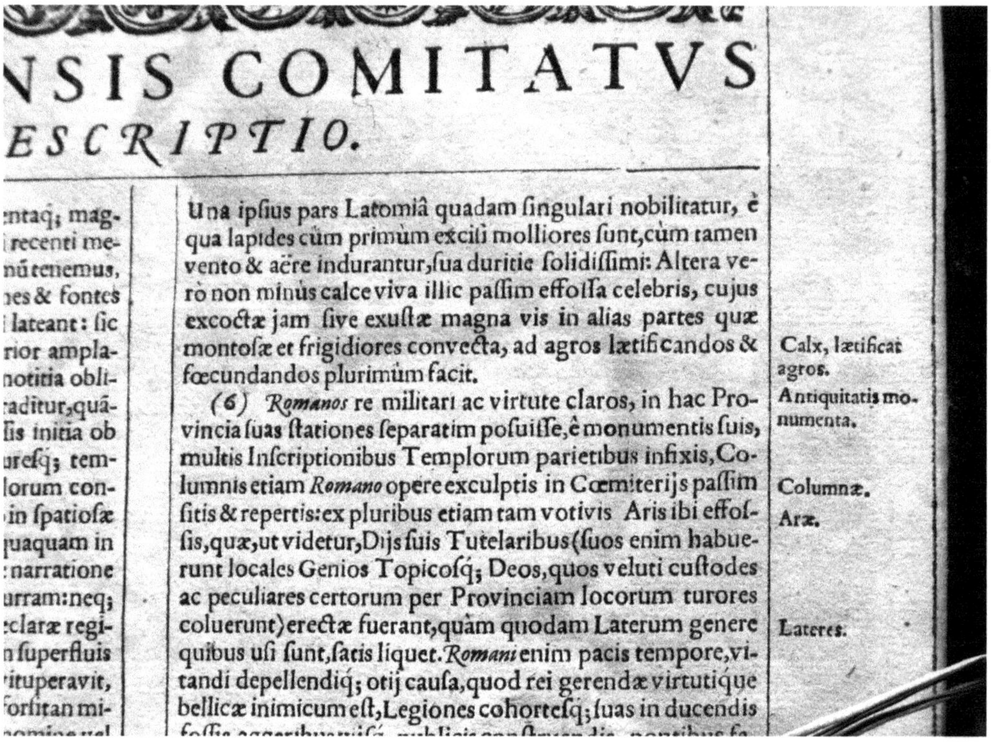

NSIS COMITATVS
ESCRIPTIO.

Una ipſius pars Latomiâ quadam ſingulari nobilitatur, è qua lapides cùm primùm excili molliores ſunt, cùm tamen vento & aëre indurantur, ſua duritie ſolidiſſimi: Altera verò non minùs calce viva illic paſſim effoſſa celebris, cujus excoctæ jam ſive exuſtæ magna vis in alias partes quæ montoſæ et frigidiores convecta, ad agros lætificandos & fœcundandos plurimùm facit.

(6) *Romanos* re militari ac virtute claros, in hac Provincia ſuas ſtationes ſeparatim poſuiſſe, è monumentis ſuis, multis Inſcriptionibus Templorum parietibus infixis, Columnis etiam *Romano* opere exculptis in Cœmiterijs paſſim ſitis & repertis: ex pluribus etiam tam votivis Aris ibi effoſſis, quæ, ut videtur, Dijs ſuis Tutelaribus (ſuos enim habuerunt locales Genios Topicoſq; Deos, quos veluti cuſtodes ac peculiares certorum per Provinciam locorum turores coluerunt) erectæ fuerant, quàm quodam Laterum genere quibus uſi ſunt, ſatis liquet. *Romani* enim pacis tempore, vitandi depellendiq; otij cauſa, quod rei gerendæ virtutique bellicæ inimicum eſt, Legiones cohorteſq; ſuas in ducendis

Calx, lætificat agros.

Antiquitatis monumenta.

Columnæ.

Aræ.

Lateres.

Figure 4.1 Part of the description accompanying John Speed's 1610 map

situation. Otley, too, had its limestone workings: the parish registers note the demise of John Wood in 1640 who had been 'slaine in Lyme delfe' (Otley Museum O/pb/1).

Evidence of early lime burning has been located in surprising sources. One such is from a witness statement in a trial held in York in 1594 to consider a case of defamation of character (Borthwick CP. G 2812) during which one particular witness was Thomas Teale of Ilkley, described as 'lymeburner'.

Equally, the Cononley (Airedale) manor court rolls offer relevant insights into lime burning for the turn of the sixteenth-seventeenth centuries (Devonshire Collection L/45/7). In October 1590, for example, a paine was laid 'that whosoever shall give or sell any stones, called "Lymstones", from which live lime (*calx viva* in the original) is burned, outside this lordship ...' would be fined 12d. per horse load. This was a common form of verdict designed to ensure that products stayed within the bounds of the manor. In October 1616 James Bradley was amerced 6s. 8d. because he had illegally taken turves off the common 'to burn in his lime-kilns, called lyme killes'. The equivalent sitting in 1617 considered a plea against Henry Walsh because he had stolen someone else's 'lymestone hamer' and, on a less trivial note, the 1618 sitting accused his kinsman, Edward Walsh, of stealing '20 cartloads of limestone being at Carleton Park head', though the intended use of the stone was not stated.

What manor court entries like these demonstrate is the extent of lime burning in the area in the post-medieval era: one may, with caution, extrapolate from them the degree to which any of it was intended for farming. Without detail provided by the likes of Camden and Speed, however, such extrapolation would remain speculative.

8. Conclusion

Contemporary views

Chapter 4 has considered the earliest surviving evidence of land improvement within the Central Pennines and the first signs of the application of agricultural lime, and has discussed soil inputs employed before liming came into common use, such as dunging and marling. The period reviewed ranges from the monastic era of 'high farming', through the tribulations of the fourteenth century and the somewhat halting economic progress achieved under the Tudors, to the middle of the seventeenth century. The line has been drawn, albeit rather arbitrarily, around the time of the restoration of the monarchy in 1660. Similar trends overlapped this divide, but the pace of change in farming and the rural economy, and the increased take up of liming after 1660, provide the justification for this cut-off point.

Before analysing the main trends identified in this chapter, a brief résumé of contemporary thinking will be provided by reference to a number of agricultural treatises written prior to 1660. By so doing it will be possible to qualify the extent to which actual real-time management practices within the research area coincided with wider, theoretical trains of thought.

Some modern writers have gone on record saying that the earliest documented reference to liming in England dates from the sixteenth century though, as they suggest, it may have been used earlier (Havinden 1974, 111). This view is broadly supported by a quote from 1616, for south-east Wales, that the 'husbandry of liming the ground for corn was first practised within the memory of the fathers or

Writer	Date	Recommendations	Locales mentioned
Walter of Henley[1]	13th c.	manure, marl	South-east
Seneschaucie[2]	13/14th	manure, marl	Unknown
Fitzherbert, John	1523	muck	South-east
Fitzherbert	1534	pigeon dung, marl	Yorkshire
Fitzherbert	1539	muck, marl, lime	Lancashire, Cheshire
Tusser, Thomas[3]	1557	dung	South-east,E. Anglia
Heresbach, Conrad	1567	dung, urine, ashes, marl, folding	Germany
Googe, Barnaby	1577 & 1614	as Heresbach[4]	General
Norden, John	1618	marl, dung, lime, watering	Welsh Marches
Markham, Gervase	1631	marl, dung, lime, ashes,urine, rags	Lancashire, Yorkshire
Fuller, Thomas	1642	marl, lime	General
Blith, Walter	1652	marl, dung, lime, folding	General
Taylor, Silvanus	1652	benefits of enclosure	General
Hartlib, Samuel	1655	marl, lime, folding, ashes, blood, shavings, rags	General
Plattes, Gabriel	1656	marl, dung, lime, folding, dregs, watering	General

Table 4.1 Soil inputs recommended by agricultural writers, pre-1660[5]

[1] See also Lamond (1890).
[2] See Lamond (1890).
[3] See Mavor (1812).
[4] Googe's book was a 1577 translation of Heresbach's.
[5] See the Bibliography for full references.

grandfathers of men yet living', which would place it in the third quarter of the sixteenth century (quoted in Emery 1967, 133-34).

Contributions from a number of contemporary writers have been examined during the current research, of which nine stressed the value of dunging, ranging from Walter of Henley in the late thirteenth century to Gabriel Plattes' work published in 1656. Other forms of soil additive promulgated in the treatises, and applicable to Yorkshire, included marling, folding of sheep, watering of meadows, and inputs such as human urine, torn up rags, ground down hoof and horn shavings, soap ashes, blood and the dregs of beer and ale. Their recommendations are summarised in Table 4.1.

The benefits of liming were discussed, usually in far more detail than for any other single additive, by seven of the writers, with the 1539 edition of Fitzherbert's treatise being the earliest. He began by saying 'Another maner of mendynge of earable lande is to mucke it, marle it, lyme it, or dunge it' thereby neatly covering all possibilities (Fitzherbert 1539, fol. 42). Later on, however, he expounded on the virtues of lime as 'he that hath Limestone ... may lyme his grounde, and that will brynge good corne, or he may sell his lyme at his pleasure' (fol. 48).

It can probably be assumed that Fitzherbert was well aware that lime was already a commodity being traded, albeit maybe only in the south of England, and its use could have extended back before 1500.

Norden (1618, 224) described lime's use in the English-Welsh borderlands where it was applied to 'great advantage' by 'poore husbandmen and farmers', and he asserted that no soil was incapable of improvement (Thirsk 1992, 27); Markham

(1631, 7) maintained that it was the 'strength and goodnesse of the Lime ... which brings forth the profit'; Fuller (1642, 117-18) believed a farmer could 'bettereth his ground, and his industry worketh miracles' if lime was applied; while Blith (1652, 133) claimed he transported lime for his estate from 30km away because it gave him 'Great Advance'.

The extent to which any of these books reached those who actually toiled on the land has been hotly debated though there is general agreement that the later examples were aimed less at the landowning classes (Thirsk 1983, passim; McRae 1992, 37-47). Tusser ran to over twenty editions between 1557 and 1638, perhaps suggesting a broad readership, and Markham openly directed his prolific output at the ordinary farmer.

Overview

Whereas the thirteenth century was a time of agricultural progress and prosperity, helped by benign climatic conditions and fostered by the monastic houses that dominated much of the Central Pennines, the following decades were the complete antithesis. Severe climatic downturn led to harvest failures and to famine; in turn, this led to epidemics among both people and cattle, and to repeated military incursions from Scotland. All this can only have impeded improvement.

The thirteenth century witnessed the first definite signs of land improvement, seen in the writings of Walter of Henley and in the *Seneschaucie*. Marling was well documented; engrossing and assarting were common practices; fleeting glimpses of liming can be had from references to lime kilns in monastic documentation. The evidence for improvement is clear, but for liming is sporadic and rather tentative.

Despite the troubles of the fourteenth century there is still evidence of investment in improvements, no doubt by those far-seeking husbandmen who saw the opportunity to improve their situation among the general gloom of population crash and labour shortage. The Scropes were actively improving on their extensive Wensleydale demesne, and there were similar signs from Otley.

As conditions slowly improved and demographics returned to a new state of equilibrium, land improvement gathered pace through the fifteenth and sixteenth centuries: marling perhaps continued to dominate the scene within the research area but liming was coming more to the fore. The evidence for liming is still rather circumstantial though the corpus of reliable references of agricultural lime production began to increase in the late sixteenth century, a view accepted by Kerridge (1967, 248) and Fieldhouse and Jennings (1978, 156), as discussed earlier.

The northern dales

Lime was being produced on a large commercial scale to the north of Richmond, at least from the 1570s, with much of it being centred on agricultural areas. It is not possible to say with certainty that some of it was going on the land but the possibility remains. Surveys within Swaledale in 1561 and Abbotside in 1613-14 provide strong evidence for some kind of land improvement and there is circumstantial evidence of liming in the former during the sixteenth century. The survival rate of contemporary documents, assuming they had ever existed in quantity, does not allow for categorical assertions of liming to be made here, though.

Craven

A similar situation pertained here, too, and conclusions must remain largely provisional. Enclosure, engrossment and improvement are well documented; there are tantalising references to kilns or limestones in wills and manor court proceedings in the sixteenth century, for Tosside, Wharfedale and Newby; and the Giggleswick bylaws of 1602 provide further clues of liming. Definitive evidence is available for Bowland, adjacent to North Craven in the middle of the sixteenth century, and it is highly likely that this was replicated in Craven. The thirteenth-century radiocarbon dates from the excavated clamp kiln at Halsteads, close to the border between Bowland and Craven and in a very remote situation, hint at the possibility of liming in the monastic era. However, the earliest quantifiable signs of liming in Craven date from 1600 in Lothersdale, and the manor court ruling made in that year was clearly formalising a practice that had been going on previously so it is perfectly justifiable to conclude that liming in Craven predates that date.

Lower Wharfedale and Airedale

Here it is possible to be rather more specific concerning the application of agricultural lime. The various anecdotal entries – such as Ryther's letter of 1589, the Cononley manor court regulations from 1590, the court case of 1594, and the Denton survey of 1596 – are clear about the large-scale nature of improvement and lime burning in general. The 1307 Extent of Otley provided stronger evidence that lime was being used on the land as early as that but, even here, caution is needed. What brings it all together in a coherent manner, though, are the observations and writings of Camden and Speed who were both unequivocal about liming and the value of lime in the research area before the opening of the seventeenth century. This correlates closely with observations made in 1595, in Pembrokeshire, that 'lyminge hath ben more used within these thirtie or fortie yeares then in times past ... and hath greatlie enriched those (lands) that used it' (Owen, quoted in North 1930, 389); and in 1616, in south-east Wales, that the 'husbandry of liming the ground ... was first practised within the memory of the fathers or grandfathers of men yet living', that is in the second half of the sixteenth century (Emery 1967, 133-34).

Having analysed a comprehensive set of data covering 350 years, it is considered reasonable to suggest that the practice of liming was established throughout the research area before 1600. Many of the instances cited are undoubtedly tenuous and viewed in isolation would not stand up to scrutiny but considered together they do add up to an emerging picture of farmers moving away from dung or marl as the primary soil additive in favour of lime. It is not claimed here that the practice was universal across the Central Pennines but the evidence provided above does suggest that liming was part and parcel of farming in the research area before 1600.

The next chapter will take the discussion a stage further by examining increased usage of lime through the later seventeenth century and the early to mid-eighteenth.

Chapter 5
Agricultural Land Improvement and the Application of Lime, mid-seventeenth to mid-eighteenth Century

1. Introduction

Chapter 4 examined the earliest evidence for agricultural lime burning in the Central Pennines and made the point that much of the surviving evidence is conjectural though there are conclusive references to the application of lime on the land, especially in Lower Wharfedale, prior to 1600. The current chapter takes the narrative forward from the middle of the seventeenth century to the middle of the eighteenth, to the point when organised enclosure of commons and wastes saw increased emphasis on improvement and greater usage of lime (or at least more evidence of it) as a means of effecting change.

Restoration of the monarchy in 1660 provided the impetus for this widespread investment in property, buildings and land throughout the research area reflecting an enhanced sense of political and economic security and opportunity, seen in the Dales and Westmorland in the latter decades of that century (Hoskins 1953, 44-48). Within south Lonsdale the changes were most apparent between 1660 and 1730, and one writer has talked of 'new (agricultural) improvements' being implemented in Tatham during the 1690s (Garnett 1987). Not all was well, however, as farm-gate prices remained at low levels across the country from around 1660 to the end of the century though the negative impact on producers was perhaps limited by reduced rentals (Hainsworth 1992, 54).

This new order also spawned wider developments in the arts and sciences with, for example, the founding of the Royal Society in 1660 and the blossoming of the Georgic movement's literary, scientific and religious 'revolution', phenomena that flourished within the Commonwealth era when 'enclosure and improvement of land gained a new moral legitimacy' (Low 1985, 354) and when estate improvement was perceived as a moral and patriotic duty (Wade Martins 2009, 81).[34] It could be argued that the later surge of parliamentary enclosures was a logical, if much delayed, outcome of the post-Cromwellian and later Stuart drive to improve and innovate, though it is difficult now to quantify the extent to which late eighteenth-century change would have happened had the Georgic movement not materialised in the ways that it did. Integral to this movement was the publication of a number of agricultural treatises which differed from earlier works by being more relevant to land managers and farmers and by having a firmer empirical base: these will be reviewed in Section 7.

The period under review witnessed the emergence of individuals who collectively became known as the 'Improvers'. While they are generally more associated with the turn of the eighteenth and nineteenth centuries, a few did make their appearance a century earlier. Notable amongst these were Christopher Crowe, of Kiplin Hall on the Swale, and William Danby, of Swinton Park on the Ure: their contributions will be assessed in Section 3.

A desire to improve led to an increase in the quantities of lime being produced, not just within the Central Pennines but also more widely. Baron Tiquet noted, in a letter sent back to France in 1738 from his journeys through Britain, that the many farmers who made use of lime in their husbandry saw their income tripled (quoted in Nef 1966, 237). This process has been demonstrated for Cumberland from a detailed perambulation, which also took in Westmorland, completed by Thomas Denton for Viscount Lonsdale in 1688. Denton had observed 'great improvement of their lands by burning of lime, which helps to fill their barns with abundance of corne' in the Cockermouth area, along with several instances where the burning of lime palpably enriched farmers (Winchester 2003, 149-50).

More generally, it has been suggested that 'new fertilisers' such as marl and lime became increasingly common during the seventeenth century (Darby 1976, 18); and that later in the following century the increased application of lime led directly to the replacement of oats and rye by wheat in the more climatically favoured parts of England (Prince 1976, 108).

Analysis of year-by-year sale prices for lime ought to emphasise variations in demand, as periods of greater demand could conceivably have been closely reflected in higher prices. However, this is not necessarily borne out by the facts: at the start of the period under review in this chapter the average price per quarter, in England in general, varied between 4s. and 6s. 8d. but remained at or slightly below that lower level for the rest of the century, falling to less than 3s. in 1710, and only rising again after 1743 (Thorold Rogers 1963, vol. 6, 480-89 and vol. 7, 407-09). These data must be viewed with a degree of caution, though, as variations in quoted sales units make direct comparison somewhat suspect.

Rising demand for agricultural lime extended to areas where the raw material – limestone bedrock - was not available and foresighted individuals and partnerships devised ingenious means of obtaining limestone from glacially-derived sources. These will be discussed in Sections 5 and 6. Much of the research in this respect has been based on extensive field surveying by this writer.

The methodology of traditional field archaeology has been employed to confirm that agricultural lime was not only being produced over a much greater spatial extent than hitherto appreciated but was also evident on a large scale further back in time. This aspect will be considered in Section 4.

[34] Virgil's influential work, *Georgics*, was first translated into English and published in 1589 (Thirsk 1992, 19).

Long-term climatic trends are likely to have had an impact on farming in the region. The Little Ice Age which reached its coldest phase in the late seventeenth century and the exceptionally cold Late Maunder Minimum from 1675 to 1696 have been demonstrated for most of northern Europe from records of fish stock landings, as different species dominate the catch depending on prevailing water temperatures (Grove 1988, 367, 415; Gauninga *et al.* 2007; Mackenzie *et al.* 2007). These conclusions have been confirmed by analysis of long-term temperature trends for north-west England (Manley 1974; Parker *et al.* 1992). Through the period under review in this chapter, only the decades 1720 to 1740 saw sustained average annual temperatures in excess of the long-term (1659 to 2007) average of 9.43^0 Celsius, while the years 1665 to 1700 were significantly below the running mean. Precipitation data are only available after 1765 so no evidence-based conclusions can be drawn about the possible influence of wetter than average conditions on farming in the research area.

The current chapter will adopt a sub-regional and exemplar focus to enable a comparison to be made of liming from one area to another within the overall research area as the approach adopted in south-east Westmorland was markedly different from that in north-east Lancashire or Lower Airedale and Wharfedale. Different localised sets of circumstances, emphases and opportunities brought about variations in approach, as will become apparent.

2. South-east Westmorland

After the Restoration landowners and tenants alike took the opportunity to make the improvements and investments that years of economic and political insecurity had prevented. This is not to say that nothing had been done in the past, as witnessed by an entry in the Ravenstonedale court rolls in 1579 concerning an indenture between Sir Thomas, later Lord, Wharton and thirteen of his customary tenants (CRO[K] WD/KILV/3/7). He had emparked land around his estate in Ravenstonedale, beginning in 1547 or 1560 (sources disagree on the date), which involved removing sixty-nine tenants from their existing properties and relocating them outside the new stone-built park pale. Wharton had tried to be fair by settling them on land 'newly improved from the waste' and subjected to 'newe Improvements', though the nature of these was not described. One such tenant, Edward Milner, received '23 acres new improvement' at Whitwall and on Ash Fell (Blackett-Ord 1986, 133-35): much of the latter, even today, is unimproved and unenclosed pasture of an inferior quality compared with much of the general area.

The late seventeenth century is seen as the period of the 'Great Rebuilding' within the research area whereby houses and ancillary farm buildings were either substantially modified or replaced by new stone structures (Marshall 1980, 511; Winchester 2006, 82) as part of a more general desire to improve.

This new-found urge was extended to land with concerted attempts at improvement being made after 1660, which continued well into the next century, and included the construction of kilns for the application of agricultural lime (Marshall 1980, 513). Peat or turf was dug to fuel kilns and their perceived economic value is suggested by their inclusion in post-mortem inventories between 1660 and 1700. The process of improvement was slow, though, and Defoe was moved to dismiss Westmorland (with Cumberland) as 'the wildest, most barren and frightful of any I have passed over' (Defoe 1726, 269-70). However, his comment needs to be seen in the light of contemporary negative perceptions of upland massifs (see, for example, Byng, 1970, 89), which have undoubtedly been exaggerated (Winchester 2006, 197). Nevertheless, this writer's archival searches have located far less evidence of early improvements in Westmorland compared with the former North and West Ridings. It is a moot point whether this merely reflects the lack of fuel resources (other than peat and possibly gorse) in the county, and the high costs of carting coal in from Stainmore and Craven, or the fact that Westmorland – along with neighbouring Dentdale and Garsdale – was occupied more by small-scale yeomen (the so-called 'statesmen' – Marshall 1972), lacking the wherewithal to invest heavily in land improvements, than by larger tenants or landowners. Indeed, it has been suggested that eighteenth-century Westmorland was essentially a peasant society with minimal market orientation (Searle 1983).

One of the few landed estates here was that of Lowther Castle, and Sir John Lowther (died 1675) in particular was a pioneering improver and prodigious purchaser of land. He himself, rather than his steward, took the decisions about which land to buy, which to enclose and which to lime (Phillips 1973, 163), not just in Westmorland but also in Wensleydale. He made a series of purchases of manors and estates in tha latter, mainly between 1653 and 1665. Sir John followed in his father's footsteps by maintaining a day book, referred to as his 'Long Vellum Book', in which he made detailed entries about new improvements (CRO[C] D/Lons/L3/1/5).

Some of the entries were very comprehensive, detailing where lime was burnt, in which fields it was applied, what the soil was like prior to treatment, and providing a breakdown of costs. Most of the entries referred to his properties outside the current research area but in 1669, for example, he noted work being undertaken in Ravenstonedale (D/Lons/L3/1/5 folio 284). He had ordered a new enclosure to be walled off from the commons at Greenrigg and then 'well lymed' because the area had 'most wettish soyle which bore little grasse'. Field investigation has located an enclosure within Ravenstonedale, called Wet Shaw, below Green Rigg (SD728 986) which has clearly been improved and is surrounded by open commons where the ground is exceedingly wet, despite being based on limestone bedrock (Fig. 5.1). Lowther did not state where the lime was obtained but there are earthwork remains of lime kilns below the enclosure opposite Elm Pot and on the limestone edge of Green Rigg, at SD7341 9833. If this is the same Green Rigg as Lowther wrote of, the latter site is more likely to have been his kiln as it was not recorded on the Ordnance Survey First Edition six inch map suggesting, perhaps, that it had already disappeared from local knowledge by the 1840s, though 'Old Quarries' were marked there.

Figure 5.1 Wet Shaw enclosure below Green Rigg in Ravenstonedale. The contrast between the improved enclosure and the open fell is apparent, despite recent neglect.

3. The northern dales

Unlike Westmorland, the eastern fringes of the Central Pennines, together with much of Swaledale and Wensleydale, were the preserve of large landed estates in the hands of the aristocracy or gentry, several of whom took a keen interest in improving their holdings, either with aesthetic or economic motives. As would be expected for large estates, the survival rate of correspondence and records is sufficient to enable a detailed picture of land improvement and the use of lime to be built up, even though at times what has survived does leave room for a degree of speculation.

The lords Wharton, of Wharton Hall in Mallerstang, owned and leased out large expanses of farmland in Swaledale and to the north of Richmond. Some of this was tenanted by the Smales family who were themselves landed gentry with their seat at Gilling and extensive lands in Gayles and Kirby Hill. In the early decades of the eighteenth century this family operated their own quarries and were actively engaged in burning their own lime as well as buying it in. Matthew Smales wrote to his son, Jack, who was running the estate from Gilling, in May 1727, advising him that horses (it is unclear if he meant any horses or their own) 'happen not to be strong enough for leeding lime' and suggesting that he 'gett Jed Mason and bargain at Melsonby for 20 Chald^{s} of lime burning' using whins (gorse) as kiln fuel (NYCRO ZQH 7/17/6). In a follow-up letter sent the next month Smales advised his son 'as to lime if have not bargaind parson Lightfoot will give you leve to burn 20 Chard att his Croft End towards the Moore which will be as near as any' (NYCRO ZQH 7/17). The letter went on to suggest his son closely observe how the kiln was managed to ensure it was 'proper to be don att her': should one read into this that Smales had been sold inferior lime on a previous occasion or that he was merely erring on the side of caution? What the correspondence does not record is how the lime was to be used and it would be premature to claim a purely agricultural use for it.

Chaytor of Croft

On the south bank of the Tees lies the manor of Croft which was the ancestral home of the Chaytor family. By the late seventeenth century Sir William Chaytor held lands between Richmond and Darlington and at Stainton, south-west of Richmond. Surviving correspondence indicates that he, too, was an active improver. A nine-year farm tenancy lease, dated 1690, included a covenant that Chaytor would 'pay at the lime kiln for as much lime as John Croft shall lay upon the close called Stockmires ... (not to exceed one chalder and a half for every acre), the leading and spreading to be at the charge of John Croft' who was the new tenant (NYCRO ZQH 9/22/4). This type of clause was to become widespread across the research area and was a prime means by which landowners, or estate stewards, persuaded sometimes reluctant and conservative tenants to lime given lands. The landowner may well have just been concerned with increasing land quality to boost the value of his estate *per se*, though the steward may have had a more mercenary intent. Tenants, on the other hand, probably had their suspicions that the ulterior motive was to raise farm rents even though they, too, must have appreciated that they would benefit directly from increased productivity and income.

Chaytor seems to have been the type of landowner with a desire to improve all round but, sadly for him and his estate, his business acumen was deficient. The estate rarely made a profit and often ran at a loss, and financial problems consigned Chaytor to the Fleet debtors' prison in 1700 or 1701 where he was to spend the remaining twenty years of his life. At first his wife kept Sir William informed of estate matters and a letter sent in his early days of confinement made clear that all was not well on the estate. Another tenant, Harry Crawforth, wanted to renew his lease, but was trying by every possible means to have cancelled the existing stipulation by which he had 'agreed with you to gitt lime as he has dun every year' to spread on the land (NYCRO ZQH 9/14/24). The letter does not give his reasons but one can only surmise that Crawforth was reluctant to put in the requisite finance and time to lead and spread the lime, as the leases dictated.

If Crawforth was uncooperative, tenant John Croft was to prove himself devious and scheming, taking full advantage of Chaytor's enforced absence and of the weakness (and early death) of the latter's two sons. Croft managed to take on for himself the role of estate steward, to his own advantage and to the detriment of the estate, though he did send regular accounts to Sir William which, among other matters, itemised expenditure on building and agricultural lime.

After the death of Chaytor's wife and sons a neighbouring landowner, at Dalton, took on the task of overseeing the estate and keeping Sir William informed and a letter, dated 1719, was full of woe concerning the dreadful state that much of the land was now in. Two recalcitrant tenants were singled out for blame, as Chaytor's friend wrote : 'I do not commend letting any tenant have tillage that keeps no draught to leed lime nor manure to ly on it, ... so the land must sertainly be impoverished' (NYCRO ZQH 9/19/84). Neither of these tenants had kept their part of the bargain to maintain the fertility of their farmland by practising the form of husbandry to which they were bound. After this date, as far as surviving correspondence suggests, all new agreements stipulated that given amounts of lime were to be applied on given fields 'according to the custom of the country' (NYCRO ZQH 9/22/10).

This latter phrase, taken together with the early date of the Chaytor agreements, is of particular significance. It suggests that liming was already an established practice before 1700 north of Richmond,[35] and Chaytor's persistence in trying to improve his estate, despite having to endure so many personal and financial adversities for so long, must emphasise the value placed by him upon the process of improvement. Furthermore, Chaytor's attempts at improvement considerably pre-dated any formal enclosure between Richmond and the Tees: the earliest award, to enclose Dalton Moor, was dated 1739. In a sense he can be seen as a pioneer and who knows what he might have achieved on his estate had he not been confined to prison.

Crowe of Kiplin Hall

The Kiplin Hall estate, east of the Swale downstream from Richmond, came into the Crowe family by purchase in 1722 and they increased the estate in a piecemeal manner from 355ha to 2000ha. Christopher Crowe inherited from his father (also Christopher) in 1749 and between then and his death in 1776 he more than doubled the estate's extent. He was, far more than Smales or Chaytor, a landowner intent on maximising the potential of his inheritance and bringing about profound changes in management practices. From 1735 day-to-day decisions were left to the steward, Robert Hutton, who maintained an account book throughout his tenure in office, though Christopher Crowe the Younger took direct management into his own hands in 1761.[36]

Hutton adopted the same policy as Chaytor, namely encouraging tenants to apply agricultural lime by giving allowances against rents, and the accounts include repeated entries of such allowances (NYCRO ZBL IV/3). Some indicate that allowances were made for building repair work, others specify agricultural use, and yet others are left vague. In 1743, for example, tenant John Rhodes was allowed three guineas for nine chaldrons of lime for his 'dung hill', meaning that he was expected to mix lime with dung prior to spreading it on the land (NYCRO ZBL IV/3/1/19). This issue, of whether lime should be applied by itself or mixed with manure, was a recurring one and we shall return to it in Chapter 8.5. At the same time as Rhodes was receiving his allowance, tenant William Wood was paid £10 for two years' liming allowances and Jacob Scott £7 for one year's.

When Crowe took in hand Hutton's responsibilities the records became more detailed, listing crop yields and profits rather than just expenditure. What is clear is that Crowe soon made a name for himself, nationally as well as locally, as an improver of considerable repute, and one who hosted Arthur Young during his tour through northern England (Young 1770, vol. 2). Crowe directly managed 133ha of arable farmland and experimented widely with new implements and new crop rotation systems; he introduced new crops to the area, notably cabbages, and found favour with Young for ensuring all his turnips were carefully hoed;[37] and experimented with different ways of manuring and improving the soil believing that it was inadvisable to use the same input for more than a few years in succession.

Young stayed with Crowe in 1768 and noted that 'as to manuring; paring and burning, and liming is their chief dependance' (1770, vol.2 221). Young described in detail how Crowe went about the task of bringing waste into cultivation (1770, vol. 2 222). The first task, undertaken in March, was to pare off a 25mm-thick layer of turf which was upturned and left to dry then burned overnight. The ashes were spread out prior to the sowing of turnips in May or June. In his estimation paring would provide 200 heaps of ash per acre at 2.5 bushels per heap, giving 500 bushels per acre to which lime was added at the rate of one chaldron per acre. This, Young was assured by Crowe, would be good for seven years on all types of land. The latter was quoted by Young as saying he found lime 'of excellent

[35] See Chapter 4.5 for a discussion of lime burning in the area from as early as 1562.

[36] I am indebted to the curatorial staff of Kiplin Hall for providing access to estate papers not generally in the public domain.

[37] Young regarded unhoed turnips as the epitome of 'very bad common husbandry' (information supplied by Kiplin Hall curatorial staff). On display in the hall is a silver cup engraved 'For the Best Six Acres of Turnips Completely Hoed Kiplin 1769'.

service in mellowing the land with the spring frosts' (1770, vol. 2 238). Young's writings often portray him as a very critical observer and the fact that he was so complimentary about Crowe must mean that Crowe was achieving creditable results with his arable and pastoral ventures. In fact, Young wrote that Crowe's work was 'greatly worthy the attention of the public' (1770, vol. 2 219).

Danby of Swinton Park

Young was equally impressed by the work being carried out on the Swinton Park estate, west of Masham, by William Danby, where generally warm and dry climatic conditions permitted intensive arable cultivation in addition to pasture. According to Young, Danby's 'principal dependence' was on lime, the effects of which were perceived to last for two years per application (1770, vol. 2 276-77). He had observed that many tenants here were apt to mix lime with dung, as on the Kiplin estate, taking lime hot from the kiln and slaking it immediately – which was an unusual procedure as most farmers let it slake slowly and naturally through the action of rain – beforee mixing it together in a dung heap. This was left to mature for three or four months before mixing more manure into the heap which was then turned over twice before being carted to pastures to be spread at the rate of about seventy cart loads per hectare. In Young's view Danby was practising 'excellent husbandry (that) ... cannot be too much commended' with results that were palpably substantial (1770, vol. 2 277).

The Danby estate was fortunate in having its own coal reserves in nearby Colsterdale. Much of it was sold, but some was retained for in-house lime production, from 1773, enabling Danby to minimise the capital costs of top-dressing arable and pastoral land and to win new land from the waste (see Chapter 6.6 for a detailed account of lime-coal output here).

The importance of Young's description of improvements at Swinton Park should not be underestimated as much of the fine detail has not survived in archival form. He contrasted the efforts of Danby and of his more forward looking tenants with those more 'slovenly' small farmers in the area who seemed, to him, ignorant of best practice even though he conceded that they did apply lime on existing land and pared and burned on new land (1770, vol. 2 283).

It is worthwhile to remember, though, that Young was not exactly a disinterested observer. His writings make clear that he considered the aristocracy and gentry indispensable to the process of land improvement that to him was so vital to the wellbeing of the nation. It was they who gave the lead and set the example to be followed (Sheeran 2006, 5). As Young himself wrote, farmland was only tenanted out at low rents because those tenants were a 'pack of slovens' with no thought for the greater good (1770, vol. 2 496).

Young singled out two of Danby's tenants for special praise, however. One, Lightfoot, was seen as a cut above the rest as he applied a 'compost of bog earth' and lime to which 'nothing answers so well' (1770, vol. 2 285). The other, James Croft, was a 'genius of a peculiar cast' and Young was full of praise for his achievements (1770, vol. 2 298).

Danby was, perhaps, ahead of his time as he not only actively encouraged his tenants to improve their own land but also provided the impetus to his employees to better themselves. Much of the land in Colsterdale was, and still is, heather moorland of minimal economic potential. Many of his coal miners lived on very small plots in the valley bottom and Danby allowed those who worked their plots well to enclose adjacent sections of moorland, rent-free for the first few years. Modern aerial images would show an intricate network of small rectangular enclosures between the River Burn and Birk Gill Beck, centred on NGR SE130 813, and around Agra slightly downstream. General enclosure in the Masham area did not come about until 1797.

James Croft was one of these miners who worked in the collieries from midnight to midday. Despite this he was prodigious in his attempts to bring new land into productive agricultural use (Young 1770, vol. 2 292-305). He had begun his efforts thirteen years prior to Young's stay at Swinton Park, which would put it in the mid 1750s. He had started modestly by reclaiming less than one hectare, first draining then paring and burning before mixing the ash with three chalders of lime to be spread across the new enclosure. He planted oats but had a disappointing harvest. Not to be discouraged he treated a further hectare, again put down to oats, and obtained better results. He limed the entire area again five times over the ensuing years, finally applying a mix of lime and soil in the sixth season before converting the ground to pasture. By the time Young saw these plots the land was valued at 20s. per acre which Danby's steward described as 'astonishing' (Young 1770, vol. 2 305).

Not satisfied with his efforts thus far, Croft had then taken in a further 3ha of very stony ground and was working on a further 3ha when Young visited. Croft's success may have been due to the greater amount of lime he applied compared to the then norm and it really is astonishing that he found the time and energy to do all this and was also able to afford the costs of purchasing the lime, charged on average at 14s. 6d. per chaldron. The estate's own kiln did not come into production until 1773, well after the time that Croft was doing this improvement. In fact, he brought the lime in from an unlocated kiln 10km distant using his own single pony. This, too, is astonishing given that a pony could carry only an average of 180kg dry weight and that Croft was applying three chaldrons per acre. The weight of a chaldron varied from place to place but it was probably around 2700kg which equates to fifteen pony loads per chaldron, or forty-five per acre.[38] Multiply this by the total area he reclaimed and top-dressed every three years and the number of loads is staggering. Added to this is the fact that each return journey of 20km would have taken all day so, leading sufficient lime just to reclaim his twenty acres, never mind further top-dressing, entailed an estimated 900 journeys: if we are to believe this incredible feat, out of the thirteen years he had been improving, the equivalent of thirty

[38] Elsewhere Young had noted that a 'galloway (pony) will draw twelve bushels of lime, five miles, with ease' (1796, 299).

months must have been devoted just to leading lime. His task was made somewhat less arduous, perhaps, by the fact that the roads around Swinton Park were 'all excellent' in Young's estimation (Young 1770, vol. 4 578). Even so, if Croft had performed all this himself, rather than possibly sending a son to the kiln, his efforts can only be seen as monumental. The steward's descriptor 'astonishing' is most apt as was Young's own conclusion that Croft was 'enthusiastically devoted to the improving of moors' (1770, vol. 2 300).

It is perhaps no surprise that other landowners in that part of the Ure Washlands were also actively taking in and improving land in exactly the way Croft had been doing, and it points up clearly the degree to which liming was valued as a soil improver prior to the main period of parliamentary enclosure.

What William Danby had encouraged since he inherited the estate in 1751 was commendable but he was also following in the footsteps of his father, Abstrupus. Abstrupus had had a clause written into every new farm tenancy agreement prescribing the manner of husbandry to be followed. A lease sealed in 1733 required the lessees to spread on arable land 'three chalders of well burnt and unfallen lime (per acre) and so proportionately a greater or less quantity of lime for a greater or less quantity of such ground to be so ploughed' (NYCRO ZS* Box 88). If any plot of land was put to the same crop for two consecutive years it had to be limed or left fallow on pain of £5 per acre and forfeiture of the lease. The Danbys were intent on improving their estate; if tenants did not willingly come on board they had to be coerced.

Further up Wensleydale other landed estates were also actively improving prior to widespread enclosure. The marquis of Winchester, who succeeded to the title at Bolton Castle in 1675, was noted for the major improvements he had made with his son and heir in mind across his estate in mid-Wensleydale and in and around Leyburn (Hoyle 2007, 102-03); and the Scropes of Danby Hall, between Middleham and Masham, were also active during the early decades of the eighteenth century.

Scrope of Danby

Lime had been burnt for common use on Watlass, or Canswick, Moor from at least the mid-seventeenth century.[39] This was affirmed by various witnesses to a court case for trespass heard at York in 1721 on land on the fringes of Danby Hall property, a case which heard evidence concerning the bounds of the moor (NYCRO ZAL 6/6/8). Various elderly witnesses attested to the fact that kilns had been operated on the moor, for example, 'for 50 years last past' and one talked of an 'old lime kiln' there. Whereas most of the witnesses merely mentioned lime being burned and sold or used by local tenants, one was more specific in stating that he could recall that many inhabitants had burned 'great quantities' of lime for a number of years and that one particular kiln 'was kept constantly burning until all the tenants of Watlass had enough lime to manure their lands'. It may be that they were liming their lands at the formal behest of the landowner, but this writer has not been able to locate any tenancy agreements from this period that would confirm the matter. They may have been liming of their own volition.

The Scropes' account books, which are extant from 1725 to 1791, have many entries for payments for leading coal to a lime kiln whose location was not given, and for burning quantities of lime (NYCRO ZPT). As with all estate records consulted, many of the entries did not state what much of the lime was to be used for though some hinted at building and the occasional entry mentioned agricultural use. In the 1730s, however, various entries were very specific. Three instances will illustrate this point: on 13 September 1731 thirty-two chaldrons were spread on 'ye High Corn Close' at a cost of 6s. 6d. per chaldron and a further twenty-six on 'ye 12 Acre Close' (NYCRO ZPT 4/1/1); on 23 April 1732 payment of 3s. was made for 'one day leading lime to the Beech Close' (NYCRO ZPT); and on 10 June in that year payment was made for fifteen chaldrons, at 7s. 6d. per chaldron, for 'ye Beck Close', a payment which this time included leading the lime from the kiln (NYCRO ZPT 4/1/1). These entries are brief but do provide useful detail about the improvements being undertaken by the Scropes and the methods utilised: in the late 1730s and the 1740s piecemeal enclosures were being walled off on the fringes of Danby Moor to the north of the hall, and were then ditched and 'stubbed' (*ie* removing roots and stumps) (NYCRO ZPT). Throughout this period large quantities of lime were burned which may well have been destined for improving the soil here. In March 1726 payment was made for two days spent spreading manure; on 9 August 1750 Anthony Buckle was paid 11s. 4d. for 'spreading eight acres and half of burnt earth upon danby moor' which is assumed by this writer to mean ashes from pared and burnt vegetation. This is similar to an entry for May 1733 when estate labourers were paid for 'setting and burning sodds' and spreading manure on one close (NYCRO ZPT 4/4/3). Finally, through the 1760s, various payments were entered for ditching and 'breasting' (*ie* paring) on newly enclosed parcels of ground.

The accounts also help to inform us about the seasonality of liming, though with a degree of caution. Payments were invariably made in arrears and those entered for June or September probably related to deliveries made in the previous autumn-winter and spring respectively. Analysis of all payments in the Scrope accounts suggests that lime was being purchased at all seasons with a slight bias towards the cooler months of the year.[40] This could suggest that lime was piled in heaps and left to air-slake until spring rather than being applied immediately, and may reflect that it was easier to transport goods when the ground was frozen.

The Scropes followed the general practice within the Yorkshire Dales of writing into tenancy agreements mandatory requirements to apply lime. We have seen above that the Danbys of Swinton Park had one generic form of wording in this respect but Scrope agreements varied over time. Christopher Hanley of Low Burton, yeoman, was covenanted to Simon

[39] The moor was enclosed and parcelled up by agreement in 1754.

[40] See Chapter 6.6 for detailed analysis of seasonal patterns of usage on the Swinton Park estate.

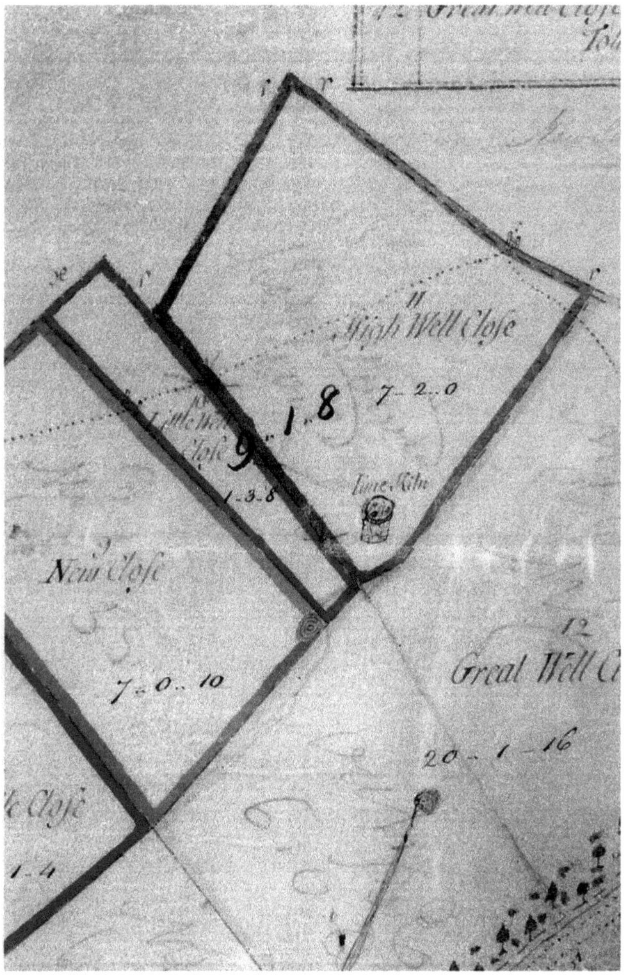

Figure 5.2 Estate map of 1772 showing a lime kiln on Hollins Farm, Ulshaw
Reproduced with the permission of the North Yorkshire County Record Office (NYCRO ZHG, Box 19)

Scrope in 1732 for an eleven-year lease of West Ground Farm and was required to treat 'with at least 8 chaldron of lime per acre when fallow' on any pasture ground which he might convert to tillage (NYCRO ZPT 4/1/5/5). By 1779 the wording had changed: John Ambler, for example, took on or renewed the lease on Hollins Farm at Ulshaw for five years and he was bound to 'keep no more than two crops from any part without fallowing and the application of two chaldrons of lime per acre at each fallowing' (NYCRO ZPT 4/1/5/6).

All the leases, though, were designed to ensure that tenants carried out the wishes of the landowner to maintain the land at the highest possible quality and, yet again, lime was central to the process. Without exception, all the estate accounts consulted during this research contain many more entries for lime burning, lime purchases or lime spreading than for spreading manure or dung or, indeed, for any other form of improvement.

The Hollins Farm lease of 1779 is of interest in another respect. The kiln from which lime was almost certainly obtained at this time for use on the Danby Hall estate and on this farm came from a kiln whose precise location, and even design, is known. A map drawn up in 1772 depicts the kiln pictorially in High Well Close field (Fig. 5.2), at NGR SE1498 8773 (NYCRO ZHG Box 19), and provides the sole evidence that these fields were indeed probably limed.

Arthur Young's comprehensive travels through the region included a visit to see Simon Scrope's efforts at Danby and he considered Scrope[41] to be a 'most accurate cultivator' as well as a committed experimenter and improver (1770, 376-85). Scrope made use of a range of soil additives, unlike the others considered so far in this chapter, with soap ashes and tanners' bark in addition to stable dung and lime.

Contemporary commentators seemed unable to agree on the best methods of applying lime to the soil and it is of interest to note that Scrope believed unfallen (unslaked) lime should remain in a large heap for two years or more prior to spreading, much longer than any other recorded practice (Young 1770, 416). We have seen that Scrope had begun enclosing around the edges of Danby Moor in the 1730s but Young noted that he had started a concerted programme of taking in the entire 360ha heather-clad moor from the waste in 1755 and had built a completely new farm within it (1770, 433-55). This was probably the farm now known as Danby (NGR SE166 875). By 1766 the transformation of 70ha had been completed, two-thirds of it under crops, the remainder 'excellent pasture'.

According to Young, all the new fields had been limed as an integral part of the reclamation process and much, if not all, of the lime was sourced within the estate. In 1764 Scrope commissioned 'some more lime-kilns' specifically for his improvements though, apart from the Hollins Farm kiln, not a single lime kiln site has been identified on the estate by this writer either on the ground or from cartographic sources.

4. Craven

Archival sources point to the use of lime within various parts of Craven during the seventeenth century. The Tatham manor court book contains a number of rulings aimed at regulating the burning of clamp kilns, such as the sitting held in April 1664 which laid a paine that 'none shall get nether toarfs or flaes nor any sods but for Limkills or fences: betweene the Spen and the Foure Score Acres yetts' (Hornby Castle Muniments. Tatham Court Book vol. 1). This rather quaint entry was aimed at those who were presumably cutting turf or peat for sealing clamp kilns near Spens Farm (SD638 668) in the Hindburn valley, where it was forbidden to do so. A further ruling, dated April 1668, forbade the cutting of turf on the Lower Common and decreed that 'none shall burne no Sod Kills on no place off the Common' (Hornby Castle Muns. Court Book vol. 2). This latter incurred a hefty payment of 39s. 11½d. per infringement, an indication perhaps that the court was determined to stamp the practice out – but why? A very similar ruling was presented at a sitting of the Tatham court in 1748 with exactly the same fine applied (LRO DDX 70/34).

[41] Young used the alternative spelling of Scroop.

Burton manor court issued a ruling along broadly similar lines in that same year but this time the motive is crystal clear. The presentment begins 'all and every persone who for the future shall dig up his soil and burn the same in Sodd Kilnes or Turff Kilns upon Bentham Moor ...' (Lancaster University, Burton manor court roll, 1748). Bentham Moor was enclosed and parcelled up in 1768 but until then it had been open commons; nowadays part of it is a golf course which is peppered with dry or water-filled hollows. These are the remains of 'marl' pits and the presentment must be referring to the digging of marl rather than just soil, though it must have been lime-rich glacial clay as marl was not burnt in kilns. The point behind the court ruling becomes clear as the presentment continued '... unless the person ... shall in every such Sodd kiln or Turff Kiln burn twenty bushels of coals'. Coal reserves were the preserve of the lord of the manor who was clearly looking after his own financial interests by insisting the tenants bought his coal.

Farm tenancies in Craven often included husbandry clauses, though a series of leases on the Craven estates of the earl of Burlington in the 1730s made no such demands (The Devonshire Collection, Londesborough MSS, Box J, x). Not all leases specified lime. In 1682 William Armitstead and John Gibson, yeomen from Stainforth, took on the lease of two closes in Cowside, above the village, for seven years and they were bound by the agreement to 'lye all the mainor' (manure) on the fields (private collection, Cowside, 1682).

The earl of Burlington received a petition in 1725 from one of his tenants, George Johnson, who occupied three farms at Bolton Bridge on the Wharfe. Johnson bemoaned the fact that the earl's agent had not fulfilled his promise to undertake a 'range of improvements including a new lime kiln' and he appealed for this to be taken into account at the next rent day (Londesborough MSS, Box J, xiv). The promised lime kiln would not have been intended for the production of mortar as one individual's needs could not have justified it, even with three farms to manage: it is reasonable to assume the lime was largely destined for improving the soil.

Surviving Clifford household accounts contain very detailed entries for disbursements for lime purchases between 1683 and 1764, and the vast majority concerned lime for mortar, though some entries were vague. However, just three payments were definitely for agricultural lime: 7½d. was paid out in June 1758 for '15 load of lime in the Pasture'; in July 1760 1s. 3d. for spreading '30 loads of lime'; and in July 1764 £1 10s. 0d. was disbursed 'for spreading lime at 44 loads' (Devonshire Collection, Clifford Household Accounts).

An indenture, dated 1751, for the privately-owned Flasby Hall estate in Malhamdale required the new tenants to apply each year for the six years of the agreement 'forty horse loads of well burnt and unfallen lime' to specified arable fields, each load to contain 'six Craven pecks' (NYCRO ZTO2). A stipulation as precise as this is a rare survival within the archival record.

It has been possible, for Craven, to supplement evidence from documentary sources with that from archaeological field surveying and excavation initiated by this writer. The archaeological work was originally conceived to test the hypothesis that clamp (or sod or sow) kilns in the Yorkshire Dales were more structured and complex than the sow kiln type-site excavated in the 1960s near Hadrian's Wall (Jobey 1966). This writer led the excavation of seven such kilns within North Craven, between 2003 and 2007, with a further excavation at Halsteads in Bowland in 2009, and was called in to advise the archaeological team working on the easement of a major gas pipeline through the area in 2006 and 2007, during which a further seven clamp kiln sites were excavated (Johnson 2006 and 2008a).

From field surveying across Craven and beyond this writer now has 102 confirmed or probable clamp kiln sites on his database (see Figure 5.4). This, plus the unearthing of seven kilns along the narrow 30km-long pipeline easement, and the not uncommon mention of sod kilns in manor court records, can be taken as firm evidence that clamp kilns were much more widespread in the landscape than hitherto appreciated. This view is strengthened by the reality of the archaeological results: none of the pipeline kilns bore any earthwork signs prior to mechanical stripping of the topsoil and, in the majority of logged sites, the surviving earthworks are visible only as vague and almost imperceptible scrapes on the surface. It is this writer's firm conviction that more clamp kilns remain to be located, and that it was a very common form of early lime kiln.[42] Field examination within the lowland fringes of the Central Pennines, such as the broad and lush Wenning Valley and Lunesdale, and inspection of Ordnance Survey First Edition mapping, show a real dearth of lime kiln sites in such areas. Yet they have emphatically been improved in one way or another and this was probably achieved, at least in part, using lime from now-hidden early clamp kilns. Later masonry-built kilns were not erected because these lands had already been improved and, even though periodic top-dressing would have been carried out, it is unlikely that the quantities needed for this would have justified the expense of building and running kilns locally.

The question of why individual clamp kilns were constructed can justifiably be raised and, apart from hints in manor court records and the very occasional reference in farm records,[43] there are no totally conclusive indications as to the usage of the lime produced. However, the context within which a kiln is found can be decisive. Of this writer's seven excavated kilns, two were associated closely with building complexes so were most probably producing lime for mortar, but the remaining five were all found in relatively isolated situations. All seven pipeline kilns were similarly isolated. In addition, the majority of logged but unexcavated sites have isolated locations.

The conclusion can therefore be drawn that the vast majority of clamp kilns were producing lime intended wholly – or at least partly – for agricultural use. To illustrate this point, two

[42] This writer was asked to lead the excavation of a clamp kiln in the Forest of Bowland in August 2009, the first such record in that area. Others here may lie awaiting discovery.

[43] See Chapter 8.6 for discussion of the work of the Garth family in Swaledale.

Kiln location	Dating method	Date AD (95.4%)	Date AD (68.2%)
Halsteads	^{14}carbon	1185 - 1280	1205- 1280 (93.2%)
Newby Cote	^{14}carbon	1440 -1640[1]	1450 - 1530 or 1570 - 1620
Feizor	^{14}carbon	1470 - 1660	1520 - 1600 or 1610 - 1650
Threapland	Archaeomagnetic	1660 - 1700[2]	
Broadwood	Archaeomagnetic	1650 - 1695	
Kilnsey	Archaeomagnetic	1620 - 1670	

Table 5.1 Clamp kiln dating results

[1] carbon dates offer different levels of probability: 95.4% and 68.2%.
[2] archaeomagnetic dating provides a range of dates for the last time the feature was fired up. It says nothing about when the kilns were first used.

Source: Johnson 2008a, 141.

examples will be chosen, one from this writer's own work and one on the pipeline easement.

On the southern flanks of Ingleborough, above the hamlet of Newby Cote, is an open area called Hagg lying at the southern extremity of Newby Moss commons. Nine clamp kilns have been recorded within a very short distance of each other, including six in a row, centred on NGR SD736 707, all of which are associated with a large but shallow quarry close by. Two of these kilns were excavated, under this writer's direction, and proven to be lime kilns while the remainder have broadly identical topographical and geophysical forms.[44] The run of six kilns is above and parallel to an early modern field wall but their flue entrances are so close to the wall that they must have gone out of use before the wall was erected. Below the wall, and parallel to it, is a pronounced ditch and bank feature which bounds an area with traces of medieval or early post-medieval cultivation ridges. During the monastic era Newby village was Furness Abbey's grange headquarters for their extensive estates in Ribblesdale and on Ingleborough, and there are several tracts of ridge and furrow and selion strips around this village and Clapham, thereby confirming that arable cultivation was once widespread.

It is suggested here that the kilns on Hagg were producing lime intended for application within the fields enclosed by the ditch and bank: we shall return to dating evidence shortly.

The second example was revealed by work on the pipeline, south of Rathmell, at NGR SD807 575. This kiln site lies within what are now improved pastures but they contain vague earthwork traces of what look like former field boundaries. This was confirmed by LIDAR investigations[45] which showed an intricate and extensive system of medieval ridge and furrow across the entire area (Cook *et al.* 2008, 33). It is quite possible that this kiln was connected with the improvement of this field system.

Dates have been obtained for three of this writer's excavated kilns found in isolated situations, for two associated with building complexes, and the Bowland (Halsteads) site (Table 5.1). Radiocarbon dates for two of the pipeline kilns, at Halton East and Embsay, both in rural situations, are remarkably similar to the Newby Cote and Feizor dates being 1460-1650 and 1440-1640 respectively (pers. comm.. Patrick Daniel, Network Archaeology Ltd).

Radiocarbon dating for the post-medieval period never provides clear cut results[46] and one could draw conflicting conclusions depending whether sixty-eight or ninety-five per cent probability is preferred. A further slight complication arises with these two sites. In both cases the dates were obtained from bone collagen from ritual partial horse burials, each assumed to have been closure rituals for the kilns, but there is no way of determining if the rituals came soon after abandonment or at some later date (Johnson 2007b). So, these two (isolated) kilns could have been in operation during the late monastic era or as late as the early seventeenth century.

Several conclusions can be drawn from these varied investigations of clamp kilns. Firstly, they were widespread features within the landscape of the southern Dales (Fig. 5.3) and many of them were sited in field rather than built contexts; secondly, it is an undoubted certainty that many more such kilns remain hidden from view; and thirdly, the dating of a sample of excavated kilns, together with the earthwork similarities to many other logged sites, point to their widespread usage before (at least) the end of the monastic period. We can safely surmise, from this writer's archaeological work and from archival evidence, that lime was being produced in Craven in simple clamp kilns for agricultural use well before and during the period under review in this chapter. The significance of the archaeological element to this research programme cannot be overstated.

[44] Space precludes description of the excavation process here.
[45] **LI**ght **D**etection **A**nd **R**anging: remote sensing technology that can detect topographical data not clear from ground or aerial observation.

[46] Because of the short time span that has elapsed, and the slow breakdown rate of ^{14}carbon, very little decay can have occurred so the standard deviation may exceed the date obtained, thereby making the latter meaningless.

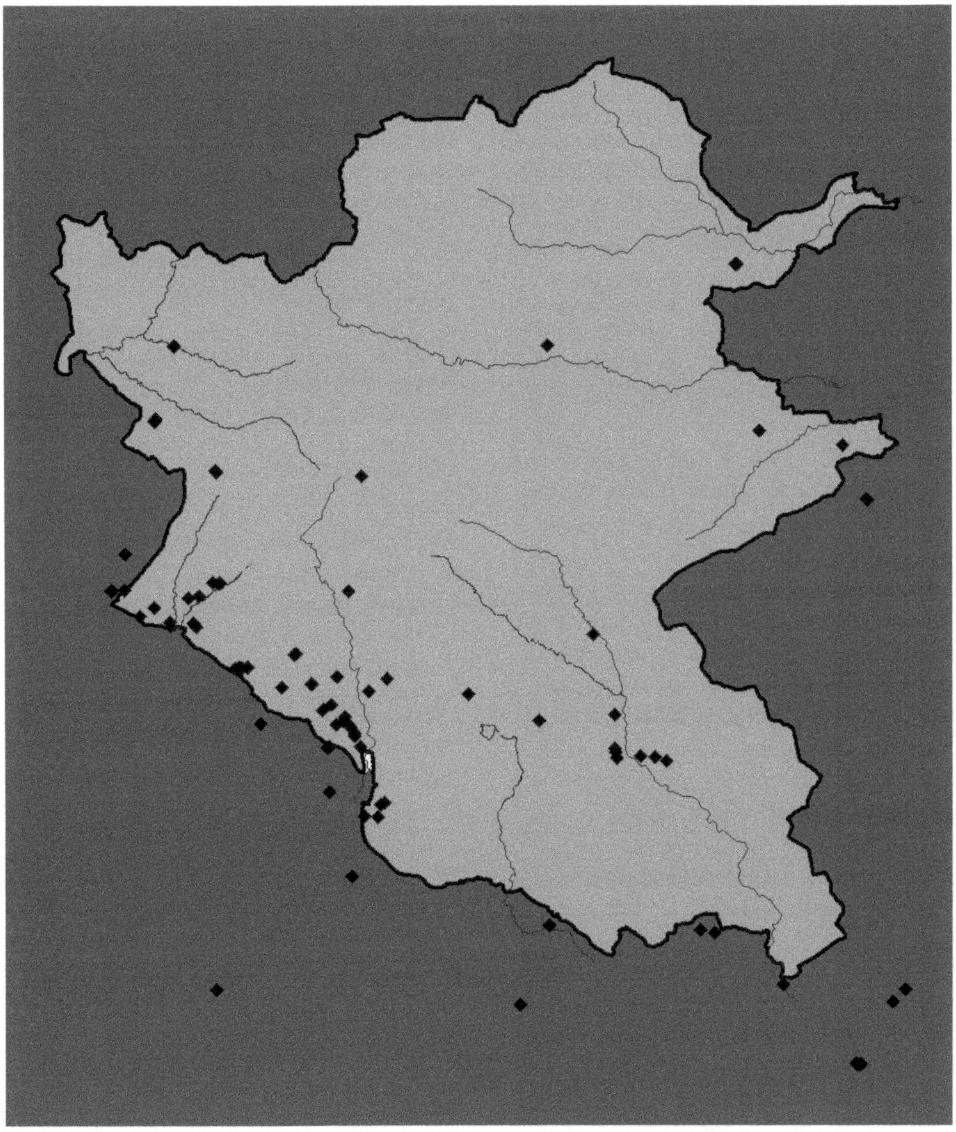

Figure 5.3 Distribution of surveyed clamp kilns in the Dales

5. Lower Wharfedale

According to one writer, in the period after 1664, farming activity 'drastically declined' in Wharfedale between Conistone and Ilkley with recorded falls in real farm income, resulting in tenancies being abandoned and rates of rural out-migration increasing, based on evidence from post-mortem inventories, though the degree of decline was not quantified (Pickles 1981, 64). This may have been induced by the climatic downturn of the Late Maunder Minimum (see Section 1), though Otley seems to have escaped the deprivations and actually saw opposite trends, so climate cannot have been the sole determinant. Other, now indeterminate, factors must have been at play.

The Royal Society's Georgical Committee, formed in 1664, was charged with seeking information across the country on the state of farming 'both to enrich every place with the aides ... and withal to consider, what further improvements may be made' in agriculture (Lennard 1932, 23-24). Responses were openly invited from 'experienced husbandmen' with all contributions being duly filed. A number of Yorkshire's farmers responded with detailed descriptions and full answers to the Committee's queries including Joseph Watkinson who, judging from the nature of his contribution, farmed in Lower Wharfedale around Otley or in Airedale (Royal Society CL. P/10iii).

Arable farming was still widespread with winter wheat and maslin (a wheat-rye hybrid) and oats planted in spring. Watkinson described the production of lime locally and its application – at the rate of ninety horse loads per hectare – and immediate ploughing-in during winter on land that was left fallow for the next two years. The treated land was first put down to wheat, followed in succession by barley and then either oats or beans prior to converting the land to permanent pasture. For application on the latter, he said the common practice was to first mix the lime with manure or dung. Watkinson wrote that he had seen 'those grounds which were covered with ling being first tilled with lime and after with dung become fine meadow'.

The general picture painted by Watkinson, and delivered in a matter-of-fact manner, was of a region where the land was well maintained and productive and where new farmland was being brought into production through a process of improvement. Whether the climatic cooling that began soon after his letter was penned caused this situation to be reversed cannot be known, but further archival evidence lends itself to the conclusion that disaster did not ensue.

The point has been made elsewhere (erroneously however, as the process has been shown in Chapter 4.7 to have started much earlier) that liming was an 'innovation' in Lower Wharfedale (and Airedale) in the late seventeenth century with 'valued and unvalued limestones' in post-mortem inventories providing the evidence (Pickles 1981, 77). Though liming was not then innovative, what the inventories do indicate is that lime was indeed being produced and stockpiled in the area and some entries support the contention that some/much of the lime was probably intended for agricultural use. For instance, the inventory of Thomas Dawson, of Morley near Keighley, proven in June 1690, included '163 loads of lime att 15d pr load' valued at £10 3s. 9d. along with 'manure in ground and att home' valued at £7 (Borthwick Prog/Exch). The two amounts are listed contiguously in the inventory which *could* suggest Dawson's intention had been to apply a mixture of lime and manure, though this is a tentative connection.

Several inventories from the period itemised manure or manure and ashes or dung coups (carts), and that of John Horsfall, proven in January 1690, valued his stock of 'Lime stones digged up' at £1 5s. along with a separate listing for manure. The wording in Horsfall's inventory is significant as it spoke of a manner of obtaining limestone for burning that, as far as this writer's research has shown, may have been unique to the Central Pennines. Specifically, the key words are 'digged up' and, furthermore, Watkinson talked in his letter of limestone being 'digged out of a soft quarry'. What was involved here was the age-old practice of digging what were called 'limestone boulder pits' whereby limestone cobbles were picked out of deposits of glacial till in shallow delfs (Johnson 2010a).

Lower Wharfedale and Airedale are underlain by sandstone and Millstone Grit and no limestone rock occurs at all. The practice was still operational in the mid-eighteenth century, with limestone being dug up and 'burnt in primitive kilns on the spot' (Speight 1899, 241), but when it started is unknown. It was certainly well-established before the late seventeenth century: in 1699 the freeholders of Bingley had petitioned the magistrates because 'for a long tyme there have been and are now great quantityes of Lyme Stones gotten promiscuously in Bingley and Micklethwait, by means whereof the ground is digged upp and very much impaired to the great decay of the yearly value' of the land (quoted in Speight 1899, 241). This confirms that the practice was already widespread.

Boulder pits were also worked at various locations on the edge of the moors west of Keighley, in Smallden Clough and Morkin Beck, at Lime Scar Hole (SE003 393) and around Slippery Ford (Johnson, 2010a).

Lower Wharfedale and Airedale, as well as the valleys of Ickornshaw and Holme Becks in Cowling, were locations for a further unusual method of obtaining supplies of limestone, known as 'river stonary' (Johnson 2010a).This writer has identified, from cartographic and field evidence, a large number of sites where limestone cobbles sourced from within river channels were burnt on the spot for lime. Whereas it has not been possible to state with any conviction what lime from the boulder pits was used for, documentary sources are more specific for stonary.

In 1620 Sir Peter Middleton devised to yeoman William Watson the right for three years to farm 'all that the Lyme stones, lyinge, beinge, and renewing, under the river browe of Wharfe ... *where the river did run*' (this writer's italics) (YAS MD 59/12/78). Watson had leased a stretch of (abandoned?) channel alongside two named meadows and it is quite possible that he wished to lime these meadows. In 1678 his descendant, John Middleton, demised to a partnership of three yeomen certain lands with implicit rights of stonary for twenty-one years on condition that the lime produced be used on the land for 'manuring and husbandry thereof' (YAS MD 59/14/103).

Manor court rulings from the same period support the contention that stonary was a major activity at least along the Lower Wharfe: paynes were laid against the washing of limestones in the 'towne beck' above Burley in 1667; against limeburner Peter Rhodes for not clearing 'rubbish' from his kiln in Newall in 1667; against leaving open the field gate into Newall Waterfield, adjacent to the Wharfe at Otley, by named limeburners or coal or lime 'leaders' (carriers) in 1684; and against the graving of sods for thatching lime kilns in Newall and Clifton (Borthwick, Burley bills).

6. North-east Lancashire

Disafforestation of the Forest of Trawden in 1507 released land for settlement and cultivation and accelerated the process of piecemeal enclosure and assarting that had been evident since medieval times. It also provided tenants with the incentive to improve the quality of their new holdings (Swain 1983, 17, 19 and 98). It has been calculated that the number of individual farm copyholdings within the manor of Colne doubled between 1527 and 1617 with almost half having been less than five Lancashire acres in size (Swain 1983, 317). Given that north-east Lancashire is based on impermeable sandstone rock and acidic soils, plots of this size could not possibly have supported a family without considerable efforts being made to improve soil quality, so these new copyholders really had no choice other than to invest in improvement.

The concept of improvement here was not new, though, and the prominent Towneley family had been at the forefront of both extending their estate and improving farmland by draining marsh and reclaiming moorland waste with marl employed as a major soil input (Raines 1878, xxxv and 7). The early dominance of marl is clear from manor court roll entries in the sixteenth century (Farrer 1897 and 1912 passim).[47]

[47] See, for example, Higham 17 January 1537, Burnley 24 February 1518 and 27 April 1542, and Colne 8 May 1561.

Further 'forest' land was made available for occupation by agreement between the Crown and its major tenants within the Duchy of Lancaster in Burnley and Colne in 1608 (McNeil and Newman 2006, 147) but a petition lodged in that year by tenants in Trawden bemoaned the fact that 'despite continual manuring and tilling' their holdings were still 'extremely barren and unprofitable' (Swain 1983, 69). Whether this was because marl was no longer available, or because it could not have the required impact on heavy clays, being a form of clay itself, is a moot point. The reality was that they were desperate.

Limestone was already in use within this area for building purposes but it had to be transported from the Clitheroe area at great expense. As early as 1304-05 the *compotus* of Henry de Lacy, earl of Lincoln, recorded the disbursement of £3 14s. for six oxen 'pro cariagio ... petre pro calce, et carbonibus Marinis pro calce comburendo' ('for carriage of stone for lime, and sea-coal for burning lime') for repairs at Ightenhill manor (reproduced in Lyons 1884).

By the early seventeenth century at least lime had also been applied on the land. The Shuttleworths began to build their new seat at Gawthorpe Hall (NGR SD805 341) in 1594, using Clitheroe lime, and soon began a concerted drive to improve the value of the estate (Harland 1856, passim). In 1612 their accounts recorded payment to 'the lymers of Clitherowe' for 1118 loads of lime specifically to 'lyme Little Cockeshuttes heade and the bancke heye' which were two enclosures (Harland 1856, 203). This is a phenomenal amount that perfectly illustrates their improving zeal, but it was just one of many purchases. Elsewhere, the accounts indicate that the Shuttleworths also mixed lime with manure: in 1605 forty-one loads were bought 'to lay in the myddings' (midden or dung hill) and in 1612 thirty-one to 'laye in the dunge hills' (1856, 168 and 199).

As the market for lime grew, and existing supplies became economically unviable, new sources of limestone were sought and sourced locally by channelling water from the moors to

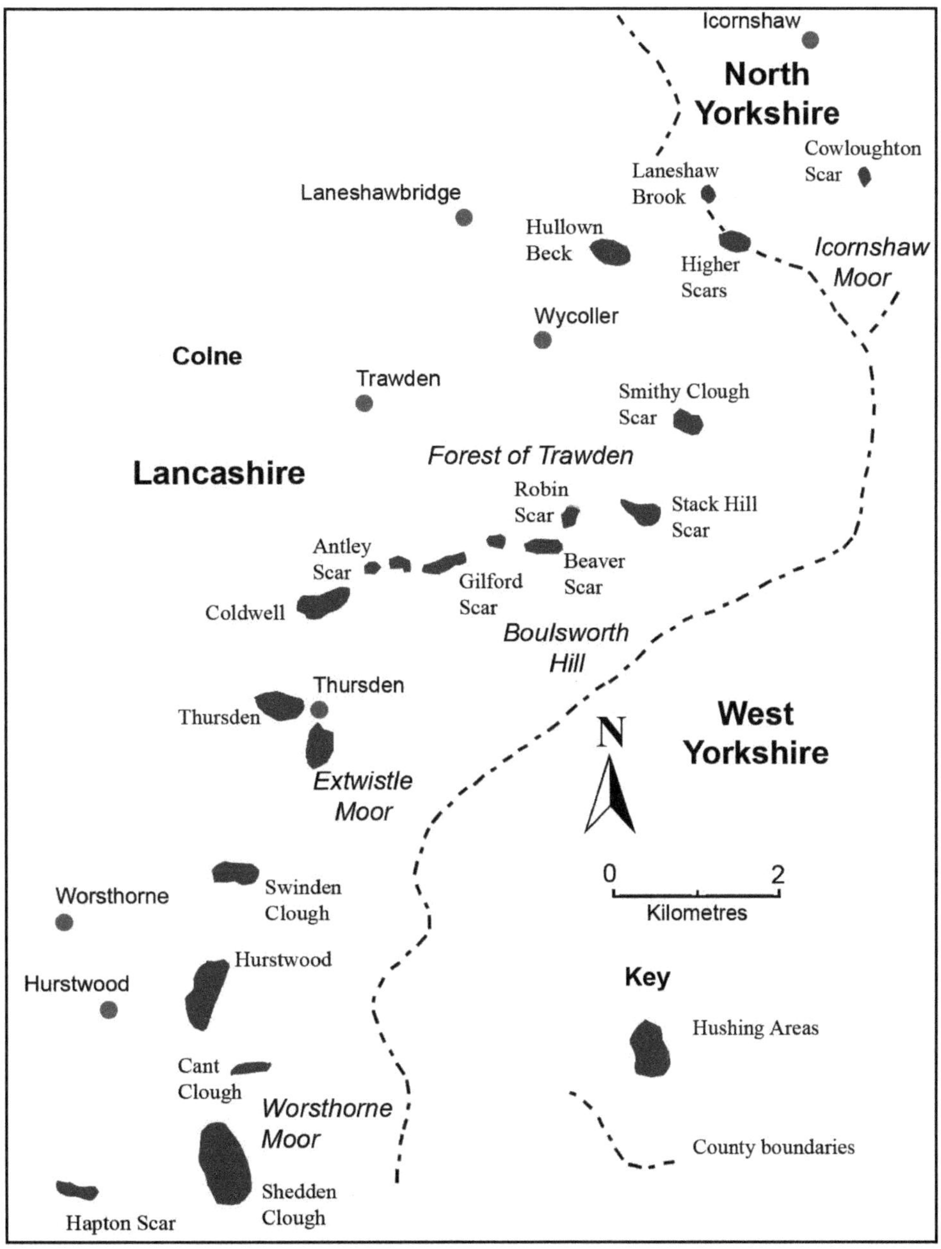

Figure 5.4 Location of hushing areas in north-east Lancashire

loosen deposits of glacial till piled up along the north-western edge of the South Pennines and thereby to flush out limestone cobbles in a process known as hushing (Johnson 2010a). Field surveying by this writer has identified twenty-five discrete areas, many associated with the toponym 'scar', where hushing was practised, in a linear belt from Cowling in the north to Hapton in the south, extending along the borders between Lancashire, North Yorkshire and West Yorkshire (Fig. 5.4). The majority have surface remains of lime kilns.

This proto-industry was controlled through the manor court system and relevant rulings have been traced from 1568 when John Forster was amerced for diverting a watercourse to his own ends (LRO DDHCl 3/51). The date broadly corresponds to the first recorded use of the word 'scar' to describe hushing grounds, in Colne in 1548 (Farrer 1897, 378). Duchy of Lancaster accounts itemised rents for the right to dig for limestone on the commons within the same manor in 1603 (NA Dl29/86/1603), and the leasing of similar rights in 1618 (NA DL5/27), and in 1632 when they were demised to Thomas Smith, a clothier described as 'of London' (NA E320/J5). Smith may have been a local man with London connections but he clearly saw a business opportunity to be had by leasing and then, presumably, sub-leasing rights to hushing grounds.

These opportunities were also recognised by the Towneleys, who also controlled extensive coal mining rights in north-east Lancashire (NA 317). They had somehow managed to secure widespread rights to hushing grounds in 1624 within Colne and Trawden, and at an unspecified date in Hapton, and this control was to provide the family with much of the basis of their later wealth, as illustrated by extant accounts for 1712-15, compiled by Ursula, widow of Charles Towneley (Ursula Towneley, private collection). The accounting period saw a net profit for Cliviger Scars of over £56 (over £4000 in current values) and for Hapton Scars of £50 despite expenses for paying 'Lime getters' at the two sites in excess of £110.

There is no direct archival proof but field examination of hushing grounds by this writer suggests that stone getting and lime burning were undertaken by individuals or small partnerships, either in the direct employ of the Towneleys (and the Parkers of Extwistle) or as their sub-tenants with the likes of Thomas Smith as intermediaries. There are, however, hints in post-mortem inventories for the manor of Colne. Between 1558 and 1600 eight per cent of all inventories contained some reference to lime or limestones but that figure rose to twenty-two per cent for the period 1601-40 (Swain 1983, 291). In the Forest of Trawden the figures were thirteen and forty-seven per cent respectively. The very fact that such a high proportion of the inventories contained valued lime or limestone emphasises the number of men involved in the trade as it is surely inconceivable, for such restricted geographical areas, that these deceased husbandmen and yeomen were not also part-time limeburners.

Several inventories itemise and give a cash value to limestones whereas others specify (burnt) lime: James Hartley of Winewall, who died in 1615, left 'lyme' to the value of 20s. (about £100 now), a sizeable sum, and John Ambler of Wycoller, who died in 1674, had '5 lime sacks' among his possessions (LRO WCW). Whether full or empty was not indicated.

Some of the lime from hushing was intended for building, and some to support the widespread tanning industry in the area, but much was also destined for the agricultural improvements outlined earlier. The extent of the industry, and the rather unusual mode of obtaining the raw material, graphically illustrate the lengths people in the area were prepared to go to in order to satisfy their need to improve the acidic soils they were stuck with.

7. Conclusion

Overview

This chapter has taken forward the discussion beyond the restoration of the monarchy in 1660 and has examined evidence showing that fundamental changes were apparent in England, not only in agriculture but also more broadly, fostered by the influence of the Royal Society and its Georgical committees. In the new political and economic climate landowners and tenants felt able to invest in buildings and land improvement, a process that occurred at different times and to a varying extent across the research area depending on localised circumstances and opportunities. In the prevailing post-Restoration climate improvement was almost a religious and moral imperative. Unproductive land must be taken in and rendered productive, improvement was increasingly given an aesthetic dimension, and the means that made all this attainable were given increased monetary and practical value. Lime was central to the whole process.

In general terms, as the period under review here advanced, the processes of improvement accelerated with a concomitant increase in the use of lime, and the ensuing century allowed landowners and managers to build upon this foundation. By the end of the period – the mid to late eighteenth century – improvement was ready to take on a radically different dimension as parliamentary enclosure and large-scale improvement transformed huge tracts of the countryside: these are phenomena to be returned to in Chapter 7.

Contemporary views

As during the period discussed in the previous chapter, various writers after 1660 published agricultural treatises, offering a range of advice on better forms of husbandry and exhorting all involved to maximise productivity. Again, to what extent these books reached the hands of the ordinary farmer, rather than merely being read by the privileged landed classes and estate stewards, is a matter for conjecture. In turn, the degree to which they actually influenced events on the land is also open to different interpretations (M Johnson 1996, 206).

Prior to 1660 thirteen agricultural writers offering advice on improved farming were reviewed in Chapter 4.8; this chapter will now look at treatises issued by ten writers, as summarised in Table 5.2.

Writer	Date	Recommendations
Reeve, Gabriel	1670	paring, draining, liming
Smith, John	1673	draining, marl
Worlidge, John	1675	paring, marl, liming, manure
Worlidge, John	1687	marl, liming
Mortimer, John	1712	paring, liming
Society of Gentlemen	1736	liming
Hale, Thomas	1756	liming, manure
Bradley, R	1757	paring, liming, manure
Maxwell, Robert	1757	marl, liming
Anon c	1760	marl, liming

Table 5.2 Soil inputs recommended by agricultural writers, 1660 – 1760[1]

[1]See the Bibliography for full reference details.

Apart from Blith, the son of a small Yorkshire farmer, all previous writers had belonged to the landed classes and had directed their advice at their own kind. Blith's views tended to hold sway during the Interregnum for the very reason that, being a commoner and a republican, he was perceived to be a man of the people but after the Restoration his books went out of favour. Of the writers summarised in Table 5.2, five (including the anonymous 'gentleman') came from the privileged classes and Bradley was professor of Botany at Cambridge; Hale and Worlidge, on the other hand, wrote with yeomen and husbandmen in mind, even though the latter was a member of the gentry. It may be that his reputation as a staunch royalist stood him in good stead and it has been suggested that he, of all post-Restoration agricultural writers, was pre-eminent as the accepted authority (Thirsk 1983, 315). Despite choosing Latin titles for his various works, he did write in a style accessible to those less well schooled than himself.

With the exception of Smith, all the writers considered here gave liming an important role in improvement; it was Smith who used the term 'Improver' to describe those intent on bettering their land (1763, 27). Reeve believed there was 'no such husbandry in the world perfectly to prepare any Heathy land, and make it nourish, receive, and ripen seeds' than to lime land which had been newly pared and burned and drained (1670, 16).

Worlidge, Mortimer, Hale and Bradley shared the belief that lime worked best if first mixed with manure after which it made 'a very excellent Improver of Lands' (Worlidge 1687, 65). (The issue of whether to mix or not has been touched on before and it will be returned to in Chapter 8.5) There was also ongoing debate about the condition in which lime should be applied: Hale (1756, 81), Maxwell (1757, 94) and 'Anon' (1760, 350-56) favoured laying it 'hot from the kiln' while the others advocated mixing lime with manure and leaving it in heaps to air-slake before spreading. It has to be said that this matter was never to be satisfactorily settled, especially so when no-one understood how lime actually worked in the soil (see Chapter 7.3 for a discussion of lime's chemical properties). 'Anon' (1760, 350), whose place of origin is unknown, made a statement that is difficult to comprehend, namely that liming was a 'new practice'. Possibly he was referring to his own home area, but it can hardly be said that liming was new across most of the country in the mid-eighteenth century.

South-east Westmorland

Extant evidence suggests that the process of improvement here was delayed relative to much of Yorkshire, owing to difficulties in sourcing affordable fuel supplies and to the dominance of small, independent farm units rather than resource-rich landed estates. Archival evidence tells of late sixteenth-century improvements on the Wharton estate in Mallerstang and field evidence emphasises the great extent, and longevity, of limestone quarrying and lime burning, and of early enclosure, on the commons and former commons of Wharton Fell and Nateby Common close by. A resurgence of economic activity after the Restoration was associated with widespread investment in farming across the board, and Sir John Lowther's improvements on his Ravenstonedale estate in 1669 are but one visible manifestation of this trend.

The northern dales

Much land in Swaledale and Wensleydale was held as demesne on several major landed estates which maintained detailed accounts and stewards' records, at least from the late seventeenth century. The Smales family at Gilling, the unfortunate William Chaytor of Croft, the marquis of Winchester at Bolton Castle, and the Scropes at Danby Hall, were all closely involved in the process of improving their estates, with an aesthetic and economic rationale, throughout the period under review in this chapter.

If Arthur Young is to be believed – and there is no reason to doubt his basic observations – this area was home to two noted agricultural 'Improvers'. Christopher Crowe at Kiplin and William Danby at Swinton Park were close contemporaries and, though there is no supporting evidence, it seems inconceivable that the two men did not engage in discussions about the best methods of achieving their aims.

On all of these estates liming was an integral and, it is suggested here, dominant element of improvement.

Craven

There are parallels with the northern dales in the Skipton and Bolton Abbey areas where the former Clifford estates were subjected to a concerted drive to improve them through the eighteenth century, with lime again being brought into the equation.

It has been possible for Craven to supplement a rather scant corpus of documentary evidence with the results of archaeological excavation of clamp kilns that place lime burning possibly within the monastic period and certainly in the immediate post-monastic. This is backed up by entries in manor court records for the post-Restoration era. The locational context of many of the clamp kilns tends to inform their use for producing agricultural lime.

Lower Wharfedale

Joseph Watkinson's carefully crafted response to the Georgical Committee, around 1664, concerning improvements and the role of liming in Lower Airedale and Wharfedale, supplements contemporary evidence from post-mortem inventories and the Burley manor court rolls, and surviving indentures from the Middleton estate. Lime was valued here as a soil input, certainly through the entire seventeenth century, and it was locally sourced in what can only be described as an innovative manner by extracting limestone cobbles from within river channels and from shallow pits dug into ribbons of glacial till on the surrounding moors.

North-east Lancashire

Innovation was also apparent here, and from a much earlier date than for river stonary and boulder pits. Limestone cobbles were obtained from till deposits along the edge of the moors by the use of controlled and targeted water flow by hushing. Hushing grounds, often termed 'scars', were carefully controlled through the manor court system and rights were jealously guarded as lime was a commodity in great demand for liming and other purposes and hushing was perceived as a lucrative business activity. The Towneley estate certainly based much of its early prosperity on controlling this, and coal mining.

Documentary sources point to the existence of hushing from at least the mid-sixteenth century and the process continued well beyond the period under discussion in this chapter. It was the coming of the Leeds-Liverpool canal around 1800 that put an end to commercial hushing operations though they may well have continued on a local *ad hoc* basis.

This chapter has demonstrated that lime was perceived across the entire research area to have been of (often paramount) importance in the business of improving land from before the Restoration and throughout the entire period under review. Evidence from a variety of contrasting and complementary sources enables more emphatic conclusions to be drawn for the decades after 1600 compared with previously.

Chapter 6 will diverge from the chronological approach to examine the symbiotic relationship between lime burning and fuel supplies in the Central Pennines, and will examine in detail the lime-coal accounts for the Swinton Park estate.

Chapter 6
Fuel Sources and the Burning of Lime

1. Introduction

Raw limestone must be calcined, or fired, in a kiln to produce quicklime and various types of kiln have been employed to this end. To recapitulate, the earliest recognised form in the survey area is the basic clamp, variously known as a sow, sod or turf kiln, and they were in use throughout the medieval and early modern eras, with the more fuel-efficient masonry-built structures that are such a common feature of the Central Pennine landscape (and of most limestone areas), appearing during the late seventeenth or early eighteenth century (Johnson 2008a; 2010b). In order to achieve calcination, fuel was required and access to an affordable source was always a prerequisite for economically viable lime production.

Calcination involves the reduction of the raw stone by thermal decomposition into its two elements, calcium oxide and carbon dioxide. The chemical formula for limestone is $CaCO_3$. When this is subjected to temperatures of 800 to 900 degrees Celsius the calcium oxide (CaO) remains as a residue once the carbon dioxide (CO_2) has been given off. In field kilns this was achieved by calcining the stone within the vertical kiln bowl: alternate layers of stone and fuel were packed into the bowl, once the fire had been lit at the base. The fire burned upwards through the kiln fill until the limeburner was satisfied that calcination had been achieved. Rising heat from the calcining zone in the central part of the bowl pre-heated and dried out fresh stone while calcined stone made its way to the base of the bowl where it cooled down to about 60^0. In some instances the entire fill was burned and then completely emptied but, more commonly, fresh stone and fuel were continually added through the bowl top as calcined quicklime was raked out through the eye at the base of the bowl, until local demand had been met. Figure 6.1 shows the key features of a typical field kiln. Farmers tended to call quicklime 'lump' lime because when it is drawn out of the kiln it is not in powder form but in lumps, much lighter in weight than the original stone but nonetheless recognisable as more or less solid lumps.

This chapter starts by examining a range of kiln fuels utilised in the research area and detailing the constraints imposed by deficiencies in fuel supply on lime production and, by association, on the adoption of lime as a soil improver. It then goes on to consider how various landed estates attempted

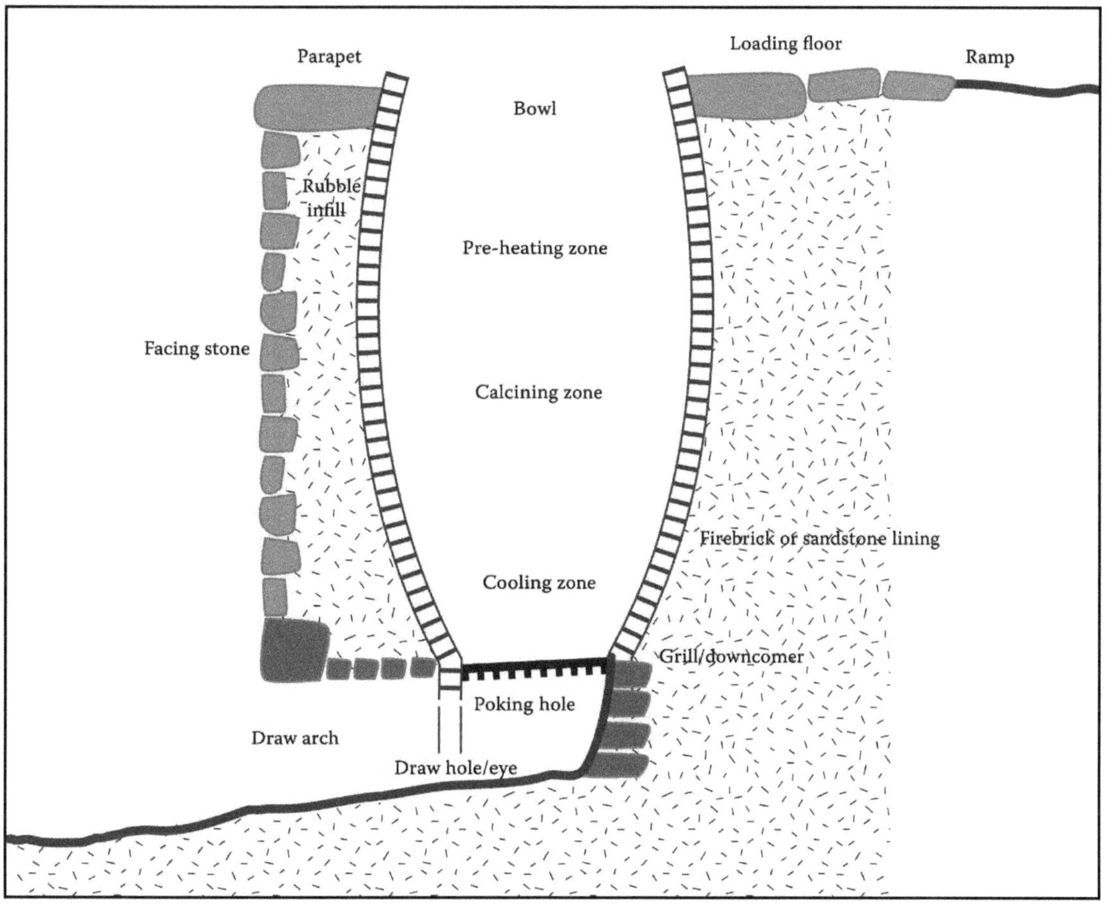

Fig. 6.1 Cross-section through a typical field kiln

to develop their own coal resources to increase the degree of vertical integration in the production and application of agricultural lime while at the same time reducing unit costs. It was in the interests of landowners and tenants to minimise production costs if the lime was destined for their own land and, equally, if some were being sold to neighbouring entities the kiln operators would have been keen to maximise their income by balancing a competitive price against production costs.

The Swinton Park estate, west of Masham, is examined in detail as a prime example of an enterprise that developed a major interest in the extraction of lime-coal, partly for external sale but, more importantly for this narrative, for improvement of the estate's farmland. Close analysis of lime-coal sales on the estate provides invaluable data concerning patterns of seasonality and long-term trends in output which can be directly correlated with the burning of lime and thus its application on the land. Such patterns can also be correlated with economic and political externalities.

2. Early sources of kiln fuel

That securing access to a reliable source of fuel was of major concern to limeburners cannot be downplayed and the field survey of kiln sites by this writer (see Chapter 3.3) has pointed up the frequency with which proximity to fuel supplies was at least as significant a locational factor as access to limestone reserves. Out of a total sample size of 1568, no less than twenty per cent of kilns were not directly associated with limestone bedrock, and a number of kilns within the total sample were sited where access to fuel was clearly a major determinant. The field survey encompassed 175 parishes of which twenty-seven per cent are known to have once seen exploitation of local coal reserves: a significant proportion.

For most of the period under review coal was the dominant fuel but a range of other types has been noted from written records and this writer's archaeological excavations. Evidence of specific kiln fuel within the medieval period has been identified from excavation of a thirteenth-century clamp kiln, in the Forest of Bowland, where significant quantities of charcoal and fragments of coal were recovered from the kiln bowl. Archival evidence points to the mining of coal in Colsterdale in 1304 (Brown 1906, XLII) and 1334;[48] in Silsden by 1310 (Spence 1992, 157); in Trawden and Cliviger from 1295, in Ightenhill in 1323-24 (Farrer 1907, 197; Frost 1982, 77); and at Tan Hill by 1387 (Page 1914, 42).

Extant Charter Rolls and *Liberate* Rolls from the reign of Henry III provide a wealth of useful material concerning the sourcing of wood for burning in lime kilns. On 24 April 1228, for example, the sheriff of Oxford was ordered to build a lime kiln at the castle there using brushwood (Cal. *Lib.* Rolls 12 Henry III); while two years later the monks of Abingdon Abbey were required to supply timber 'needed in the lime kilns' for constructing the town walls at Oxford (Cal. Charter Rolls 14 Henry III, April 1230). Other royal writs were issued throughout Henry's reign requisitioning brushwood or timber for lime kilns for work at various royal properties. Two instances, from many, will suffice: payment of forty marks was authorised for the purchase of '600 pieces of wood' for kilns at the Tower of London (Cal. *Lib.* Rolls 13 Henry III); and the Keeper of Windsor Forest was commanded to supply and deliver eight tree trunks to the castle there 'to make a lime kiln' (Cal. *Lib.* Rolls 45 Henry III).[49]

Evidence has been provided earlier (see Chapter 4.3) to show that lime was being burnt in the lowland fringes of the Central Pennines in the medieval period, and the assumption has to be made that timber or brushwood (possibly including gorse) was not just the preferred fuel but was probably the only source then locally available. The lowland valleys of the Wenning and Lune to the west, and of the Lower Swale and Ure to the east, were sufficiently well wooded to supply the necessary quantities of timber fuel needed whenever kilns were fired up. Crossley (2005, 106) has demonstrated that, if chopwood had been the source of fuel, lime burning required wood twenty to twenty-five years old: to maintain the requisite supply would have required vast areas of coppiced woodland. For most of the research area this was impracticable.

The accounts of Richard le Skenner, park keeper at Ightenhill in the manor of Blackburn, record payment of 4s. 10d. in 1323-24 as wages for 'men making a kiln, collecting stones and brushwood for burning lime' (Farrer 1907, 197). The same set of accounts also itemises the purchase of '8 qtrs of sea coals dug and collected' at a cost of 16d (see Section 3 for a definition of sea-coal). We can deduce from these two entries that the brushwood was intended as kindling fuel within the kilns with the coal being the calcining fuel. Whether this also applied to the royal kilns is debateable as the Rolls, with one exception, record brushwood *or* timber, suggesting that the former was the sole kiln fuel.

Archaeological evidence can shed light on how early modern lime kilns within the survey area were fired. Seven clamp kilns were excavated within North Craven and dated to between the fifteenth and late seventeenth century: all provided samples of fuel (Johnson 2006 and 2007). Three contained fragments of coal whereas all had charcoal deposits, some of which were identifiable by species: alder (*Alnus*), ash (*Fraxinus*), oak (*Quercus*) and willow (*Salix*) were represented suggesting either that these limeburners considered a range of species suitable or that they made use of whatever was close at hand. One sample (*Salix*), recovered from a valley bottom site below Chapel le Dale near Ingleton, had been cut down in summer which, by association, indicates that the lime from that burn had been produced in the same summer and, further, that the lime had been spread on the ground soon after as it is not possible for the lime to have been stored for any length of time given the open situation of this kiln. In addition, it was not

[48] 'Agreement between the Abbot and Convent of Jervaux, and Sir Geffrey le Scrop, that theAbbot and Convent should get coals in Colsterdale, dated 7 Edwd. III, AD 1334'. Quoted in Fisher (1865, 66-67).

[49] The term 'making a lime kiln' does not refer to the construction of the fabric of a kiln but to the refilling of an existing structure, or the creation of a clamp kiln. See, for example, NYCRO ZS*, August 1696.

then general practice to store lime elsewhere prior to taking it to the land in question.

Further documentary confirmation of dual-fuel practices comes from a set of vouchers for a specific lime kiln burn in 1759 at West Witton in Lower Wensleydale (NYCRO. ZPL 5). Out of a total cost of £12 13s. 2d., 6s. was spent on cutting and leading wood to the kiln with £6 for purchasing and leading coal. The wood was clearly for kindling and the coal for calcining. To complicate matters here, though, 2s. was spent on 'two loads of peats' for the kiln but the accounts do not indicate how this was used: it certainly would not have been used to seal the kiln from the atmosphere as this would have been a masonry kiln and they were not sealed.

Peat was utilised as a kiln fuel within the survey area, as elsewhere. In a letter sent to his master in 1694, the steward at Levens Hall Estate near Milnthorpe reported that he was 'now leading peats for another lime kilne' and that, if the weather held, he would also 'get coals for it' (Bagot and Munby 1988, 96). In the following month he followed this up by saying he had ready 'two stacks' of peat (Bagot and Munby 1988, 100). However, the letter sheds no light on how peat and coal were to be used in the same kiln firing. Whyte (1979b, 202) has shown that peat fuelled lime kilns in Banff in north-east Scotland and along the Solway coast in the seventeenth century, while detailed descriptions of how peat was used in lime kilns were published at the turn of the eighteenth and nineteenth centuries. Ure (1795) concentrated on kilns, in Dunbartonshire, for which seventy cart loads of peat were required per burn; and Radcliff (1803) wrote a prize-winning essay on the potential use of peat in kilns in Scotland. None of these sources indicated if peat had been used by choice or because there was no viable alternative.

Around the same time Dodgson (1805), a farmer in eastern Cumbria, had supplied the Board of Agriculture with detailed specifications for burning lime with peat. His reasoning was clear: peat was available close by, cheaper than coal by a factor of 2.5, and he claimed that peat performed better than coal, being able to complete the calcination process within one day rather than coal's three while producing lime of a more consistent quality. Peat would have been much cheaper, but doubt has to be cast on his claims that the burn would have been completed more quickly and that it produced a better quality end product. According to Oates (1998, 130-31) peat could never be a satisfactory kiln fuel given its low thermal efficiency and slowness of burning whereas wood with its 'long, lazy flame' was ideal. Nevertheless, peat was indeed used as a kiln fuel within the research area.

In fact wood is still used as a kiln fuel in parts of the developing world and is considered to produce 'some of the best quality limes' as its 'long even flame penetrates the stone readily' giving uniform calcination (Wingate 1985, 43).

Later in the nineteenth century, however, it was reported that in parts of modern west Cumbria peat had largely been superseded by coal as a kiln fuel partly because the latter was so accessible locally and partly because the 'trouble of securing' peat was no longer considered worthwhile (Dickinson 1850, 69). This is in stark contrast to observations reported in east Westmorland where peat was still in use as the main kiln fuel because coal was unobtainable (Pringle 1794, 339). Farm leases in the parish of Horton-in-Ribblesdale, for Top Farm in 1727 and Dubcote in 1823, both included the right to delve for coal on Pen-y-ghent as well as rights to dig peat for use as kiln fuel (Horton History Group 1984, 43).

Two centuries earlier James Bradley had been amerced 6s. 8d. in the Cononley manor court sitting of October 1606 (Dev. Coll. L/45/19). He had dug and taken away turves on the common 'to burn in his lime-kilns, called lyme-killes' contrary to the regulations. At this time the kilns in question would have been clamp kilns, whose outer layer was sealed with clay or turf, but the entry here very specifically refers to burning the turf. The word turf may have been used here to mean peat, though turf was itself traditionally used as a fuel source in rural Yorkshire into modern times (Hartley and Ingilby, 1975, 74). Turf was easier to cut and dig than peat, it dried more quickly and burned more brightly and at a higher temperature than peat.

Even Arthur Young, the forward-looking improver, was not averse to using peat as a kiln fuel. He purchased an upland estate near Blubberhouses, probably in 1792 (see Chapter 7.3), which included a kiln and he noted that burnt lime cost 4s. per chaldron to produce including the cost of peat for fuel (Young 1796, 298).

Turf, or peat, was a common kiln fuel in Ireland though in parts dried gorse was preferred (Lucas 1960, 64-65), and we have earlier seen evidence of this within the Central Pennines. It was certainly widely used, and valued, in lime kilns elsewhere in the country, especially where other sources were not available (Robinson and Cooke 1962, 22).

Documentary and archaeological evidence, therefore, illustrates the use of wood and peat as important kiln fuels from at least the early thirteenth century for in excess of 600 years. In most such instances it can be inferred that peat was often the only fuel locally available and that wood, with exceptions, was used for kindling rather than as the main source of fuel. As time passed coal became more readily available replacing wood and peat as the dominant kiln fuels, with a surge in the price of timber around 1270 leading to increased use of coal as kiln fuel (Brimblecombe 1975).

3. Sea-coal

When coal was first employed as a kiln fuel is not known but documentary evidence does confirm its use since the last decade of the thirteenth century. The use of the term sea-coal (*carbonibus marinis* in the medieval period, or later 'smith's coal' [Stat. Realm 6 Henry VIII, 1514-15, and 34-35 Henry VIII, 1543]) in lime kilns is recorded from the monastic period. Sea-coal was not necessarily sourced from coastal pits: the term was initially used for coal mined along the north-east coast, and exported widely within England and across the Channel (Blake 1967), but later on was used generically for any coal to distinguish it from charcoal.[50] The *Compotus* for Bolton Priory in Wharfedale records payment in the accounting year 1300-01 of 17s. 6d. 'in carbonibus marinis ad calcem ardendam' ('in sea-coal for burning lime'), the lime to be employed within the demesne. A further payment of 3s. 10d. was made for carriage of sea-coal to the abbey's lime kiln at an unspecified location ('in carbonibus marinis cariandis ad turriol calcis').

Nef (1966, 11) described coal as the 'new fuel' of the Middle Ages. Given the reality that to produce 1 tonne of burnt lime required four wagon loads of wood, which broadly equates to the trunks of two mature trees, it is not surprising that usage of sea-coal spread very rapidly once resources were opened up. It is also, perhaps, not surprising that the authorities soon came to realise the income-generating opportunities presented by trade in sea-coal. The burgesses of Newcastle upon Tyne were granted full rights to 'dig and have mines of coal ... and make their profit thereof' in 1357 (Cal. Charter Rolls 31 Edward III), and those of Beverley the right, in 1483, to exact a toll of 2d. per chaldron on lime landed in the port there (Cal. Charter Rolls 22 Edward IV).

Trade in coal from north-east England to London and other east coast ports flourished for several centuries, with much of it being destined for lime kilns – as well as for domestic consumption and tile and pottery kilns – in the capital. It has indeed been said that limeburners and smiths were the two 'most frequent customers for the medieval coal merchant' (Blake 1967, 5). Coal was consumed to the point where proclamations were issued in 1307 banning the use of sea-coal in lime kilns in Southwark because of the 'intolerable smell (which) diffuses itself ... and the air is greatly infected, to the annoyance of (people) ... there dwelling and to the injury of their bodily health' (Cal. Close Rolls 35 Edward I). Brushwood or charcoal was to be used instead. Sea-coal is essentially bituminous coal (though, as said, the term sea-coal was used generically) and it was of generally inferior quality noted for emitting noxious fumes and large amounts of soot so its use in lime kilns, at least in urban areas, was resisted and frequently opposed.[51]

In the rural parts of the survey area, where timber resources were lacking, estates with reserves of sea-coal exploited them for domestic use and for burning lime. The estates of Henry de Lacy, 2nd earl of Lincoln, in north-east Lancashire, were operating coal mines by at least 1294[52] (Lyons 1884, xxii); and the estate *compotus* records receipts for sales of sea-coal in Cliviger, Colne, Trawden and Ightenhill. The de Lacy properties passed to the Duchy of Lancaster on the earl's death in 1311: a petition was made to the Chancellor of the Duchy by Richard Tempest in 1576 concerning the getting of 'coles called seacoles' within the manor of Barnoldswick, though the outcome is not recorded (Ironbridge, Box 8, RS 796.10) As discussed in Chapter 5.6, coal was still being dug in these manors, including for use in lime kilns in the hushing grounds, into the early nineteenth century. It was the ready availability of coal prior to the canal age that made lime burning in north-east Lancashire such a long-lasting and productive activity: indeed, it could be argued that limestone hushing only developed because of cheap local coal supplies.

Nef (1966, 105) emphasised the strong link between fuel supplies and lime burning. Even in the eighteenth century, he wrote, lack of coal and the expense of carting it left farmers in many areas having to 'perform their husbandry without any fertilizer' to the obvious detriment of their output. Long before this, however, the cleric, antiquarian and topographer John Leland had made acute observations on his travels through Richmondshire in 1539 and 1544. He is worth quoting (Toulmin Smith 1964, 140):

> Ther be some vaynes of coles found in the upper parte of the west mountains of Richemontshire, but they be not usyd for incomoditie of carriage to the lower parte.

Yet, he noted that most coal burned in Richmond was sea-coal imported from Durham, obviously much more distant. His comment, however, does neatly summarise part of the link made earlier between access to fuel, the extent of lime burning and the use of agricultural lime, a matter returned to in the next section.

4. The development of local coal reserves

Coal occurs naturally throughout the research area, though not uniformly so (Fig. 6.1). Apart from small Basement inliers along the North Craven Fault (see Chapter 1.5) the entire area is underlain by Carboniferous strata. At the top of the stratigraphic sequence are the true, high quality, Coal Measures but they only outcrop widely in north-east Lancashire and at Ingleton. Elsewhere the coal was sourced from the Yoredale Group or Millstone Grit Series within which thin bands of generally poor quality coal were extracted from adits, shallow bell-pits or deeper shafts. Many of these were necessarily located on moor tops as the Yoredales and grits are found near the top of the sequence. Even though the quality was poor, these coal and coal shale seams were deemed suitable for lime burning (Phillips 1837, 12; Kelly 2008), and their high levels of ash and sulphur were not seen as a problem for lime burning (Ironbridge, Box 13, RS 1026). Indeed, the symbiotic relationship between small-scale coal pits and lime kilns has been noted by Raistrick (1971, 44) and Simmons (2003, 96-97).

Not all this type of coal was of low quality, though. Sale particulars for the East Scrafton Estate in 1853 stressed the 'valuable seam of coal (equal in quality to the best Durham coal) ... and the best lime' to be had on the estate, which may well have influenced Lord Bolton, of Bolton Hall in Wensleydale, to purchase it (NYCRO ZBO 1).

[50] The term may have originated because coal was exported from the north-east by sea or, more likely, because the earliest coal was gathered up along the coastline there (Nef 1966, 452-53). By the thirteenth century all coal was referred to as sea-coal even if it was mined inland.
[51] Lewis *et al.* (1995) have shown that pollution from lime kilns, for example in thirteenth-century York, was responsible for lung diseases and sinusitis.
[52] Bolton priory obtained its sea-coal requirements from de Lacy's pits.

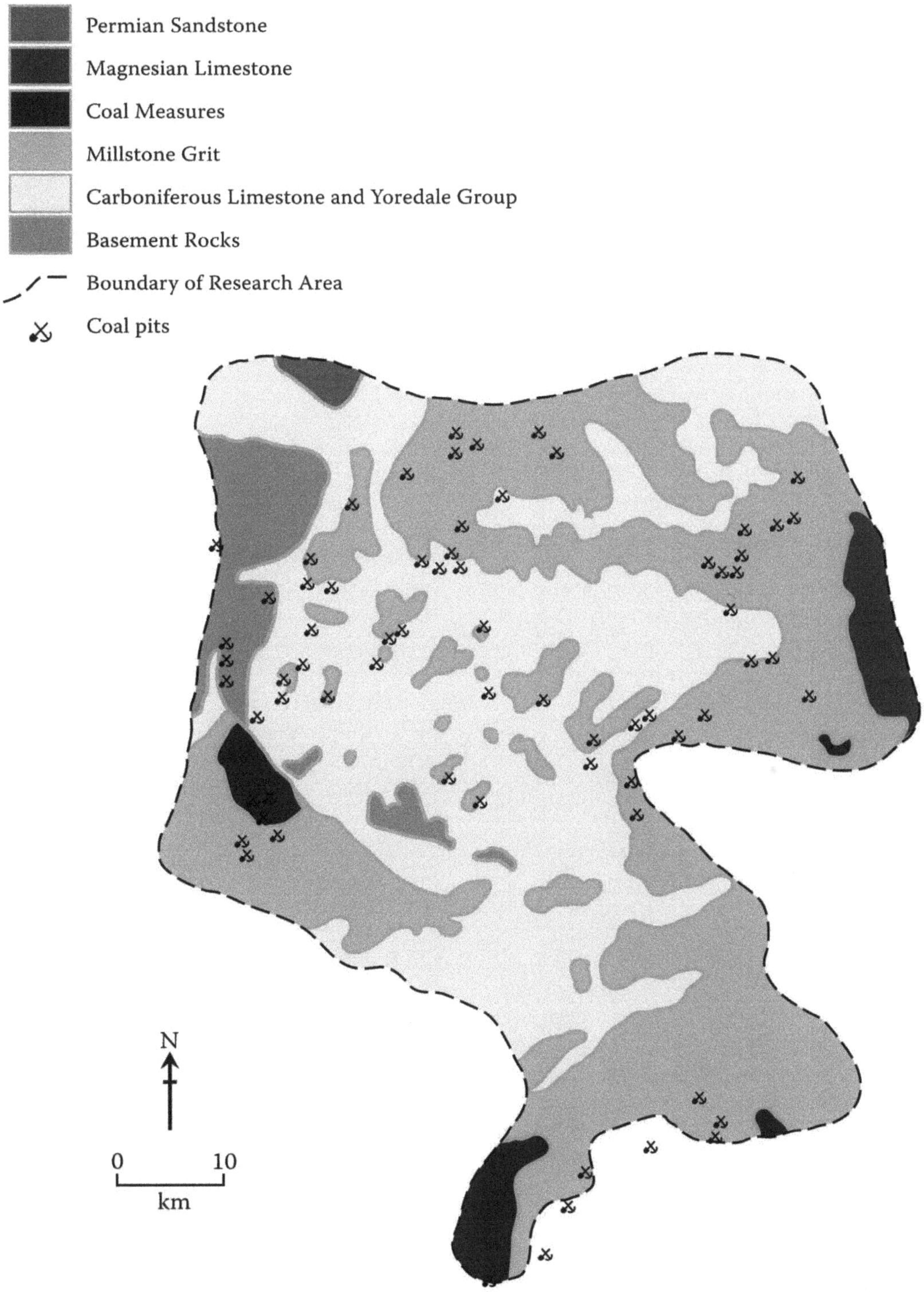

Figure 6.2 Location of collieries in relation to simplified solid geology

More than eighty discrete collieries have been securely identified within the research area with operational dates spanning six centuries: the earliest known mining was at Tan Hill above the head of Swaledale in the thirteenth century and the last known in 1938, also at Tan Hill (Gill 2008).

Many landed estates within the area leased licences to local prospectors or worked their reserves in-house, and careful regulation of coal mining within the manor court system ensured that infringements added to manorial income. As we have seen earlier, the Burton in Lonsdale court sitting of 9 November 1748, for example, laid a pain of 20s. against any limeburner who:

> shall dig up his soil and burn the same in sod kilns ... upon Bentham Moor ... unless the person so doing shall in every such ... kiln burn twenty bushels of coals. (Hornby, court roll).

The point behind this ruling was not a desire to collect the occasional infringement fine but to apply pressure on limeburners to buy their fuel from the lord of the manor's pits.

As early as 1518 the lord of the manor of Colne was engaged in similar income-generating measures by decreeing that a fine of 3s. 4d. would be levied on 'any person out of Craven (who) do break the king's soil anywhere within the township of Colne, to obtain coals' (Farrer 1897, 267). The court roll does not state whether inhabitants of the manor would be similarly amerced but it could indicate that limeburners from Craven were travelling widely to obtain coal supplies, despite the risks involved. In turn, this suggests that their fuel needs could not be met from within Craven even though there were operative coal pits. It could also suggest that lime production was at a high level in early sixteenth-century Craven.

The next section will develop the theme of how fuel supply constraints impacted on lime production and either caused those involved in the industry to look far and wide for supplies, or prevented agricultural lime burning at all. The link between *local* coal supplies and the adoption of liming is clear: the greater the distance coal had to be transported, the higher were production costs for limeburners and this in turn acted as a severe check on the ability of farmers to afford lime.

5. Lime-coal and fuel supply constraints

As explained in Chapter 3.3, the Central Pennines are well endowed with surviving sets of farm and estate accounts and day books, many of which were meticulously kept by estate factors and stewards or in some instances by freehold farmers. Data which have proved especially useful in building up a picture of lime-coal production and consumption include combined leases of estate-owned coal mines and lime kilns; farm leases granting tenants the right to dig for coal specifically for burning agricultural lime; lime-coal purchases by estates; and lime-coal sales by coal mining operations.

A representative sample of such estates, farms and collieries is summarised in Appendix 2 with their location being shown on Figure 6.2. They are distributed across the entire survey area with all major dales being represented thereby reinforcing one of the central tenets of this book, namely that agricultural lime usage was spatially widespread throughout the Central Pennines and surrounding areas. The temporal distribution of lime-coal mining is equally broad, extending from the very end of the seventeenth century to the third quarter of the nineteenth, which also supports the tenet that lime was being used on the land much earlier than was hitherto generally accepted. Similar evidence prior to about 1700 is scant though the records for the Swinton Park estate (see Section 6.6) suggest that similar lime-coal links existed even earlier.

Some estates purchased lime-coal, for agricultural lime, from collieries quite some distance from where the kilns were located which serves to emphasise the importance attached to the practice of liming the land. Given the parlous state of roads in the area and the expense of leading coal, whether by cart or packhorse train, added to the costs of buying coal and burning lime, land managers clearly perceived the expenditure to be a worthwhile long-term investment.

The Scrope family at Danby Hall between Leyburn and Masham purchased lime-coal for use in their estate kilns, and farm account books covering the period 1725 to 1786 have been examined in detail (NYCRO ZPT and ZPT 4/4/3). Lime-coal was bought in on a regular basis with every month of the year being represented at some point in the accounts.

Some of the Scrope vouchers were for coal for producing lime mortar, others were for agricultural use in reclaiming Danby Moor and Dalton Moor, though some did not specify end use. Some indicate the quantity purchased – either in cart loads, bushels, corves or simply loads – and most name the colliery from which the lime-coal was sourced. Regrettably, payments to cover carriage costs are rarely entered in the accounts and when they do appear they cover unspecified periods of time, so it has not been possible to estimate transport costs except for a single entry, dated 12 November 1754, 'leading lime kiln coal at 5d. per corf £3 0s. 2d'. Even here, though, the data are of limited value as the colliery was not indicated.

Distances between mine and kilns were often considerable. The Danby estate purchased lime-coal at different times from five collieries with West Pits above Braithwaite Hall in Coverdale having been the preferred choice. Table 6.1 shows the location of each colliery, the distances between mine and Danby Hall, and the first known year during which lime-coal was sourced from each one. Distances given are estimates based on the most likely direct route for horse-drawn carts. West Pits was the closest so transport costs would have been lower but the mine was situated 300m high on Braithwaite Moor with an arduous descent to the valley below. Given the disproportionate distance from the Colsterdale pits and the difficult upland terrain between Colsterdale on the River Burn and the River Ure, the choice of this source defies explanation on purely economic grounds and no reasons are suggested in the estate accounts.

Colliery	Location	Distance	Date
West Pits	SE125 845	5km	1730
Cover Pits	SE118 862	6km	1731
Middleham	SE?	c. 7km	1732
Colsterdale	SE124 824	14km	1752

Table 6.1 Sources of lime-coal for the Danby Estate
Source: NYCRO ZPT and ZPT 4/4/3

The Willis family farmed land along the Ure to the north of Bainbridge at Yorescott and Skellgill. Edward and then James Willis kept a regular day book from 1822 to 1850 with meticulous entries for sales and purchases (NYCRO TD 124). There is frequent mention of purchases of lime-coal for

1 Hall Intack, Ravenstonedale
2 Langstone Fell
3 Morley of Marrick
4 Danby, Stainton
5 Bolton Hall
6 Wyvill of Constable Burton
7 Scrope of Danby Hall
8 East Scrafton Estate
9 Horton in Ribblesdale
10 Fountains Fell
11 Raikes Farm, Threshfield
12 Towneley Hall
13 Hornby Castle Estate
14 Kettlewell Trust Lords
15 Swinton Estate

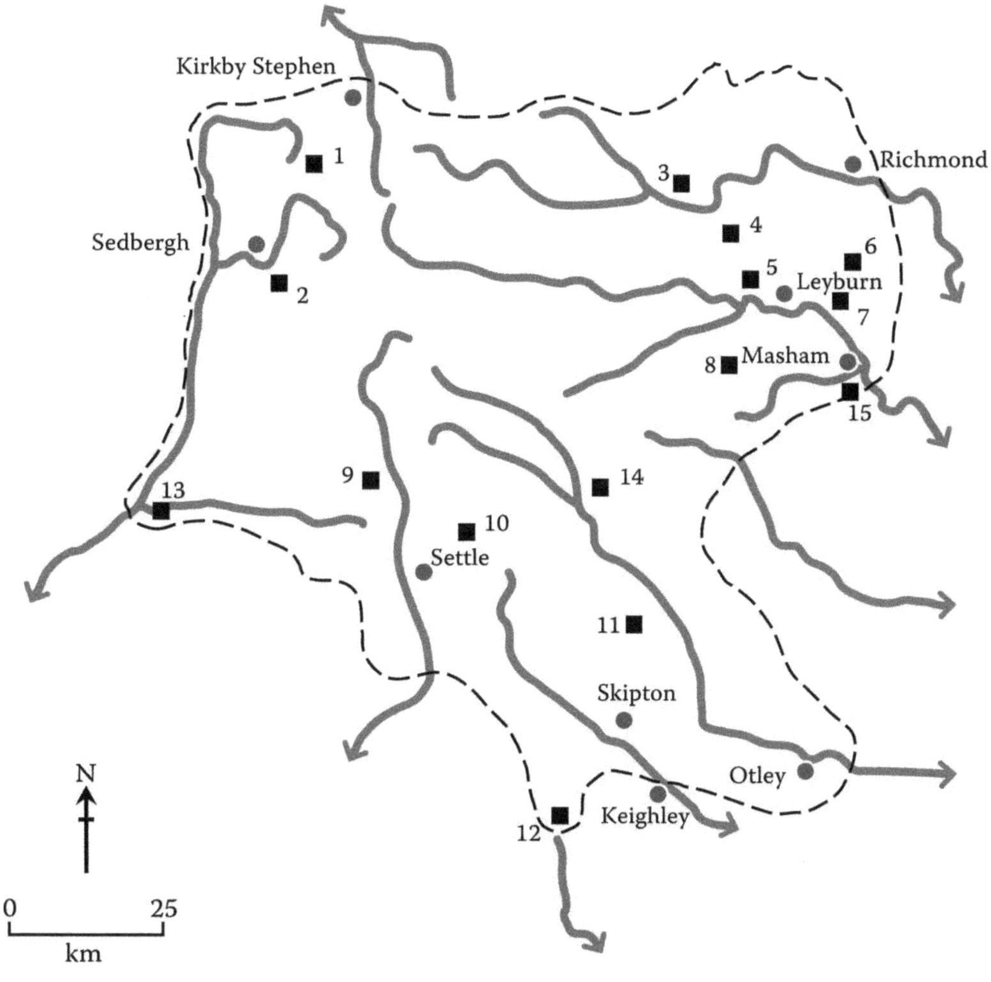

Figure 6.3 Distribution of estates with direct lime-coal associations

their own kilns, some of the lime intended for construction and some for agricultural use, sourced from six collieries in tributary valleys of Upper Wensleydale. Distances here are at the top end of the Danby scale, as shown in Table 6.2. The Willis's transport costs were inevitably high but, before the coming of the railway through the dale in the 1870s, they had no choice. Each of the five pits was situated at a high altitude with rough tracks connecting them to the main valley road, adding to the total price of coal delivered to the kiln. Bringing coal in was indeed a costly enterprise but if they felt burning lime was necessary they had to bring it in, and the mines they used were the nearest.

Large estates and farms had the financial wherewithal to absorb high transport costs but this did not apply to smaller tenant farmers and landowners. In 1834 the landed proprietors of the school at Muker in Upper Swaledale decided the time had come to bring into profitable use pastures they owned by a process of draining and liming, producing lime in the township kiln. The main constraint facing them was the cost of procuring lime-coal as the nearest coal pits were at Tan Hill 12km-20km away, so they agreed to petition the lord of the manor for permission to get coal from the moors contiguous to their own land in return for 'a small duty per load' (NYCRO Z 760). The outcome was not recorded.

Colliery	Location	Distance
Mossdale	SD822 913	13km
Garsdale	SD783 896	19km
Storth Gill	SD865 866	12km
Bank Gill	SD853 853	14km
Cotterdale	SD839 955	12km
West Pit	SD855 957	12km

Table 6.2 Sources of lime-coal for Yorescott and Skellgill
Source: NYCRO TD 124

Thirty years later the outcome of another petition also went unrecorded, at least in surviving records. High Gingerfield farm (NZ151 022), on the Zetland Estate north of Richmond, was tenanted by the Wilson family and they spent many years trying to square one particular circle (NYCRO ZRL 11/3). They had successfully negotiated the right to build a kiln on the farm in 1864 and, in an unspecified year, entered into discussions for permission to obtain 1000 corves of lime-coal from pits at Hudswell and Hipswell, to the south of Richmond, belonging to the Marske Hall Estate. They clearly did not gain satisfaction as, in 1873, Wilson complained bitterly that his offer of £2 per ton at the pit was reasonable especially as he would have to pay 10s. per ton for carriage on top of the purchase price. In the following year he wrote a pleading letter to John Hutton of Marske Hall explaining that he had a kiln and ample limestone on his farm but 'coals are so dear this cannot be done', meaning he could not bear the expenses incurred in burning lime.

Contemporary accounts bear out the difficulties faced in obtaining fuel supplies and the amount of traffic dominated by lime- and fire-coal. On a prolonged tour through Craven another John Hutton observed the:

> number of small carts laden with coals, and each dragged by one sorry horse, that we met, was astonishing. Many of the smaller farmers ... earn their bread with carrying coals ...from the pits at Ingleton, Black-Burton ... to Kendal, and the neighbouring places, for fewel, and burning lime in order to manure their land (Hutton 1781, 12).

Prior to this, poor quality coal had been eked out of thin seams high on Whernside. According to the authors of the county report for Westmorland submitted to the Board of Agriculture (Pringle 1794, 292-93), burnt lime was exported from Craven through Kendal to the central Lakes 'at great expense' because the cost of importing coal was considered prohibitive. The use of agricultural lime in Westmorland was markedly delayed by a lack of affordable fuel, except in the north-east – in Ravenstonedale and Orton, for example – where limeburners brought in coal from Stainmore even though that source was more than 30km distant, and the cost of lime-coal 13s. per ton inclusive of transport (Robinson 1803, 417; Curwen 1932, 106). Coal of variable quality was also worked on the moors above Barbon and Casterton from the 1600s to the 1800s, a trade closely linked to lime burning (Kelly 2008, 21). A deposition, registered in 1843, made the point that coal reserves above Barbon and Casterton had never enjoyed a high value, with those in Barbon being of 'inferior value' and then only used by the limeburner who operated three local kilns as well as the coal pits (CRO [K] WD/PP, Box 8). A pamphlet written in 1791 to attract investors to the proposed Lancaster-Kendal canal claimed that farmers would save £5 per ton on lime transported by boat and would no longer have to cart it 'twenty miles from Kellet' for 'manure' to those parts where it had not been burnt owing to a lack of local coal (LRO DDPa, Box 1).

The crucial link in Westmorland between cheap coal, lime and land improvement was also emphasised by John Hodgson (1820, 11) who expressed the view that 'till coals can be more easily obtained, it (lime) can never be extensively admitted into agriculture here'. Certain areas within the south-east of the county did manage to find suitable coal locally. In Mallerstang farmers wishing to burn agricultural lime dug 'small coal' on the moors (Hodgson, 1820, 19). It was not suitable as fire-coal but acceptable for lime kilns. Further south, in Casterton, a similar layer of poor quality coal was 'wrought' for producing agricultural lime but this was dying out by his time (Hodgson 1820, 234).

The same pattern can be recognised from the seventeenth to the early nineteenth century throughout the research area: where coal was more or less at hand, lime was burnt in quantity for applying on the land.

6. Colsterdale colliery lime-coal accounts

Coal had been mined in Colsterdale from monastic times.[53] In 1334 the abbot of Jervaulx in Wensleydale was granted the right by Sir Geffrei le Scrop to dig for coals for abbey use on any of the 'wastes of Colsterdale' (NYCRO ZS). The Danby family of Swinton Park to the west of Masham operated coal mines there from the seventeenth to the nineteenth century (NYCRO ZS*).[54] The earliest reference dates from 25 May 1609 when Christopher Metcalfe was paid 10s. by estate steward Miles Danby as part payment for burning lime intended for the rebuilding of Leighton Hall on the estate. Sir Abstrupus Danby, who had recently taken over the estate, was rebuilding Swinton Park itself through 1696, carting hundreds of loads of coal from his pits to the lime kiln specially built at Swinton Park for use in making lime mortar. Much of the lime, however, was also intended for agricultural use on the estate.

The estate accounts provide remarkable detail of lime burning operations intertwined with aspects of social history. Between 1695 and 1700 Ralph Horner was paid regularly for burning lime at Danby's kiln with individual settlements ranging from five to 173 bushels. Over the six-year period Horner earned

[53] The word 'colster' means someone who is involved with coal.

[54] Unless otherwise stated, all references in this section are ZS*, a large and uncatalogued set of archives.

rather more than £35 for burning 1360 quarters at 6d. per quarter. He also found time to get the limestone and was paid by the cart load for 'gathering limestones', though the weight per load was not stated. This work added £12 to his income for the period. He clearly did not have the time to gather all the necessary stone as he was helped out by several other people. For whatever reason, the quantity gathered was paid by the load when men performed the task but by the day when women or children were involved. Ralph's wife was paid, at 4d. per day, for fifty-eight days work gathering limestone in 1695 and 1696 and his son for fifty-seven, at the same rate. She did no such work between July 1696 and April 1698: was she pregnant perhaps for at least part of that time, or was her husband able to cope with the work load in this period? The surviving accounts do not permit these questions to be answered. After April she no longer appeared in the accounts; her son was not listed after April 1696. From time to time there are miscellaneous entries in the accounts, such as payments for 'filling stones' at the kiln, work again undertaken by a family partnership of father, mother and son; for 'making up the Lime Kilne anew'; every few months for the purchase of fresh scuttles for use at the kiln; and, of course, for lime-coal purchases before Danby's own collieries came into production.

Danby eventually sold lime-coal on the open market and to this end agreements were signed between Sir Abstrupus and a partnership of miners, John Ascough, William Abbott and Robert Wilson, to supply '6000 corves of good and saleable lime kiln coals such as shall content to (his) customers ...'. The lime-coal accounts for 1797 show that one chalder was made up of fourteen corves. A further agreement, signed on 25 February 1724, committed the partnership to bringing 10,000 corves to the surface while a third agreement granted a second partnership consisting of Aron Abbott and James Blaids to mine a further 10,000. The records do not identify the destination of all this lime-coal, though a number of large-scale kilns were operational in the area around Masham, Well and Snape. An indication of the importance of lime-coal mining is given by contrasting lime-coal sales with those of fire-coal: from June 1723 to June 1724 8896 corves of fire-coal were sold as opposed to 12,507 corves of lime-coal.

The estate owned, in partnership with Lord Ailesbury, two mines that concentrated on lime-coal, both in Colsterdale. Brownbeck Colliery, sometimes referred to as the 'Limecoal Colliery', was the more profitable of the two (Table 6.3). The colliery accounts list annual profits, derived from sales revenue less costs of mining the coal and costs of maintaining the mine and operating surface plant, such as it was. Though the accounts do show considerable variation in profit throughout the period, they were on the whole making a good contribution to the viability of both mine and estate.

Brownbeck's first heyday was relatively short-lived, largely confined to the 1750s, with a single-year revival in 1778, and the accounts offer no explanation as to why there were marked fluctuations in sales from year to year. The lack of surviving records directly attributable to Brownbeck, between 1779 and 1805, is unfortunate and leaves a major gap in understanding

Year	Annual profit	Year	Annual profit
1746	£188 17s. 8¾d.	1755	£216 17s. 2¼d.
1748	79 5s. 6¾d.	1760	91 8s. 8¾d.
1750	229 3s. 7½d.	1764	55 7s. 9¾d.
1751	301 10s. 8½d.	1778	220 0s. 1d.
1753	363 5s. 8¼d.	1779	68 19s. 7d.
1754	413 7s. 9¼d.	1805	30 17s. 6d

Table 6.3 Brownbeck Colliery, Colsterdale, annual profit margins

sales trends at the end of that century, especially as it was a period when there was an almost missionary zeal to improve land in the nation. This colliery is named again in the mid-nineteenth century in terms of lime-coal sales but, for the twenty-one months listed between February 1850 and April 1852, sales only occurred in seven of those months. In 1850 847 corves were sold, over two months, while 1727 corves were despatched in 1851 spread over five months. None was sold in 1852. More tellingly, perhaps, are sales prices: in 1850 lime-coal was sold at 10d. per corve; in 1851 the price had been halved. It has to be assumed that the colliery manager was fighting a losing battle to keep the colliery going, not helped by problems posed by nature. From March 1842 to February 1843 Brownbeck produced nothing as water poured into the pit, causing great expense in pumping it out. It did not survive after 1852.

Gollinlith Colliery was even less successful in the long-term, partly because the coal was of an inferior quality and partly owing to difficult mining conditions. Quantities extracted quickly rose from 391 corves in 1780-81, the first recorded year, to 25,185 the following year but then steadily declined to 14,782 by 1794-95 with a collapse to only 3538 the year after that. Gollinlith closed down at the end of 1796 having made a profit of only £42 4s. 7d., too small to be economically viable. Production was transferred to other pits on the estate, though the name Gollinlith reappears in the accounts in the 1840s. The years for which accounts survive re-emphasise its economic fragility. The first year in the later series, 1842, saw a loss of over £84, just returning to profit the following year to the tune of around £7, with a further loss of over £8 in 1844, and barely breaking even in 1845 with an operational profit of just over £1. As will be seen in Chapter 8.3, agricultural lime sales were decreasing by this time as alternatives came on the market and, combined with the colliery's physical inaccessibility, it could not survive. Gollinlith met the same fate as Brownbeck, for the same reasons.

During the third quarter of the eighteenth century, when the coal mines were still productive, the Danbys were intent on improving the profitability of their estate by winning new land from the open moors. Their agent, William Morton, was at the forefront of the campaign urging on tenants and keeping his master, often in residence in London, fully abreast of developments in a regular series of detailed letters. As an example, Morton wrote on 7 December 1773 with a further development to report (NYCRO ZS):

> But as I thought it better to make a little further trial of the limestone they have found I directed Sam (Ascough) to get up as much stone as would conveniently burnt in a common Sod Kiln, by which means we should not only judge the quality of the Lime, but form a better idea of the substance of the quarry, accordingly he set about it, and I was up there on Saturday to see some Lime drawn out, which indeed seems to promise great things for Colsterdale and its neighbourhood; The Lime is good in quality ...; you will consider of the proposed method of working it, that all necessary preparations may be made against the season for burning lime comes on

Morton was so pleased with the quality of lime from the sod kiln that he ordered the building of a masonry kiln nearby in 1774, at a total cost of £12 6s. 4d. Both kiln and quarry were rented on an annual basis to Thomas Wintersgill for 1s. per year. This kiln was worked hard, particularly as the surrounding area was in the process of being enclosed and improved through the 1770s, and by 1825 it was in need of a major rebuild. Thomas Clarkson was employed as mason to build the new kiln with the help of ten other men who put in a total of 171.5 days of work on the new kiln and access road. Mathew Wintersgill was one of two men employed to 'shift the earth at the kiln bottom'. Henceforth he was retained to superintend the working of this kiln. However, the kiln (which still stands) only worked for four years (Table 6.4).

The kiln could not operate at a satisfactory profit because, as the accounts relate, 'the above works discounted as it was found not to answer, the Lime Stone not good, and a great expense in getting it'. Neither of these contributory factors is surprising: the quarry concerned was cut into a very small outcrop of limestone in what is basically a sandstone landscape,

Year	Expenditure	Revenue
1825	£39 18s. 0d.	-
1826	£ 7 9s. 6d.	£48 4s. 1½d.
1827	£77 1s. 5½d.	£39 15s. 9d.
1828	£11 16s. 4d.	£17 8s. 0d.
Total	**£136 5s. 3½d.**	**£105 7s. 10½d.**

Table 6.4 Colsterdale lime kiln accounts, 1825-28

and the stone is fractured and awash with impurities; while the kiln itself is perched near the top of a long and steep slope on the valley side. The access road reaches the kiln down a series of zigzags and transport of coal in and lime out cannot have been easy, especially in wet weather.

How the estate obtained its supplies of lime after 1828 is not recorded but, in May 1842, William Wells was paid £14 for 'building the kiln, as per bargain'. A new kiln, called the Maines kiln (Fig. 6.3), had been built close to the Ure 4km north of Masham. It, too, had a chequered life as the first four years' accounts testify, as shown in Table 6.5.

No coal was sold to Maines in 1846 though in its final year, 1847, 1204 corves were sold. The figures emphasise Maines' lack of viability from the outset and it could be argued that this should have been apparent even before the decision was taken to build it. There were well established banks of selling kilns in the general area – at Well and Snape – and the estate already had a kiln at Ilton, so the Maines venture may have been effectively redundant at inception. It is also possible that it was brought into production when demand for lime was already beginning to fall off, a hypothesis which will be explored below.

The detail in the overall colliery and lime accounts is of importance to this research process in that it illustrates in a variety of ways patterns of lime usage: a number of themes will

Year	Sales revenue		Expenses	Net profit (loss)
1842	£77 14s. 5d.	Labour on kiln	£84 17s. 7d.	
		Coal purchases	50 19s. 4d.	
		Ale, lime leaders	**2 6s. 3d.**	
			138 3s. 2d.	(60 9s. 9d.)
1843	136 10s. 6d.		129 1s. 11d.	7 8s. 7d.
1844	166 8s. 2d.		175 0s. 2d.	(8 12s. 2d.)
1845	111 7s. 7d.		108 11s. 2d.	2 16s. 2d.

Table 6.5 Maines lime kiln accounts, 1842-45

Figure 6.4 Maines lime kiln, built 1842. Its size and technological features suggest it was intended for large-scale production.

be investigated. The relationship between lime-coal sales and lime production will be discussed as they have a clear symbiotic relationship. Lime burners reacted to customer demand, boosting or cutting back on lime production to suit the current economic situation. To keep their accounts in order, and to avoid having capital tied up in unsold lime or unused coal at the kilns, they were ordering lime-coal almost on a just-in-time basis. The successful commercial lime burner had to be adept at reading the market, anticipating demand in the near future, and purchasing sufficient coal to burn the amount of lime they had orders for.

Limeburners were also at the mercy of the limestone getters, those who quarried the stone. Virtually all lime kilns in the Masham area had an adjacent quarry operated by the lime kiln owner or lessee so the lime producing business here was vertically integrated. Problems in supplying stone to the kilns arose when the quarrymen could not work. If inclement weather prevented them from quarrying or breaking up stone to the required size, the limeburners could do nothing.

This all meant that lime-coal sales fluctuated according to what was happening at the kilns and to what lime customers were ordering. Table 6.6 shows sales of lime-coal, from all Danby collieries, in ten-year classes.

Nearly half of all sales were achieved in the twenty-year period 1796 to 1815, with a marked drop after 1835. Various contributory factors can be identified to account for these trends, some local and some national, though any conclusions drawn must remain tentative in view of the absence of records from before 1796.

Ten-year class	Total sales (corves)	% of total sales
1796-1805	179,248	23.27
1806-15	171,221	22.23
1816-25	162,978	21.16
1826-35	149,055	19.35
1836-45	93,242	12.10
1846-51*	14,585	1.89
Total	770,329	100

Table 6.6 Lime-coal sales from Colsterdale 1796-1851
*Coal production ceased in 1851

The correlation between lime-coal sales and lime production on the one hand and parliamentary enclosure on the other was a factor here (this will be explored in Chapter 7.4), and insight into the state of the local market can be gained by comparing the sale price for lime-coal through the period in question (Table 6.7).

The doubling in price between 1723 and 1796 may reflect inflationary pressures in the general economy during that

Year	Price per corve	Number of corves sold
1723	4d.	6000
1796-1805	8d.	17,925
1806-09	9d.	18,585
1810-12	10d.	14,507
1813-17	1s. 0d.	17,388
1817-19	10d.	17,923
1819-22	1s. 0d.	14,896
1822-23	10d.	13,320
1824-27	1s. 0d.	17,906
1850	10d.	2385
1851	5d.	1727

Table 6.7 Sale price for lime-coal from Colsterdale, 1723-1851

century (the value of the pound slumped by sixty per cent between 1700 and 1800) or a response by the estate to increased demand for lime-coal. The gradual increase from 8d. in 1796 to 1s. in 1813 was a direct response to increased production costs and inflationary pressures rather than exploiting a seller's market; the drop in sales from 1810-12 needs to be explained by other, as yet unknown, factors. The price fluctuations between 1813 and 1827 are difficult to explain but the final increase to 1s. from 1824 may well have been in response to increasing demand. Beyond 1827 annual figures are not available in the accounts except for 1850 and 1851 and the plummeting sale price then was a direct result of the falling sales noted earlier.

These colliery accounts accounts provide some information about lime-coal customers by purchasers' names and also, in a few cases, the name of the kiln they operated. It is, therefore, possible to extrapolate trends in lime sales from the pattern of their purchases, as shown in Figure 6.4.

The trend is clear: over the fifty-four years for which there are records the mean number of customers was statistically slightly over five (5.24). In the period from 1796 to 1833 only in four years was the number of customers less than the mean while after 1833 the mean was not achieved in any year. The mode is six and this was achieved or exceeded in twenty-five years up to 1828 but not once after that. These statistics support the conclusion from the annual sales figures that there is a clear positive correlation between the number of customers and the total sales of lime-coal through the period under review. Up to 1828 in particular, sales were buoyant, and more commercial lime burning concerns were exploiting a rise in demand for lime and more individual limeburners, most probably part-time farm-based, were also buying into the much-promoted idea that the best way to bring newly enclosed moorland into productive usage was by the application of lime. This buoyancy goes against the national trend, though, as the 1820s are perceived as a time of agricultural depression: this theme will be returned to in Chapter 8.2.

Of the customers who are named in the monthly and annual lime-coal accounts, it is possible to match them to four named kilns. It has already been seen that the lessees of the Danby Estate's own kilns at Colsterdale (from 1825-28) and Maines (1842-47) were purchasing lime in varying quantities for use on estate lands; the other named kilns were at Well and Gebdikes.

When the kiln in Colsterdale was built in 1825 Thomas Clarkson was the one contracted to carry out the work. He is presumably the same person who had been buying lime-coal from the estate since 1817 for an unidentified kiln. He remained a regular customer until 1832 when Mathew Clarkson's name begins to appear in the registers, possibly Thomas's son. He bought lime-coal on a regular basis until 1851. The kilns at Well and Gebdikes were worked by Thomas Hauxwell and George and John Lee. Hauxwell's name is entered in the accounts more or less continuously from 1803 with his first actual purchase being 1365 corves two years later. His annual purchases were substantial though variable (Table 6.8) but he seems to have nearly come to grief in 1819 when he took delivery of only one consignment of 308 corves, and there is an annotation in the accounts about Hauxwell: 'I fear (the outstanding account) will not be recovered as he is unable to pay'. He was in arrears with his payments. Two years later, however, he seems to have bounced back taking 833 corves over the year. Up to this point the accounts do not state which kiln he was working but the

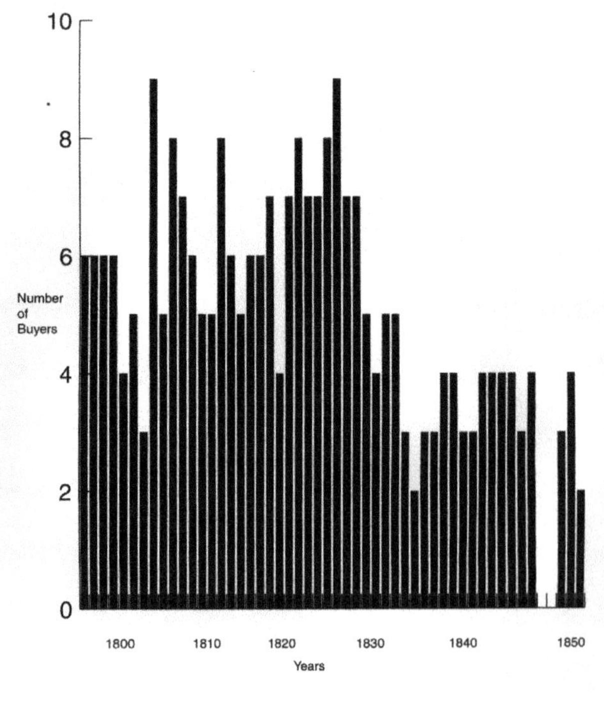

Figure 6.5 Number of lime-coal customers, Colsterdale, 1796-1851

records for 1822 and 1823 mention that he was taking delivery of lime-coal at both Gebdikes and Well kilns.

Comparison of Hauxwell's purchases with general sales of lime-coals (Table 6.8) points up an anomalous situation from 1818 to 1822. General sales were very buoyant through those years yet he was virtually bankrupt: perhaps he had other ventures elsewhere that impacted in a negative way on his lime burning business. The colliery accounts, however, throw light on how he managed to recover from 1821, and John Topham was purchasing lime-coal on his own behalf, for an unnamed kiln, up to 1821 but in 1822 and 1823 he is listed jointly with Hauxwell at Gebdikes.

The Lees must have operated as a partnership but it is not known how they were related. In some accounts George is named as the payee while in others it is John (Table 6.9). It is of interest to note that the Lee partnership experienced similar difficulties to Hauxwell in the years 1819-21, though the former's almost immediate recovery can only be described as spectacular.

Nonetheless they, too, saw their business decline to the point of extinction from 1832. Whatever local factors were at play in the Masham area, the effect on their purchase pattern was in line with all other local lime-coal customers, as illustrated in Figure 6.5. The possibility of all customers having transferred their purchases to other sources of coal can be firmly ruled out as there were no alternatives.

Because the lime-coal accounts itemise sales on a monthly basis, it is possible to identify seasonal patterns of sales which can be extrapolated to seasonal patterns of lime burning and usage (Table 6.10). It will thus be possible to identify the months when lime was being applied to the land, even if some of the lime produced at the kilns was intended for mortar or other non-farm end uses.

Late spring and early summer (April to June) account for over half (54.27 per cent) of all sales in the period under review with April being the busiest single month overall, despite the fact that sales were not achieved in that month every year. The four winter months, as might be expected, were quiet months accounting for only 4.62 per cent of total sales. Quarrying the stone was subject to the vagaries of the weather; transport of coal from colliery to lime kilns may have been hampered by often impassable tracks; building work and therefore demand for mortar and limewash would have been minimal; and the ground would have been alternately frozen and waterlogged militating against any thought of applying lime on pasture or crop land.

Year	Corves purchased
1803	1365
1808	2873
1809	3952
1810	2053
1812	4241
1813	3814
1814	3692
1815	3322
1816	3502
1817	3111
1818	2851
1819	308
1820	0
1821	833
1822	2864
1823	3534
1824	458
1825	0

Table 6.8 Lime-coal purchases by Thomas Hauxwell, 1803-25

Year	Customer	Purchases (corves)
1818	John	3410
1819		316
1820	George	26
1821		853
1824	John	11,915
1825	George	6967
1826		4207
1827		5686
1828		2002
1829	John	1904
1830		2095
1831		3050
1832		543
1833		0
1834		0
1835		0
1836		0

Table 6.9 Lime-coal purchases by George and John Lee, 1818-36

Month	Total sales by Month	% total sales	No. of years with sales (n=42)
January	8080	1.39	7
February	4108	0.71	6
March	37,475	6.45	13
April	147,257	25.34	39
May	91,550	15.76	42
June	76,490	13.16	42
July	65,079	11.20	42
August	51,987	8.95	42
September	41,354	7.12	40
October	43,014	7.40	36
November	7483	1.29	12
December	7151	1.23	12
Total	581,028	100	

Table 6.10 Sales of lime-coal at Colsterdale by month, 1796-1851

If a delay of a week or two between buying coal and selling lime is built into the equation, the data in Table 6.10 would suggest that lime was being burnt on a large scale continuously from March to the end of October or early November.

Comparison of Table 6.10 with Table 6.6 provides a rather anomalous picture: the years 1796-1805 and 1816-25 were the decadal periods with both maximum sales (23.27 and 21.16 per cent respectively of the 1796-1851 total) and the highest number of months when winter sales were registered (seventeen months in each case), yet the decadal period 1806-15 achieved 22.23 per cent of total sales but had only eight months when winter sales were noted. The years 1796 to 1805 were encompassed by the French Wars: the impact of these wars on farming – and therefore on the application of lime – will be discussed in Chapter 7.3. Climatic factors can help explain the lack of winter sales between 1806 and 1815 as temperatures were below the 1659-2007 average of 9.43°C in seven of those years. On the other hand, though, precipitation levels were only above the long-term average in 1810-11.

Peace in 1815, with the opening up of trade, wrought havoc across many farming communities in the country and, together with a disastrous summer in 1816, one might have expected these events to have impacted negatively on agricultural lime sales. The data from Colsterdale, however, indicate buoyant trade in the decade 1816-25.

7. Conclusion

This chapter has discussed the importance of obtaining reliable sources of fuel for firing lime kilns, sources that were economically viable both for estate-based lime production and for commercial operations. Without this, the costs of producing agricultural lime would have prevented many estates and tenant farmers from fully benefiting from land improvement. Indeed, it has been shown that the widespread use of agricultural lime was delayed by fuel supply constraints, particularly in Westmorland.

As evidenced by documentary sources and archaeological excavations, a range of fuels were used in the period under review with wood and peat having been dominant prior to the identification and development of local coal reserves during the late medieval era, mirroring practices elsewhere in the country. In some areas, however, where coal was absent or import costs made its use prohibitive – as in parts of Westmorland – peat continued to be used into the early modern period. Initially kilns used what was widely known as sea-coal, but early exploitation of coal in the study area was focussed on coal shales or thin bands of carbonaceous strata within the Yoredale Group rather than on actual coal from the Coal Measures. The former were of inferior quality and highly polluting but were deemed adequate for lime burning and, in any case, were all that was available except in north-east Lancashire and around Ingleton.

Documentary evidence has been used to demonstrate the lengths certain land managers went to in order to obtain coal for their lime kilns. Those who were wedded to the improvement ideal purchased lime-coal, often at considerable expense, from collieries between 12km and 20km distant, simply because no alternative was available.

Estates fortunate enough to possess sizeable coal resources tended to exploit them to the full. Sales of lime-coal were logged separately from fire-coal and the larger estates developed their own lime kilns to produce agricultural lime, as well as lime for other end uses, for use on demesne and tenanted land. Estate managers saw the financial benefits of vertically integrating the process: quarrying limestone on the estate, mining their own coal and building lime kilns for the use of the estate and its tenants. Foremost among these landed estates were the Towneleys of Towneley Hall east of Burnley and the Danbys of the Swinton Park estate to the west of Masham.

The existence of monthly lime-coal production and sales accounts, from 1796 to 1851, for Swinton Park estate collieries in Colsterdale has enabled detailed analysis of lime usage across the Masham area. Sales of coal to lime kiln operators closely paralleled the production of lime; in turn, this was closely correlated with sales of lime and thus with its application on the land. If farmers were not minded to apply lime at any given time, they would not buy it, thus the limeburners would not produce it nor purchase lime-coal. Each element in the supply chain was inextricably linked to all the others.

Thus, it has been possible to identify patterns of seasonal use of agricultural lime in the Masham area, with November to February having been the only months when its use was minimal. Long-term application trends have been similarly identified with clear peaks and troughs visible in the archival record. Usage rates were high in the years 1796 to 1805, a decade coinciding with the early years of the French wars when a 'dig to survive' mentality spawned massively increased farm activity. Usage was correspondingly low between 1806 and 1815, despite this also having been a time of blockade with reduced imports of food, so this particular set of data is anomalous and may be explained by climatic and/or unidentified stochastic factors. Equally anomalous are the strong lime-coal sales data for the years after the declaration of peace in 1815 when farming came under acute pressure. This will be explored in Chapter 8.2.

Having provided an analysis of the close connections between the use of agricultural lime and the availability of affordable fuel supplies, the following chapter returns to a chronological theme and considers liming during the period of parliamentary enclosure when there was a determined nation-wide drive – one might even say crusade – to bring about improvement of existing (unimproved) agricultural land and transformation of the waste into productive farm land.

Chapter 7
Agricultural Land Improvement, Liming and Parliamentary Enclosure

1. Introduction

Chapter 6 discussed the relationships between lime burning and the supply of fuel, and focussed on lime-coal production on the Swinton Park estate in Colsterdale. A long run of production accounts made possible detailed analysis of seasonal and periodic fluctuations in lime-coal output from 1796 to 1851, which can be directly related to variations in patterns of liming. This chapter considers the role of liming over that period, the time of parliamentary enclosure.

A range of stimuli impacted on improvement in general, and liming in particular, during this period and these will be examined in turn in Section 3:

a. the growth of agricultural societies,
b. the spread of periodicals aimed at farmers and landowners, and an upsurge in the number of farming manuals,
c. the establishment of the Board of Agriculture,
d. a developing scientific appreciation of the effects of lime on soil and plant growth,
e. the influence of the Improvers, a body of individuals intent on modernising farming and bringing more land into productive use,
f. the impact of the French wars on trade and pricing structures in farming,
g. improvements in transport, especially canals and turnpike roads.

The decades after 1760 are associated with a concerted drive to 'tame' the countryside through the privatisation of stinted commons and fell pastures in the process of enclosure. Many of the 'new' ideas were tested on a grand scale where formerly barren, or waste,[55] ground was walled in and coerced into a semblance of productivity. Where liming played a part in this process – and it did not always – it was but one element in a succession of reclamation strategies: paring and burning, ditching and deep ploughing were also integral to the process.

This chapter will make use of examples from across the research area as a means of identifying improvement measures; will isolate those areas where it is possible to demonstrate a correlation between enclosure and liming; and will investigate the prime movers behind liming.

2. Climatic trends

The effects of climatic downturns on farming within the research area in earlier times have been considered. In some instances the impact was seen to have been negative, in others farming seemed to have gone on much as before. For the period under review now, there were times when farming was adversely affected by natural phenomena, to the extent that contemporary diarists felt moved to note exceptional rainfall events and cold spells.

Long-term data (for 1659 to 2007) indicate that temperatures were below the average of 9.43°C in far more years than they were at or above the average value: between 1760 and 1817 below average values were recorded in thirty-four years and only in 1779 and 1781 were they significantly above average (**www.metoffice.com**). The coldest spells occurred in the late 1760s, in 1784 and in 1816-17.

The latter dates can be readily explained by exogenous natural forces. The Icelandic volcano Laki erupted through 1783-84. A plume of ash, dust and sulphur dioxide mixed with rain to form sulphuric acid. This covered much of Europe north of latitude 40° throughout July and August 1783, directly leading to widespread harvest failure, repeated storms in the autumn and a very long and severe winter (La Rochefoucauld 1988, 30, 98, 107; Highwood and Stevenson 2003). Its impact was felt across Britain well into 1785. Tambora, in Indonesia, erupted in 1815 with climatic deterioration felt across the world for over a year. In Britain the following year came to be known as the 'year without a summer' and the impact on farming has been regarded as disastrous (Grove 1988, 415).

Precipitation levels fluctuated wildly: the driest single year between 1766 and 1980 was 1788, with 1779-86 having been drier than the norm, whereas 1768 was the wettest. Twelve years endured above-average rainfall between 1760 and 1817, with 1789 one of the wettest on record (**www.metoffice.com**; Wigley *et al.* 1984).

The degree to which these climatic downturns entered the popular psyche and impacted on farming in the research area will become apparent as this chapter unfolds.

3. Catalysts for improvement

Agricultural societies

In 1800 England had thirty-five agricultural societies, by 1845 there were between 360 (Tarlow 2007, 35) and 700 (Goddard 1990, 166; Wilmot 1990, 9). Possibly the first attempt to establish such an organisation was the foundation of the Society of Improvers in Edinburgh in 1723, though it was wound up in 1746 (Marshall 1929-30, 50). Those societies which were established within the research area are summarised in Table

[55] The word 'waste' in this context has been described as a 'weasel' word because much of the so-called waste land was not waste at all, in the true sense of the word (Tarlow 2007, 45).

7.1.[56] It is significant, perhaps, that only two are known to predate the period when Britain was at war with France. This was a time when every effort was made to boost land productivity, an aim agricultural societies subscribed to.

They were perceived as a means of disseminating information among the farming community as widely as possible, and many instituted annual shows and prizes to recognise best practice and thereby encourage others. Lord Somerville, President of the Board of Agriculture, wrote to prominent landowners requesting their assistance in setting up regional societies because the Board could thereby 'render the farming world ... the most essential advantages' by disseminating information down to the lowest level (LRO Towneley, DDTO, Box UU, 1800).

Their value has been debated widely. Some commentators have taken the view that local societies were little more than social clubs with a tendency towards self-congratulation and vanity among the leaders (Macdonald 1975, 139), with the argument expressed that they gave the landed classes social cachet (Wilmot 1990, 6). An alternative view sees them as disseminating information to those who were illiterate and therefore beyond the reach of agricultural manuals and periodicals; they were, indeed, clubs for a 'new class of person' (Hudson 1972, 3, 113). Of more practical value, perhaps, in disseminating information to the ordinary man were the localised farmers' clubs which possessed none of the frills and pomposity of some of the more formal societies (Macdonald 1972, 33).

Local societies may have had a meaningful role to play *locally*, but the RASE was probably seen as an elite body, dominated by the landed classes, and thus more relevant to prosperous farmers and to landowners than to the ordinary small farmer. Membership figures for RASE in 1854 may illustrate this point; in the whole of Yorkshire there were only 296 members, in Lancashire 150 and in Westmorland a mere fifteen, out of a national membership of 5260 (Wade Martins 2004, 141). Alternatively, these relatively low numbers could be explained by the fact that, in broad terms, the North suffered less than southern areas in times of agricultural recession during the period under review here and it may have been the case that northern farmers and landowners felt no need for the support the Society was able to offer.

A contemporary survey of the state of farming in the North Riding bemoaned the facts that not a single society existed there and that the Richmond Agricultural Society had already died out (Tuke 1794, 94). Tuke put this down to innate conservatism among the farming classes who, he said, were suspicious of improvers trying to foist new-fangled ideas on to them. It should be borne in mind, though, that he was an improver with a specific propagandist agenda to follow.

[56] The first national organisation in England, the English Society of Agriculture, was founded in 1838 and became the Royal Agricultural Society of England (RASE) in 1840.

Societies did undoubtedly play a part in developing new farming techniques – in selective breeding, in promoting scientific advances, in demonstrating best practice – which helped to bring farming into the 'modern' age of Victorian Britain. However, as will be seen in Chapter 8, the highs and lows of nineteenth-century farming lead one to doubt their long-term impact and the extent to which they fully permeated through the farming hierarchy.

Society	Founded	Other recorded dates
Appleby & Kirkby Stephen	1841	1890
Ashton	1838	
Bentham	1865	1840s – 1855
Craven		
Kendal	1799	1867
Kendal Farmers Club	1861	
Lancaster		1810
Lunesdale	1839	1862
Orton	1860	1867 closed
Ribblesdale		1840s – 1855
Richmond	1777	1794 defunct
Royal Lancashire	1767	
Royal North Lancashire	1847	
Shap		1861
Westmorland	1867	
Wharfedale	1796	
Yorkshire	1837	

Table 7.1 Known agricultural societies within the research area

Agricultural literature

Fox (1979, 54) asserted that the period 1780 to 1850 witnessed a five-fold increase in the volume of agricultural manuals and double that rate for the number of farm-oriented periodicals in print, compared with previous years. This writer has examined fifty-four manuals, published between 1760 and 1839, compared to eight between 1660 and 1760 (see Chapter 5.7) and thirteen prior to that (see Chapter 4.8). A breakdown by decadal periods, from 1700 to 1900, confirms that more were published in the 1790s (sixteen titles) than at any other time (see Table 3.4). In terms of periodicals though, the comparison is rather meaningless as none of the nine titles identified by this writer pre-dated 1780 (see Chapter 3.3).

The question was raised in earlier chapters whether manuals and periodicals reached the lower strata of farming society. The same issue can be raised for the period now under review, and the authors of several manuals made their position abundantly clear by dedicating their works to aristocrats. Kent (1775) was open in titling his book *Hints to gentlemen of landed property*; Winter (1787) dedicated his to the duke of Beaufort; and Parkinson (1798) listed thirty-six aristocrats among his subscribers with the prime dedication being to Thomas Coke, earl of Leicester. Titles such as these may have had a very different readership, and arguably less overall impact on grassroots farming (Macdonald 1975, 138), than an anonymously written book addressed to the 'ordinary class of practical farmers' (Anon d 1776). However, a rather more cynical view might see such dedications being reflections of these authors' social deference or even sycophancy.

Perusal of the contents of the libraries in any great house will testify to the large number of such volumes, but the evidence from lesser farmers is largely absent owing to a dearth of relevant records. One, undated, entry in an account book for Thomas Smailes' farm in Wensleydale is a rare example (NYCRO. OUTFAC): listed as purchases were eight volumes of the *Journal of Agriculture*; *The book of the farm* (1855), presumably by Henry Stephens; and two books by (J F W) Johnston, *Dictionary of the farm* and *Elements of agricultural chemistry* (1877). Smailes, however, cannot be described as a

Author	Date	Application of lime	Benefits of liming	Draining	Paring & burning
Hitt	1761	Y	Y		Y
Home	1762	Y	Y		
Dickson	1765	Y	Y		
Mills	1767		Y		
Henry	1771		Y		
Peters	1771		Y		
Kent	1775	Y		Y	Y
Anon d	1776	Y			Y
Clarke	1777	Y	Y	Y	
Hunter	1777	Y	Y		
Dossie	1782	Y			Y
Turner	1784	Y		Y	
Winter	1787	Y	Y		
Bailey & Culley	1794	Y			
Pringle	1794	Y			
Tuke	1794	Y	Y		Y
Somerville	1795	Y	Y		
Donaldson	1796	Y	Y		
Headrick	1796	Y	Y		
Parkinson	1798	Y	Y		Y
Banister	1799	Y	Y		
Verral	1799	Y	Y		
Dickson	1805	Y	Y		
Potts	1807	Y			Y
Kames	1815	Y	Y		
Rees	1819	Y	Y		
Sinclair	1819	Y	Y	Y	
Johnson	1834	Y			
Loudon	1839	Y			
Sproule	1839	Y	Y	Y	Y

Table 7.2 Summary of manuals published, 1760 – 1839[1]

[1]See the Bibliography for full reference details.

small farmer: he was descended from the Smales, the lower gentry family met in Chapter 5.3.

The most widely known periodicals were the *Annals of Agriculture*, compiled by Arthur Young between 1784 and 1815, and the *Journal of the RASE*, issued from 1840. Articles in the former were, arguably, of more practical value to the ordinary farmer, especially the regular sections of 'Queries' by which questions were asked and problems raised – and responded to by those with relevant experience. Volumes three and four for example, published in 1785, contained eight short pieces concerning how and when to apply lime to greatest advantage, all written in straightforward language offering sound advice. Yet, the *Annals* rarely exceeded a circulation of 350 (Mingay 1975, 18; Horn 1982, 320-21), though what cannot now be determined is how widely each copy was circulated locally by subscribers.

The Agricultural Magazine (published 1799-1811) and *The Farmer's Magazine* (1800-37 in Scotland, thenceforward in London; see Anon e 1800) offered equally practical advice on all aspects of farming, some of which provide contemporary evidence of then current practice within the research area (see below). Also worthy of mention are compilations in book form of supposed farmers' letters offering wide-ranging advice (Anon f 1771).

Some manuals were general all-inclusive works while others were more specifically concerned with methods of improvement, but only six fell fully into the latter category. Their emphases are summarised in Table 7.2.

The clear conclusion to be drawn is the predominance of liming during the period of parliamentary enclosure and the enormous emphasis put on it by agricultural writers: while they disagreed on details of the benefits of liming, and on optimal methods of application (see Chapter 8.5), every writer acknowledged the role lime played in land improvement during this period. Verral (1799, 108), on the other hand, cast doubt on lime's value as a soil additive, believing that the rate at which lime was applied was 'injudicious' with its benefits out of all proportion to the expense involved; and Henry (1771, 110) was of the opinion that lime would ultimately 'impoverish every soil it mixes with'. Peters (1771, 143), however, was scathing about those 'absurd' people who doubted its value, while Home (1759, 39) maintained that those farmers who rejected liming did so out of ignorance because 'they did not know how it acted'. Such contrasting opinions only serve to emphasise the unscientific basis of their arguments.

Board of Agriculture

Proposals to establish a national body to promote change in farming were first mooted in 1790 by William Marshall (Rowe 1972, vii), though Sir John Sinclair, a landowner from Caithness, was the one who steered them through parliament in 1793. He was appointed first president of the Board of Agriculture and Internal Improvement, with Arthur Young installed as its first secretary.

The Board's aims were laudable but it was impeded from the outset by a lack of finance, by disputes between Marshall and Young, and by a lack of interest from some board members, many of whom had no farming experience (Virgoe 2005, 98). One of its main achievements was to commission county reports looking into the current state of farming – eighty were submitted, all within a remarkably short period of time. Their quality is variable, the whole process has been described as ill conceived, and not all compilers were sufficiently experienced to pass judgement (Rowe 1972, xiii).[57] Whether the reports achieved anything substantial is debateable; nevertheless, they do form a valuable archive of contemporary practice and thinking.

Four reports cover the research area (see Tuke 1794 for the North Riding; Rennie *et al.* 1794 for the West Riding; Pringle 1794 for Westmorland; and Holt 1795 for Lancashire. See also Dickson 1815 for Lancashire). In addition, Marshall (1808) compiled a useful volume of critical abstracts of the county series with his discussion of Yorkshire being based on his own survey of the county, published in two volumes (1788 and 1796).

Each county surveyor was given the precise brief 'to set out as quickly as possible' on the survey and to report on thirty-five wide-ranging questions common to all the surveys (Brit. Museum TY 7/140). Questions included what manures were used, whether draining, paring and burning were practised, and if any agricultural societies existed. Six questions specifically referred to enclosures and five to improvements.

Sadly, perhaps, the board was unable to live up to expectations and did not have the resources necessary to implement its aims by building upon the county reports. Secretary Young was blind for his last nine years in post (1811-20) and by then the board was in a 'sorry state of decline' (Gazley 1973, 693): in 1822 it was dissolved. The relevance of the board to this argument is not whether it achieved anything of lasting value but the insight the county reports offer into farming at the time, as will be seen.

Developments in chemistry

Land managers had been liming for centuries knowing that it had a positive effect on plant growth but without understanding why; they still had no conception of the chemistry behind it. At the start of the period under review in this chapter, Home (1762, 69) believed that lime attracted oils from the air which were then dissolved by contact with soil. It may seem to us fanciful and naïve but he was trying to make sense of the chemical reactions taking place in an age when science was unable to supply the answers. In the same vein, a later correspondent hoped to satisfy his readers by saying lime acts as a 'provocative' to the soil allowing it to release its natural fertility (Anon f 1771, 124-25). A manual claiming to be all-knowing merged these two theories into the equally vague hypothesis that lime 'attracts oils, acids, and salts, from

[57] Marshall (1808, 247) cuttingly described Holt, surveyor for Lancashire, as having the skills 'of an inexperienced tourist'.

the earth and atmosphere' (Winter 1787, 30). They all knew lime was causing something to happen but, as a discipline, chemistry could not explain what or why.

It was the earl of Dundonald (Cochrane 1795, 42) who made headway with the chemistry but even he was less than precise in saying caustic lime destroys the natural adhesion of organic matter. His explanation of the 'how' is unconvincing. Perhaps the most honest eighteenth-century explanation came from Donaldson (1796, 211): he offered various explanatory theories but concluded by writing that 'some how or other' it worked. The Bath Society was also exercised in trying to understand the chemistry of lime in a series of experiments begun in 1805 but, effectively, these came to nothing (Hudson 1972, 36).

Humphry Davy (1814), arguably the most prominent chemist of his day, gave a series of lectures under the auspices of the Board of Agriculture, on the chemistry of agriculture, and it was he who finally began to tease out how slaked lime reacts with minerals in the soil, and how the application regime was so crucial to ensuring that the lime was effective rather than harmful to plant growth. However, he did not fully explain the processes and still could go no further than to scrape the surface, but his was undoubtedly a major contribution and provided the foundation for later advances in the scientific understanding of soil processes. Real understanding, however, came from work sponsored by the RASE after 1840 under its motto 'practice with science' (Wade Martins 2004, 13). This is illustrated by the comment made by Wilson (1849, 186) that the age-old saying 'lime enriches the fathers but impoverishes the sons' reflected earlier ignorance of how lime worked. Lime had been perceived as a fertiliser in its own right and farmers were wont to apply too much too often, thereby 'burning' the soil. By the 1840s, however, its true role was fully appreciated.

The Improvers

As discussed in earlier chapters, improvement was a phenomenon of the countryside for centuries though the rate and scale of change increased incrementally from the mid-seventeenth century. By the middle of the eighteenth the very idea of improvement had become ingrained in estate management. Agricultural and landscape improvement were perceived as being inseparable (Whyte 2002, 84). Farming, in particular, was not just integral to the education of, but also a 'locus of innovation and experiment' for, the landed classes (Tarlow 2007, 41). Following the Georgic ideal, improvement had gained a moral and aesthetic dimension, so it was not merely farmland that needed to be changed. The concept was extended to houses, non-productive parkland and woodland areas: it was part and parcel of developing an 'improving prospect' within the countryside. It also enhanced a landowner's status locally and nationally, fed his paternalistic tendencies and his desire to be seen at the forefront of modernity (Williamson 2007b, 5-6).

The accession of George III, in 1760, was described by Young as an improvement watershed (Gazley 1973, 594). There is no doubt that this king held farming close to his heart, and he was a regular correspondent to Young's *Annals of Agriculture* writing under the pseudonym 'Ralph Robinson' (Marshall 1929-30, 56). He regarded farming as the 'greatest of all manufactures' (quoted in Wade Martins 2004, 9), but to what extent he was able to influence farmers in distant places remains conjectural. At least he gave farming a certain cachet.

Some have regarded the landed classes as central to improvement (see, for example, Mingay 1975, 16),[58] and Young certainly did, as seen in the previous chapter, while others tended to see the majority of large landowners as irrelevant and even incompetent with genuine improvers having been few and far between. Macdonald (1975), for example, believed it was the (few) committed and forward-looking farmers in any given area who were instrumental in progressing farming and not the large landowners. It was the former group who passed down knowledge to those less educated and free-thinking.

It is understandable that the former view, namely placing the aristocrats and gentry at the forefront, became so entrenched. Travellers and diarists of the day were from those classes and their itineraries took them from estate to estate. It is possible that they set out with preconceived notions and built up a picture of change that fitted their own agendas. Even Young, who was sceptical of the landed classes' commitment to improvement, and critical of government's role, was not averse to furthering his own interests. When he stayed with William Danby on the Swinton Park estate he was most effusive in his praise for what Danby had achieved (see Chapter 5.3). We cannot now determine which was cause and which effect but, in 1777, Young took up an appointment supervising improvements on a new landed estate in Ireland: it was Danby who had secured him the post (Gazley 1973, 109).

No single improver was more prolific in his output than Young, and no one was more forceful in urging liming than him (see Young 1768; 1771; 1773; 1804; 1805). He had a lifelong and almost missionary zeal to enclose the 'waste' and to bring ever more land into productive agricultural use. Why emigrate to the 'swamps and forests' of America, he asked, when they could move to the 'moors and heaths' of this country; why not make inroads into 'ling, whins, and fern' and transform the land into crops and clover so that 'bread and beef will be plentiful' (Young 1773, 38-39)?

Yet as a hands-on farmer Young had little to show. His Irish venture lasted but a year, the two farms he managed on his own family's behalf did not prosper, and his one real chance to prove that a 'barren waste' could be improved was virtually still-born (Gazley 1973, 26, 311).

Sometime after 1790 he was told of a large estate consisting of 'fine and very cheap' but mainly barren moorland that was on the market; he was partly attracted to it by the existence on the estate of a lime kiln (Jennings 1992, 332). He bought

[58] A close correlation in Westmorland between the rate of enclosure and aristocratic desire to improve has been suggested (Whyte 2000, 83).

2220ha,[59] and leased more, and set out his vision to gradually enclose it by draining, paring and burning, and liming, laying it down to a rotation of crops before finally turning it to permanent pasture. His plans included parcelling the estate up into a series of new farms to be sold to finance further reclamation. The estate extended across the Pennine watershed between Greenhow and the Skipton-Pateley Bridge road, and encompassed Pock Stones Moor and Rocking Moor. His plans came to nothing, however, because in 1793 he was offered the post of secretary to the Board of Agriculture. He could not refuse that opportunity to influence farming change nationally so the estate was put straight back on the market, though it took over a year to secure a sale (Young 1796).

Perhaps Allen and Ó Gráda (1988) were justified in their view that Young was not always correct in his ideas on effecting change, that he was a propagandist guilty of pushing 'landlord ideology', and even mistaken in assuming that he lived in an age of improvement. In their rather revisionist view the late eighteenth century was an 'age of stasis' rather than progress in farming (1988, 116), to which the likes of Young contributed little of substance. There is an opposing analysis of Young's place in farming history, which recognises there is some validity in criticising him for his practical failings, but nevertheless sees him as personifying the 'pioneering spirit of change' associated with that era (Chambers and Mingay 1966, 74).

The impact of war

In 1803 open hostilities between Britain and France were finally declared and both parties employed economic warfare in their armoury, though the two nations had effectively been at war for a decade by then. As war progressed, and especially after 1806 and 1807, Britain was subjected to a concerted economic blockade designed to starve the nation into submission (Derry 1970, 31). Britain was forced to rely on its own resources and one effect of this was an added impetus to bring more land into production, either by improving existing land or by reclaiming the waste, and to seek advancements in technology. In addition, farm-gate prices soared and remained high for the duration of hostilities. The response was to develop marginal land that, in normal circumstances, might never have been considered for productive use. Overall, agricultural production doubled in volume between 1795 and the end of the war in 1815 (Derry 1970, 47). Set against this, however, is an estimate that perhaps only a tenth of all enclosed upland was ever successfully improved (Chapman 2004, 87).

As early as 1803, Sir John Sinclair had commented that, though the nation had embarked on a military campaign, it should also 'attempt a campaign against our great domestic foe ... the hitherto unconquered sterility of so large a proportion' of the country (Sinclair 1837, 111). The waste lands must submit to the 'yoke of improvement'. It is arguable that the country had not needed a Board of Agriculture or Improvers to promote improvement: the strategic need to produce more food, and the patriotic slant given to improvement (Simmons 2003, 141), may have been enough to bring about widespread change. Cynically perhaps, the enhanced profits to be made from higher farm-gate prices may have sufficed in themselves.

Developments in transport

At the end of the eighteenth century packhorses were still a major means of transporting raw limestone or burnt lime within the Central Pennines, and the network of causeways across the moors from Airedale and Lothersdale to Calderdale was still in general use in 1798 (Keighley BK 79/205). As late as 1865 strings of 'limegals' (packhorses carrying lime) were still operating southwards out of Clitheroe (Clarke 1900, 57) though, from around 1800, this form of transport was increasingly unable to cope with the demand to move lime or stone from here to there. It was slow, expensive, and subject to disruption in adverse weather conditions. It was, in short, a severe constraint on development and improvement, despite the spread of turnpike roads which had somewhat eased the situation.

It was the expanding canal network that enabled commodities to be moved in bulk at relatively low cost. One local estimate, dated 1792, suggested that a saving of nearly £8 per hectare on the cost of lime could be achieved if lime was brought in by canal rather than by packhorse (LRO DDPa/Box 1). Examination of extant promotional material for projected canals in northern England shows that most schemes placed lime (and coal) at the top of the list of profit-making commodities likely to be carried (see, for example, WYAS B, BCC/14/3, 1768; and see Hadfield 1972; 1973; and Hadfield and Biddle 1970, for overviews).

Table 7.3 summarises relevant schemes within the research area. Not all came to fruition but, of those that did, company records indicate that lime was indeed a prime commodity. Without canals it is doubtful if large-scale enclosure and

Scheme	Proposal date	Route
Parkfoot[1]	1760	Settle - Burton-in-Lonsdale
Leeds & Liverpool[2]	1773	Leeds – Barnoldswick
Bradford[3]	1770	Bradford - L & L
Settle[4]	1780	Settle - L & L
Lancaster[5]	1791	Preston – Kendal
Swale Navigation	1767	From the Ouse confluence
Teesside-Bedale	?	via Boroughbridge

Table 7.3 Canal schemes in the research area citing lime as a potential major commodity

[1] YAS. MS1186
[2] Hustler 1788; Killick 1897
[3] WYAS (B) BCC; DBS2.C
[4] LRO. DDPa/Box 3; WYAS (W) WYL 162. GA/2/61.4; YAS MD 335/1/10/6/15
[5] NYCRO. ZAG 279

[59] Young paid £4400 for the estate, the equivalent of £132,484 today (based on Bristow 2001, 224).

improvement within the gritstone regions could have been achieved: hushing and river stonary could not have satisfied demand.

Towards the very end of the period under review here, expansion of the railway network facilitated the development of industrial-scale limeworks, especially in Craven and Wensleydale, and made agricultural lime a more affordable commodity. Rather than leading to an overall increase in lime consumption, however, the most visible impact was to bring about the demise of the ubiquitous field kiln.

The foregoing account has painted a picture of various forces combining to encourage enclosure and improvement of old and new farmland. Some were exogenous (weather, the Board of Agriculture, war) while others were localised responses to wider processes (agricultural societies, local improvers, canals). According to Young and Tuke, most improvement was at the behest of the landlord, and the ordinary farmer was, if not hostile, then uninterested and often suspicious.

From one perspective, the improving zeal pushed cultivation (of grass or crops) beyond sensible and sustainable limits and much enclosure was (arguably) misguided and doomed to failure. Ignorance of the chemistry of lime sometimes led to harmful application strategies which, in turn, led some to refute its value as a soil additive. If Sinclair was correct, it was twelve years of Napoleonic wars that fostered change in farming more than any other factor, because naval blockades reduced the volume of food imports so domestic production had to be increased to feed the nation.

Discussion will now turn to the reality of the situation within the research area and will examine if this summary has any validity here.

4. Enclosure and liming

The rationale and mechanics of enclosure are well documented within historical literature, both nationally and locally.[60] The purpose of this section is to investigate the causal link between enclosure awards and hard evidence of lime burning initiated by enclosure in the research area.

Table 3.2 showed that twenty-two awards (fourteen per cent) out of a total of 150 across the entire area had made direct provision for the burning of lime for use on new enclosures. A further forty-five awards (thirty per cent) had made provision for the quarrying of limestone for unspecified purposes. The greatest proportion of awards were enacted between 1760-80 (twenty-four per cent), 1801-20 (twenty-four per cent) and 1841-60 (thirty-three per cent).

Exploitation of limestone permitted by non-specific awards may have included liming but in the absence of firm evidence any link must remain speculative, as the stone may have been destined for building repairs, new walling, road repair or, if burnt, for mortar. In some cases the degree of speculation is reduced to a certain extent by evidence from elsewhere, as with the Masham award of 1797 (NYCRO NRRD no. 3).

This award encompassed an area of 2070ha between the Pennines and Masham town. It enclosed Gollinlith and Potts Moor, as well as more lowland tracts within the parishes of Ilton, Fearby, Ellingstring, Swinton, Masham and Healey. A further area in the valley bottom of Colsterdale had been privately enclosed and improved in the 1770s. It has already been explained (Chapter 6.6) that the Swinton Park estate's agent, William Morton, was keen on land improvement and on liming, and it may be valid to suggest that there was a strong linkage between this enclosure award and lime-coal output from the estate's collieries. In the fifty-five year period for which accounts are extant, it was the years immediately after 1797 that saw the greatest output and sales. As was shown in Chapter 6.7, much of the coal was being delivered to kilns around Well and Snape and it would have made perfect economic sense for carters to deliver coal one way and bring lime back on return. Once the area had been improved from the moor, however, there would have been a reduced need for lime hence, conceivably, a direct cause of decline in lime-coal sales and the ultimate demise of the mines.

The Masham award set aside two 'common quarries', but without specific provision for lime burning, though field surveying has located a kiln in one of them. Whether the kiln predated the award or was built as a result of it is unknown but it goes beyond mere speculation to suggest there may be a causal link, given the Danbys' faith in liming.

As illustrated by Table 3.2, field surveying has located lime kilns within the areas encompassed by thirty-eight enclosure awards, excluding those where specific provision was made for lime burning. This represents one quarter of all awards. Table 7.4 abstracts key data for these awards.

Overall, there is no meaningful correlation between the size of an area enclosed and the number of kiln sites found. What these data do show is that lime kilns were extensively found across the Central Pennines; what cannot be determined, however, is the correlation between date of award and date of building. The kilns could just as well post-date as pre-date awards; it is not possible to determine if these kilns were coeval with the awards as no specific lime burning provision was included.

A further complication arises. Table 7.4 cannot take into account kiln sites lying very close to areas enclosed, for example on the lower edge of open moorland. Field evidence tends to suggest that many of these kilns provided lime for the adjacent enclosures but they may have been producing lime for use in other areas, especially if such kilns were accessed from below by cart tracks. It is not beyond the bounds of possibility that the number of kilns serving but outside enclosed areas, and not provided for in the awards, could be significantly higher. However, this must remain purely speculative.

[60] For general discussions see Williams 1970, Turner 1980 and 1984, Dent 1983, Wordie 1983, Williamson 2000b, and Kain *et al.* 2004. For material relevant to Westmorland see Searle 1993, and Whyte 2003b, 2005 and 2006.

Township	Award	Date	No. of kilns located
Thornton Steward	Town Fields	1694	1
Thornton Watlass	Watlass Moor	1754	4
Hudswell	Hudswell Moor	1755	0
Clapham	Clapham Thwaite	1758	3
Dalton	Dalton Moor	1760	3
Arncliffe	West Moor, Clowder	1766	4
Halton East	Halton Green	1768	2
Bentham	Bentham Moor	1768	marl pits
Litton	Various	1769	8
Newsham	Commons	1781	1
West Witton	West Witton Moor	1781	6
Leyburn	Leyburn Moor	1785	5
Horton	Selside	1791	8
West Tanfield	Thornborough Moor	1799	0
Conistone	Various	1803	12
Horton	Horton Scar, Harber	1804	3
West Burton	W. Burton Pasture and Common	1805	4
Walden	Bents and West Pasture	1807	6
Grinton Grinton	Crackpot Pasture	1809	0
Bainbridge	Worton Pasture	1810	1
Grinton	Birks	1813	1
Bainbridge	Various	1816	24
Thornton in Lonsdale	Various	1819	15
Horton	High Moor, Moughton, Pen-y-ghent	1821	2
Hawkswick	Hawkswick Moor, Clowder	1824	2
Buckden	Various	1831	6
Starbotton	Various	1831	8
Abbotside	Camshouse Pasture	1837	1
Abbotside	Abbotside Common	1837	4
Coverham	Caldbergh Pasture	1839	3
Marrick	Marrick Moor	1841	1
Hawes	Snaizeholme Pasture	1847	3
Horton	Horton Moor	1847	5
Kirkby Stephen	Kirkby Stephen Intake	1850	2
Waitby	Waitby Common	1854	1
Hebden	Hebden Moor	1857	1
Grinton	Harkerside Common	1872	7
Carlton-in-Craven	Carlton Common	1873	4

Table 7.4 Abstract of key data for enclosure awards with lime kilns located in the field, but with no provision within the award[1]

Source: field surveying by this writer.

[1]The table excludes kilns in hushing, stonary and boulder pit areas.

Very few kilns can be accurately dated. Most field kilns were built to fulfil a local need and whoever caused one to be put up would probably have seen no need to make a specific note of its erection or intended purpose – for liming or mortar. As will be seen, though, some dating evidence has survived.

One case in point concerns a kiln that was erected long after the surrounding area was formally enclosed. In 1816 4570ha within the Hawes area were enclosed by act of parliament (NYCRO QDD.D, Book P), and twenty-four kilns have been located in the field. One of these was the subject of an agreement, in 1857, between Elizabeth Routh, of Gayle near Hawes, and John Moore, yeoman, of Fossdale Farm above Hardraw. Routh agreed:

> to allow the sum of five pounds to the said John Moore for his building a new Lime-kiln in Green Bank (*sic*) Pasture; and the said Elizabeth Routh further agrees to allow the said John Moore at the end of his first year of tenancy the sum of twenty pounds, the said John Moore having previously laid out that amount in well draining and liming at reasonable prices certain portions of the lands hereby agreed to be rented; and the said Elizabeth Routh further agrees after the year 1860 to allow the said John Moore ten pounds towards draining and liming for a period not exceeding ten years from that date (NYCRO Z2).

It has to be presumed that Greenbank Pasture was either not improved at enclosure or not sufficiently improved to provide Routh with the necessary net return. The record does not indicate how long Moore had already been in occupation though he had clearly been active for some time liming and draining. Neither does it indicate whether this new kiln (illustrated in Figure 7.1) was a replacement for an older structure or to avoid having to buy lime in. Ground surveying does confirm, from the size of the adjacent quarry, that a substantial volume of stone was removed and (presumably) burned in this kiln, which ties in well with the length of the agreement.

As mentioned above, twenty-two awards made provision for the burning of lime (Table 7.5). In addition, the oldest possible example is from 1694 when Town Fields in Thornton Steward, near Middleham, was parcelled up (NYCRO ZPT4/1/4). This gave a Mr Thornton the duty to maintain two gates, one being the 'gate into ye same field at ye end of his Kiln': the context suggests this was a lime kiln.

The wording of lime-burning provision in enclosure awards varied to an extent. In almost all cases the provision was for getting stone 'for burning into lime' or 'to burn the same into lime' in kilns to be erected in allotments set aside as public quarries, either to be publicly owned, eg Bainbridge 1849 (NYCRO QDD/I no. 9) and Casterton Moor 1816 (CRO[K] WQR/I/14); or built by whoever wished, eg Barningham 1807 (NYCRO ZQA I/4/1) whereby William Johnson was given that right, and Studfold Moor 1771 (Horton PC). Without exception the lime was for the use of allotment holders or township inhabitants only, with none to be sold.

Some awards did not stipulate what the lime could be used for; others stated it could be used for building, repairs or on the land; yet others said it could be used in accordance with past custom, such as Muker 1843 (NYCRO TD 53). Some made actual mention of liming. The award for Bellerby High Pasture and Bellerby Moor, dated 1772, included provision for liming for 'manuring their antient enclosed lands or their said new allotments' (NYCRO PC/BEL); that for Casterton Moor in

Figure 7.1 Greenbank Pasture lime kiln, Burtersett (SD8910 8897)

Township	Award	Date	No. of kilns located in public quarries
Hetton§	Cow Close, Ox Pasture	1771	0
Horton	Studfold Moor	1771	1
Bellerby §	Cragg Common	1772	1
Gayles §	Gayles Moor	1774	2
Grassington §	Old Pasture, New Pasture, Botton	1792	2
Horton §	Dub Cote Scar	1795	3
Orton §	Various	1779	3
Barningham	Cow Close	1807	2
Carlton in Coverdale	Carlton Common	1815	1
Casterton §	Casterton Moor	1816	4
Carperby	Various	1819	2
Redmire	Redmire Pasture	1819	1
Tunstall §	Leck Fell	1825	2
Threshfield	Various	1827	1
Muker	Thwaite Common	1837	2
Muker	Angram & Great Sleddale Commons	1843	0
Muker §	Satron Common	1848	1
Bainbridge	Cragg Common	1849	2
Thornton Rust	Thornton Moor	1855	1
Dent	Various	1859	7
Garsdale	Garsdale Common	1859	3
Hawes	Greenscarr Pasture	1859	1

Table 7.5 Enclosure awards with specific provision for lime burning
§ awards with specific mention of (agricultural) liming.

1816 said they could get stone at Bullpot to burn into lime 'for the improvement of their respective lands' which meant liming (CRO[K] WQR/I/14); the Orton 1779 award said the lime could be used for 'tilling of their several lands' (CRO [K] WQR/I/73 and WPR/9/Z11); while that for Gayles in 1774 phrased it as 'for the manurance of their lands' (NYCRO NRRD AX No. 54). They all meant the same, namely that limestone could be obtained from public quarries to produce agricultural lime as part of the process of improving newly enclosed land.

The award for Settle Banks, High Scarr and Scaleber, east of Settle, in 1759, adopted a different tack, at least for Scaleber Pasture where a kiln has been located on the ground, even though the award did not specifically mention one (NCHRG). Grants to new enclosures within Scaleber included the right of passage through others' allotments 'for the purpose of leading lime and tillage only either on horseback or in carts or carriages'. The kiln was sited at the far edge of one of the new sub-divisions of Scaleber and it is possible that the other parties insisted on this clause being written into the agreement to safeguard their own interests, even though most of the new enclosures were directly accessible from a public road.

Archival evidence permits comparison between the date certain kilns were built, or agreed to be built, and the date of the corresponding award. Two examples will illustrate the point: for one, the two dates tie in perfectly; for the other, not. In August 1836 James Willis, of Yorescott Farm near Bainbridge, paid tenant John Atkinson 10s. towards the cost of erecting a new kiln (at SD9285 9130) in Grange Gill and 8s. for burning 6 chalders of lime (NYCRO TD124). The enclosure award covering this area was dated 1837, so Willis may have been acting on what had been agreed by the parties to the agreement prior to its formal implementation. In November that year Willis paid for burning 113 loads, but his accounts do not record what it was all to be used for.

The second example involved Wisphill kiln above Grassington, a kiln that cannot now be identified but was high up on the moors. John Peart took out a one-year lease on lands which contained the kiln in 1806 yet the relevant enclosure award was dated 1792 (Skipton Raistrick 557). This mismatch can be interpreted in various ways. It could be that the land enclosed in 1792 was not worked on by its owner or previous occupier and that Peart had different intentions. Alternatively, he may have wanted the kiln for non-agricultural purposes; or, indeed, the kiln may have been irrelevant to him if all he wanted was the land, possibly for field sports. Without being able to tie the place-name down to an actual kiln (there are several possibilities), no conclusion can be drawn.

Four kilns, out of the total field survey of over 1500, have datestones. One, in Melbecks in Swaledale, bears the date 1865 but this area was never subjected to formal enclosure; one in High Abbotside (SD8850 9213) is dated 1841 and stands on the edge of land enclosed in 1837, so there is a possible connection here if improvement followed on from enclosure after a slight time lag. A kiln at St. Simon's Chapel in Coverdale (SE0868 8494), dated 1786, is adjacent to an area enclosed by an award dated 1839, so there can be no direct connection here as the time difference is too great; no corresponding documentation has been located.

Another kiln (SE1134 8577) near Braithwaite Hall, in Coverdale, is dated 1795 but sits within land enclosed by an award also in 1839. Archival evidence informs its commissioning (NYCRO Z 775; ZS* Middleham). The Middleham Estate belonged to Thomas Wood and he had been closely involved in improving the value of his properties for several decades prior to 1795, and archival records provide useful detail about improvement strategies on this small estate. In May 1775 his agent, William Head, noted that, even though The Limekiln Close was 'much the best arable field', supporting oats, he intended to have it dressed with lime in 1776 at the rate of '3 chaldrons per acre' and then planted with turnips. These were to be eaten off by sheep, thereby providing 'free' manure, before putting the field down to barley the following year and to permanent meadow in 1778. All other arable land on the estate was to be similarly treated with a three-year rotation of wheat or maslin (a wheat-rye mix), clover or rye-grass, and oats or beans prior to sowing to permanent pasture. There was a disagreement between Head and Wood in 1776, Head expressing in a letter his feelings that it was unreasonable to expect tenants to commit themselves to a twelve-year rotation regime as that would have required too great an investment from them (NYCRO ZS Middleham).

In 1795 a valuation of the estate was undertaken and this gave details of a tenancy agreement for Ewe Close Farm, which incorporated Lime Kiln Close. Wood reserved to himself the right to 'dig lime stone and to burn lime for the use of all or any of his tenants' and to install forty roods of drains per year on Cover Banks. The valuation also included a tenancy agreement for Braithwaite Hall itself, signed in 1795. Again, Wood reserved to himself the 'liberty of building a lime kiln on East Hanghah' (now Hanghow Pastures) to burn lime for the use of all his tenants. This is the kiln bearing the date 1795.

Estate records testify that 'proper' draining and liming remained key elements of tenancy agreements here at least until 1811, when relevant records cease (NYCRO Z 775, 1811). If the conditions were not met, tenancies were to be terminated.

Wood was by no means a major player but he stands out as a determined improver. With Head, he ensured that it became accepted practice for tenants to expect inducements if they clearly wished to 'try their ability and disposition for agricultural improvements, several of them having expressed a strong desire for it' (NYCRO ZS* Middleham 1795). One unidentified tenant had even offered to encourage other tenants to adopt new practices designed to make the land 'sweeter and wholesomer'.

He obviously had his work cut out though as one of the sitting tenants at Braithwaite, called Ryder, had 'neither the capacity nor inclination to make any improvements from his present slovenly mode of management' despite the fact that the other tenant there, Sigsworth, was the one who had agreed to go along with his landlord by building the Hanghow kiln. However, he demanded his price, a one-off rental allowance of £60.

This reference to a 'slovenly' tenant, and to a tenant who was prepared to go along with improvement if he could gain something to his advantage, fits in neatly with the contemporary views expressed by the likes of John Tuke and Arthur Young that ordinary farmers would not change their ingrained methods, and that improvement depended on the landed classes. On the other hand, Wood was indeed a landowner, but not a large one, and the unnamed progressive tenant belies Tuke's and Young's prejudices. As will be seen below, the smaller players were actively engaged in improvement during the enclosure period, and the negative views expressed by the likes of Young and Tuke were too generalised and unsympathetic.

5. Evidence of liming

The West Riding county report summarised the state of farmland, especially in Craven and Lower Wharfedale (Rennie *et al.* 1794, 13, 109; Brown 1799, App. 1). Though much remained unimproved and therefore 'of little benefit' (Rennie *et al.* 1794, 13), some pasture land was described in glowing terms, especially in North Craven. These perceptions are given extra credence by comments made by travellers on their grand tour at the turn of the eighteenth and nineteenth centuries. In Lunesdale, for example, the grassland was described as 'the

Figure 7.2 Improved meadow land in Greenfield

richest I have ever seen' (Walker 1792, 49); while Housman (1800, 195) observed how fertile and verdant much of Craven was, a conclusion also reached by the artist and musician Charles Dibdin (1802, 262).

Travellers such as these were at the mercy of the elements and their diaries included entries when particularly adverse weather conditions affected their progress. Hodgson (1820, 7) provided detailed climatic statistics for south-east Westmorland in 1792, emphasising how unusually wet that summer had been, and Byng ([1970], 89) wrote from bitter personal experience of crossing the watershed from Bainbridge to Ingleton, in June 1792, during which 'there fell many storms of rain' without warning. As seen earlier, that decade proved to have little to cheer the farmer yet this was a time when improvement was very much in progress.

The farm accounts for Ingmire Hall near Sedbergh, for 1787, detail rental allowances for 'lime or manure' set at £1 per annum for five years (SDHS, Ingmire); and the accounts for the manor of Garsdale, which owned and leased out lime kilns at Longstone Fell, east of Sedbergh, noted the granting of rights to tenants to 'get limestone there for the improvement of their lands' (SDHS, Garsdale). As was inevitably the case with such account books, no detail was given of which land was involved, and the entry is terse. Nevertheless, this does help inform the general picture of improvement and liming.

According to Arthur Young, Thomas Elliot, of Fremington in Swaledale, was 'one of the greatest improvers of moors' in Yorkshire (Young 1770, 212). Elliot was not a landowner but a tenant, since at least 1742, on the Smales' estates in Swaledale (NYCRO ZQH 7/46). No doubt he was bound contractually to do as bid at Fremington, but on lands that he rented elsewhere he was able to implement the changes he thought necessary. He leased a 925ha-estate in Greenfield, at the head of Langstrothdale, much of it being moorland, and actively enclosed it in a piecemeal manner. He pared and burned the sward, and open-drained, then limed before planting turnips for two years prior to sowing the land to permanent pasture (Young 1770, vol. 3, 213-16; Fieldhouse and Jennings 1978, 156). According to Young, the value of non-limestone pastures here had increased from 1s.6d. to 8s. per acre and of limestone pastures from 5s. to between 12s. and 20s. By the time of Young's visit, Elliot had improved 90ha. The impact of Elliot's improvements can still be seen today: Greenfield stands out as a series of hay meadows and improved pastures, as shown clearly in Figure 7.2, surrounded by upland that was considered so unproductive that it was mostly given over to blanket afforestation after World War II. The documentary record does not exemplify the reason for Elliot's downfall but, by 1803, the estate was on the market and he had been declared bankrupt (NYCRO TD31). Possibly he had over-extended himself financially in his attempts to transform this valley.

Improvements elsewhere were coming to the attention of interested observers. The county surveyor for the North Riding described in positive terms work being done on the Marske Hall estate where, for example, one farmer applied a mix of 'kelp-ashes, sea-weed, slam from the alum works, and lime'[61] (Tuke 1794, 52) which says much for his sheer determination. It may also support the notion that knowledge of new ideas and concepts was spread across the country and through the

[61] Slam is a term used for the refuse from alum processing.

farming hierarchy by slow diffusion. Clarkson (1814, 321) also visited Marske where, in his opinion, improvements 'worthy of the observations of the agriculturist' had been promoted by the estate's policy of giving annual premiums to those tenants whose efforts at improvement had been most successful. It was no coincidence that this landowner was president of Richmond's Scientific Society which awarded prizes for good practice. John Hutton of Marske Hall gained a further favourable mention in the travel journal of Charles Fothergill in 1805, being described among other compliments as an 'agriculturist' – not just a farmer or landowner but someone who was intent on improving his estate's agricultural condition (Romney 1984, 158).

Similar improving work was being commented on in Westmorland. Orton Moor, newly enclosed in 1779, was subject to the same sequence of processes as in Greenfield, and along the upper reaches of the Lune, lime, 'till lately ... seldom used as a manure', had by then become the main soil input (Robinson 1803, 110). In Kirkby Stephen Robinson had noted similar successful improvements, based on liming and manuring (1805, 215). An estate valuation for 1803 found meadow and pasture land at Wharton Hall in Mallerstang to be 'very fine' and much improved by liming thanks to the efforts of the current tenant (CRO[C] D/Lons/L12).

6. Farm leases

The influence of lease conditions on tenants' attitudes to investment has been pointed up for Ravenstonedale where, in the early nineteenth century, terms of as little as seven years acted as a disincentive to improvement because tenants could not believe that their investment would be recouped in such a short time (Bouch and Jones 1961, 230). If that was so in Westmorland, the willingness to invest must have been even lower in the West Riding where it was common for leases, stipulating liming, to be renewed annually (Rennie *et al.* 1794, 15-16). Unless tenants had been guaranteed renewal it is difficult to comprehend why tenants would tie themselves to such stringent leases. The value of long leases had been pointed out at a much earlier date by Thomas Hitt (1761, 21-22, 49) who recognised the benefits to tenant, landowner (and country) of giving leases of twenty-five years in return for improvements, increased output and increased rents.

An anonymous Scots farmer, on a tour of England, later drew attention to the stifling effect of short-term tenancies that he could see had led to 'errors of the Yorkshire lime husbandry' because tenants were not permitted to use their own judgement (Anon g 1800). This was not a phenomenon unique to Yorkshire; a correspondent had noted similar problems in Ayrshire some decades later (Anon h 1839, 386).

There was a view current at the time county reports were being compiled that the very inclusion of liming clauses was fundamentally flawed (Somerville 1795, 66-67). Many agreements forced tenants to apply lime at too regular intervals regardless of soil conditions, and applying lime in excess of the soil's lime requirement led to crops being scorched. This was done out of sheer ignorance of the chemistry of liming: it is easy to see how a landowner or agent could assume that failed crops resulted from too little being applied rather than the contrary.

The documentary record provides many examples of liming clauses in farm tenancies across the research area. An admeasurement for the Flasby Hall estate in Malhamdale, dated 1763, listed tenancy conditions in place at the time (NYCRO ZTO 5/2). Any incoming tenant was bound on all ploughland 'to lay 40 customary horse loads of unslecked lime upon an acre' with the lime having been 'got and burnt upon ye premises and coal isn't far to fetch'.[62] The coal, presumably, was brought down from the collieries on Fountains Fell.

The Westmorland estates of the earl of Thanet followed a standard format: for instance, in 1767 John Thompson was required to apply 'at least *thirty* horse loads of well burnt lime', the amount having been handwritten on the printed agreement (CRO[K] WD/Hoth/Box 26). The format remained much the same as late as 1815, though agreements in this year tended to stipulate forty Carlisle bushels of unfallen lime.

In 1779 George Fawcett had accepted very stringent conditions when he took on the three-year lease of Hall Intack Farm in Ravenstonedale though he was allowed 20s. to purchase coal to fuel the farm's kiln in Limekiln Close which he was to fire at his own expense for applying the lime on arable ground (CRO[K] WDX/94). Shortly after this date John Waller of Winton required his new tenants, *inter alia*, to 'till and lime (the demised land) in a husband like manner' though no amounts were specified (CRO[K] WDX/668).

In that same decade tenants around Dent and Sedbergh were subject to very similar demands. In 1778 the Garths inserted into an agreement for land at Sedbergh that each tenant must lay out 50s. annually 'in manure or lime for the land'; while the lease of Bridge End Farm (SD720 861) at Kirthwaite in upper Dentdale, in 1787, required Richard Shaw to apply thirty loads annually (Cleasby 1994).

Robert Foster inherited Hebblethwaite Hall, north of Sedbergh, and, though he was clearly not subject to any external demands, he immediately planned a process of improvement in Penny Farm Gill (SD698 931) starting with the construction of a new kiln in 1782, an access road to it and a new bridge (SDHS MF 80). Foster recorded in his day book how much lime had been burnt and on which pockets of land it was applied.

Across on the Swale Christopher Crowe's descendant, George, at Kiplin Hall, was still intent on building up the value of the estate he had inherited and a series of tenancy agreements, running from 1778 to 1836, laid down precise rules for managing the various farms. They changed little throughout the period (NYCRO ZBO IV2/1). The earliest was for Layland House (now Laylands Farm, SE261 987) whereby

[62] Two fields were under crops: Lime Kiln High Close and Lime Kiln Low Close.

John Eden took on a rolling three-year lease, which provided him with the security to expect a return on his investments. He was to 'load lay and spread upon every acre ... one and an half chaldron of well burnt Sod Lime', together with compost and manure, and to follow a prescribed six-year rotation of arable and fallow. In later leases the term 'sod' was replaced by 'clod': the meaning of the latter term is not clear as the word clod implies a lump of soil without turf. It could suggest that the lime was burnt in clamp kilns, in which it was common practice to strip the covering turf and burnt lime together, but this would not be using clods; or that the lime was mixed with earth prior to spreading, as was also common practice (see Chapter 8.5).

Simon Scrope III of Danby Hall, descendant of the Simon Scrope discussed in Chapter 5.3, had similar clauses written into his farm leases, though here the recurring quantity to be applied was two chaldrons per acre each time arable land lay fallow (*eg* NYCRO 4/1/14/10 in 1771 and ZPT4/1/5/6 in 1779). One lease, for a farm at Stainton in Swaledale, stipulated three (NYCRO ZPT4/2/10). Possibly the soil here was more acidic and had a greater lime requirement, or had been neglected in the past. The Wyvills, of Constable Burton Hall in Wensleydale, wrote identical clauses into their tenancy agreements during the 1790s (NYCRO ZAW 31).

It is, in a sense, remarkable that such similar conditions were laid down in the different counties from the 1770s to the 1810s. Whether word of current best practice circulating from one landowner to another and filtering down to lesser landowners was the reason, or whether the common ground had its roots in the legal profession, is another unknown quantity. The fact remains that estates as far apart as Swaledale, Ravenstonedale and Craven imposed broadly the same terms. A small sample of leases from Craven, all in the 1790s, will complete this comparative survey.

In 1790 a twenty-one-year lease for East Pallener Farm (probably now Painley Farm, SD841 500) in Horton in Craven required the tenants to lay 180 bushels or sixty horse-loads of 'well burnt lime' per acre of 'sand land' prior to sowing with grass, but they would be subject to a penalty of £70 per acre if they failed to carry out these demands (NYCRO ZXF 4/6/1.i). Such a punitive clause has not been found elsewhere during this research process. A coeval lease, for an unnamed farm, stated the requirement in a different manner: here, the lessor was to spend £100 to purchase lime but the tenant was to transport it from kiln to farm and spread it at his own expense, but with a rental allowance of £5 (NYCRO ZXF 4/6/1.ii).
By an agreement for Hezlerton Farm (now Nether Hesleden, SD887 746) in Littondale, sealed in 1793, the landowner, John Foster, was bound to supply 1000 loads of lime at his expense to tenant, William Baynes (NYCRO ZXF 4/5/1). Baynes, however, was to pay interest of 1s. in the pound and to transport and spread the lime, as directed by Foster. The agreement bears a handwritten note, dated 1798, that £2 6s. 9d. had been paid in interest for 1513 loads, so the implication is that liming here had been undertaken in phases.

On Lord Walsingham's estates at Blubberhouses and Beamsley, from the 1790s, tenants were bound by a common agreement to ensure that all arable land should be 'limed and managed' according to the stipulations laid down by that agreement, though the actual details have not survived (NRO WLS LXVI/210/4).

A markedly different approach had been taken on the Bolton Hall estate, as explained in a memorandum to Lady Bolton by her agent, William Sadler (NYCRO ZBO IV.8). He pointed out that, though considerable areas of pasture were capable of improvement 'by the application of lime', he felt it unreasonable to expect tenants to meet the cost themselves, given current tenancy practices, so he advised paying for the lime out of the fund that had been accumulated specifically for improvement and recouping a proportion of the outlay from them after three years, by which time the lime would have had chance to take effect. Perhaps more importantly, the tenants would have seen the benefits for themselves. Lady Bolton's response, sadly, has not survived.

7. Charles Towneley: improver extraordinaire

The Towneleys had held estates around Burnley since 1295 (Bourne 1979, 3) and successive generations built up the estate and its mining and lime burning interests. Charles Towneley (1737-1805), who inherited in 1758 (Smith and Smith 2004), was without doubt an improver in the Georgic, aesthetic sense of the term. Though unwell for much of his later life, he employed John Carr to remodel much of Towneley Hall, and John Nash to transform the parkland (Law 2006, 39-54), and built up an impressive reputation as an antiquarian collector and connoisseur (Vaughan 1988, 436). Indeed, he was instrumental in helping to establish the collections that were to substantially increase the early holdings, and size, of the British Museum (Bourne 1979, 22).

As shown in Chapter 5.6, improvement by liming was already well established in north-east Lancashire using lime obtained by hushing. The comment was made in 1798 that large quantities of lime (and coal) were coming down off the moors on the backs of 'wretched starveling horses' to feed the demands in the lowlands below (Whitaker 1800-01, 347). Around the same time the value of local hushing grounds to improvement was illustrated by an advertisement for New House Farm (SD908 356) in Extwistle which stressed there was 'plenty of Lime-Stone ... and great Improvements may be made at an easy expence' (*Blackburn Mail* 1795).

To implement his ambitious plans, Charles Towneley brought lime in from Clitheroe using his own cart (LRO Towneley, DDTO, Box II, 1785), and from local hushing grounds. In a letter to his uncle, John Towneley, Charles wrote that he had purchased a string of twenty Galloway ponies – together with their drivers – and had 'already commenced my liming operations at Cowside and other places' (LRO DDTO, Box OO, 1804). Towneley was no absentee landlord and not one who managed his estate solely through a steward, and when in London he received regular correspondence from his steward.

Towneley was very much hands-on in overseeing day-to-day improvements and, for example, took in-hand the formerly leased-out Jacob's Farm and turned it into a model farm for the estate, renaming it Towneley Farm (SD856 306) (Law 2006, 4).

Exactly when he started improving is not recorded but he was certainly active in 1769-70 when a 'clear' profit of £191 had been made over the two seasons from '20 acres plowed in the park' less payment for liming at 8s. per acre (Brit. Museum, Towneley, TY 1/21). Estate records and accounts during Towneley's tenure are very detailed and frequently embellished with his own hand-written notations, and it is clear he was genuinely interested in all aspects of farm and estate management.

He made allowances to tenants for 'lime set' on their land (LRO DDTO, Box CC, 1781-86), and gave instructions about what was to be done and where. In 1795 he ordered 'dung compost manure or lime to be put on the worst parts of the meadows and the pastures' that were 'over run with sedges, rushes and swamp' (LRO DDTO, Box II, 1795); and he corresponded with, or visited, other estate owners to exchange notes and ideas. A letter sent to Edward Standish of Standish Hall near Wigan contained Towneley's own detailed notes and calculations about improvements he had made using a 'compost of earth and lime' and drainage (LRO DDTO, Box UU, 1796). According to Towneley, his once-sceptical neighbours now declared 'they ne'er seed the like' and he urged Standish to spread the message among the unbelievers.

Towneley was also passionate about the benefits of draining and was instrumental in furthering methods devised by Thomas Elkington who, by 1795, had successfully drained nearly 9000ha on farms and estates in Lancashire and the Midlands (LRO DDTO, Box UU, 1795). In fact, he was responsible for the writing of a manual on Elkington's techniques and for bringing them to the attention of the Board of Agriculture's Committee on Draining (Brit. Museum TY7/1064; LRO DDTO, Box J, 1795; LRO DDTO, Box UU, 1795; Parliamentary Papers XL 1794/5).

As might be expected, Towneley and his steward also noted prevailing weather conditions if they impinged on estate work: through 1789 there had been terrible rains and floods (LRO DDTO, Box II, 1789); in 1790 it had been impossible to obtain lime from Clitheroe because severe winter weather had prevented lime burning 'for one month' (LRO DDTO, Box II, 1790); and in 1797-98 there had been so many wet days (Brit. Museum TY1/7).

Arthur Young did not record a visit to Towneley Hall but, if he had, he would surely have accorded him his catch-all accolade 'He who is the BEST FARMER, is with me the GREATEST MAN' (Young 1770, vol. 1, xv). Towneley's efforts and achievements, given his many other commitments and bad health, are remarkable.

8. Conclusion

This chapter has examined the significance of liming within the wider process of improvement during the period of parliamentary enclosure after 1760. It was a time when a diverse range of stimulants conspired to promote improvement in the quality of existing farm land and the taking in of 'new' land from hitherto unproductive areas. It was a time when vast areas – especially, though by no means exclusively, uplands – were enclosed within walls or hedges, thereby removing common rights and encouraging private investment to, hopefully, bring about increased productivity and profit.

Some of the stimulants can be seen as deriving from positive happenings, such as the evangelical role of the Improvers in furthering the idea and ideal of improvement, or the establishment of the Board of Agriculture with its programme of county surveys. Whether individual improvers or board members collectively were overly optimistic is a moot point, but the key issue is the philosophy imbued in their efforts: they were determined to transform the nation by modernising farming and removing all those elements (such as 'waste' land and the indigent or 'slovenly' classes) that impeded progress. Developments in transport, with an expanding turnpike network and the development of canals, were equally positive stimulants, and it could be argued, for much of the research area, that these were essential in the improvement process. Intellectual growth, too, had a marked impact in bringing about greater appreciation of the chemistry behind liming.

Tied in with all this was the setting up of agricultural societies and farmers' clubs to spread the message and a burgeoning of the amount of available literature on every aspect of farming. Again, one can question their overall impact. Were they, along with the Improvers, central to the process of change or largely irrelevant and on the margin, being aimed at and read by the landed classes and the armchair farmer rather than by the 'ordinary' farmer who was probably illiterate anyway?

It is impossible now to judge the extent to which these positively-derived stimulants would have achieved their laudable aims because the picture is clouded by a negatively-derived factor: the French wars. There can be no doubt that years of economic blockade and warfare generate a 'dig for victory' or perhaps more realistically a 'dig to survive' mentality, so change in farming and expansion onto marginal land may have happened without the positive stimulants. Certainly, the rate and scale of improvement, and of liming, expanded despite the dreadful weather conditions that prevailed through many of the early years of conflict.

Central to the improvement process is the matter of who were the prime movers, whether the landed classes or the tenants and small farmers. The likes of Young and Tuke promoted aristocratic and landlord ideology; this implies that only this class had the wherewithal (and the intelligence, they might have unfairly added) to bring about widespread change. Those lower down the farming hierarchy, it was said, were too entrenched in conservative values, too resistant or even hostile to new ideas, too incompetent, and 'slovenly' in their demeanour. They needed someone prodding and cajoling them into action.

Fortunately, perhaps, this loaded perspective does not stand up to scrutiny in the Central Pennines. There were indeed large estates where the landlord was promoting improvement, as for example Crowe at Kiplin, Danby at Swinton Park, Scrope at Danby, or Towneley in north-east Lancashire. However, the process was apparent on a number of smaller estates across the area where the landowners could hardly be described as gentry, let alone aristocracy. Colonel Wood at Middleham and Coverham, Elliot in Greenfield and around Fremington, the Horton in Craven and Marske estates, were all small players but nonetheless at the heart of improvement. Then there were the ordinary yeomen who appear in the record and who recognised the value and potential of improving their farms, even though they were more often than not bound by clauses in their tenancy agreements. If they were not prepared to accept improvement and liming constrictions, though, would they have taken on the leases?

There is a very clear positive correlation between liming and enclosure in the research area, suggested by evidence from the awards themselves, by this writer's field surveying of lime kilns, and by examination of lime-coal accounts (see Chapter 6.6). A quarter of all the 150 areas subjected to awards have lime kilns on the ground, fourteen per cent of all awards made specific provision for lime burning, though only nine actually specified liming; and thirty per cent of awards set land aside for quarrying limestone for unspecified purposes. It is inconceivable that many of these quarries had not fed the kilns that were built at some time, and that some of the lime was not destined for improving the surrounding enclosures. However, because it is not possible to date many kilns, we do not always know whether liming was being carried out as a direct result of enclosure agreements or because the new allotment holders had taken independent decisions before, during or after enclosure.

When field evidence, enclosure awards, estate and farm records, and contemporary accounts are all taken into consideration, the overall picture emerges of a period when the improvement of existing land and the taking in of the waste accelerated compared with before 1760. Farm leases were employed to ensure liming, and other management practices, were undertaken (though some stipulations were restrictive and stifling); and inducements were offered to those tenants who showed enthusiasm.[63] The extent to which improvement and liming were going on here belies the views expressed by Allen and Ó'Gráda (1988, 104) that the late eighteenth century was an age of stasis when the idea of improvement was a myth. It was reality.

This chapter has considered the role that lime played in the improvement process, and the relationship between liming and enclosure, after 1760. Chapter 8 will take the discussion forward from the cessation of hostilities in 1815 to the end of the nineteenth century. It will examine a range of forces, positive and negative, that impacted on the extent of liming and eventually brought about its (temporary) demise as a prime soil additive.

[63] Horn (1982, 314) made the point that eighteenth-century landowners could have influenced innovation among their tenants had they made more use of leasing covenants, but this view does not stand up to scrutiny in the current research area where such covenants were common.

Chapter 8
Land Management and Liming, 1815 to c. 1900

1. Introduction

Chapter 7 examined evidence of liming, and of land improvement more widely, in the Central Pennines during the era of parliamentary enclosure between 1760 and 1850, specifically looking at the strong relationship between liming and enclosure. Documentary evidence was considered to illustrate how estates had furthered the idea of improvement by responding to external stimuli and by means of clauses inserted into farm leases. A picture was built up of widespread improvement across the region, some at the behest of the landed classes, some undertaken by the lower gentry, or by tenants. The impact of the Napoleonic wars in stimulating agriculture and driving improvement was stressed.

The current chapter backtracks a little by picking up the picture at the end of hostilities in 1815, a year which proved challenging to farming in more ways than one owing to climatic and economic forces. In farming terms the nineteenth century was one of extreme contrasts. On the negative side, distress after 1815 was mirrored by further distress in the early 1830s and by a severe and deepening depression that began in the late 1870s. More positively, though, were advances made across the board in the industry during the so-called 'High Farming' period mid-century.

Developments in science and more liberal trade regimes were the catalysts for the introduction and widespread take-up of 'new' soil inputs: guano, imported from South America, artificial chemical additives and pulverised limestone began to erode the formerly dominant position occupied by lime. By the end of the nineteenth century lime was no longer the obvious choice for a farmer wishing to take in new land – assuming any remained to be taken in – or to top-dress existing pastures. According to one source in Yorkshire, written in the 1920s, very little liming had been undertaken there over the 'last forty years' (Ruston 1924, 739), taking decline in usage back to the 1880s. However, a similar conclusion had been made two decades earlier, putting decline back to the 1860s (Board of Agriculture 1907, 1). It has also been suggested more recently that the late nineteenth-century depression in farming effectively killed off liming as farmers felt unable to meet the costs involved (Havinden 1974, 129), but this view conflicts with evidence from the Yorkshire Dales where commercial limeworks enjoyed a buoyant market in agricultural lime beyond the end of the century (Johnson 2002).

Liming was still being practiced in the Central Pennines and there is ample documentary evidence to question the negative perceptions alluded to in the previous paragraph. In Wensleydale, at least, draining and liming were still being practised through the century with kilns having been built wherever limestone was available and lime having been applied 'very liberally to the meadowland' (Fieldhouse 1980, 193). True, the extent of liming did decline as the century progressed, alternative inputs did make serious inroads, and the late-century depression had caused retrenchment in marginal areas, but a picture will emerge of liming still playing an important role.

Two particular instances will help illustrate this assertion. In upper Swaledale successive generations of farmers, of the Garth family, maintained day books from 1795 to 1911 which contain a wealth of material on daily and seasonal farming practices, including their faith in liming up to 1879. The Farrers, on the Ingleborough Estate in North Craven, were engaged in extensive improvements from at least the 1830s and beyond the end of the century: liming appears in the accounts almost to the end.

2. The post-war legacy

The global impact of the Tambora eruption of 1815 was alluded to in Chapter 7.2, but it was only one of several serious eruptions in that decade (Post 1977, 4-5). Their impact on climate patterns was immediate with markedly increased precipitation levels and reduced temperatures through 1816 and 1817. Harvest failures were an inevitable consequence, as were famine and outbreaks of diseases like typhus and 'relapsing' fever.

This could not have come at a worse time. Demobilisation threw thousands on to a restricted job market (Hueckel 1981, 192), and the sudden end of economic blockade and trade restrictions saw a surge in cheap imports of grain. Furthermore, an above-average grain harvest in 1813 had swamped the domestic market with cheap grain which depressed farm-gate prices (Hall, 1987, 24). To reverse the situation, the protectionist Corn Laws were enacted in 1815 placing restrictions on grain imports, unless the price of home-grown grain was high (Gazley 1973, 654; Ward, 2004). Predictably, this legislation led to higher grain (and staple bread) prices which added to the nation's general woes. The overall economy was adversely affected but the farming industry was 'seriously depressed' and subject to utter calamity (Post 1977, xiii and 35). This confirms Prothero's opinion that peace had brought only 'beggary' and 'unexampled' misery to the farming community (1912, 317). Cheap imports had hit grain producers but not stock farmers; however, these were hit in two ways by the climatic downturn. Massive mortality was recorded across northern Europe in sheep and cattle as a direct result and failed hay harvests were an indirect but equally devastating consequence of cold and wet summers.

One, albeit delayed, effect of the accumulation of problems after 1815 was a notable drop in the number of small farming units, particularly in Westmorland where many smaller yeomen succumbed and gave up farming altogether, leading to an estimated fall in the number of yeomen of 38.5 per cent between 1829 and 1849 (Beckett 1982, 97; Walton 1986, 227-28). Consolidation of small units was the result but it did enable more efficient and cost-effective land management, though this may have been the resumption of a long-term trend that had begun before the Napoleonic wars. In Beckett's view (1982, 103), one result of the post-war situation was that most small farmers here sank back into their pre-war lethargy and resumed age-old practices, giving up all thought of improvement. However, the current research has shown that this was not the case in those parts of the Central Pennines where evidence has survived.

A contemporary account, penned during a journey through Westmorland, remarked on the number of carts seen on the road 'mostly laden with lime' which that traveller concluded was the main top dressing used hereabouts on summer pastures (Moore 1817, 188). His remarks were made in August 1815, probably too early for any post-war economic downturn to have made itself felt.

No conclusive proof has been found to inform meaningful comparison of the extent to which different parts of the research area were affected by the post-war slump, though it has been posited that Lancashire fared better than much of the country (Phillips and Smith, 1994, 157). As will be shown, the concerted efforts of the Garths, Farrers and others lend credence to the view that the Central Pennines were perhaps not so adversely affected. On the other hand, the picture could be distorted by the nature of the documentation that has survived: for a farm or estate to have kept detailed records might suggest that it was progressive and well-managed and thereby better placed to ride out the problems which undoubtedly affected it.

There is the occasional documentary reference to distress within the research area. In September 1816, for example, the clerk to the Property Tax commissioners, William Stackhouse of Giggleswick, wrote to his masters on behalf of a farm tenant who had been dispossessed for non-payment of rent. The tenant in question, rearing cattle on a farm at Swinden south of Settle, was unable to find the rent owing to the recent 'great depreciation in the value of livestock' (Stackhouse, private collection). An indication of just how long the depression had lasted is provided by a further submission, dated 1828, whereby Stackhouse reported that James Calverley, a farm tenant at Feizor to the west of Settle, was pleading for a rent reduction because of 'bad times' in farming (Stackhouse, private collection). Meanwhile, in Westmorland, it was reported in the local press that many farmers were so crippled by debt as rents, tithes and taxes exceeded their income that they were forced to apply to the Poor Laws (*Westmorland* Gazette 13 November 1830). Successive editions of the paper contained large numbers of farm 'sale' or 'to let' notices. Perhaps the true extent of post-war distress in the industry is masked by a dearth of local evidence.

Set against these realities, as demonstrated in Chapter 6.6, was the high volume of lime-coal sales in Colsterdale after the war ended. Sales remained buoyant between 1816 and 1825, even during winter months, despite adverse weather and economic conditions. As lime-coal sales can be directly correlated with lime burning operations – and by association with lime sales and usage, too – this buoyancy would seem anomalous, especially given that average winter sales during the war years of 1806-15 were not high. No rational explanation can be found for this situation but it is conceivable that the farming economy in the Masham area remained strong enough to weather the setbacks, or that unidentified factors were at play.

If 1816 was an exceptionally cold year, 1836 and 1837 were even worse (Post 1977, 7). The immediate post-war problems were long-lasting and took on various guises: for example, wool prices in 1821 were barely half those of 1819 (Beckett, 1982, 108). Though no local archival evidence has been found to support the assertion, such collapses must have had a devastating impact on farmers in the Central Pennines who were so dependent on sheep for their income.

The early 1830s had witnessed equally disturbing occurrences for cattle farmers, and disease (the first recorded outbreak of foot and mouth) hit Westmorland hard in 1839-40 (Curwen 1926, 70-71). For these problems there are surviving first-hand accounts. Richard Rigg, who farmed at Applegarth in Middleton (SD626 846), wrote a lengthy account in his pocket book of the 'epidemic' among cattle in the area during the 1830s, though he showed no signs of having suffered like so many of his compatriots (CRO[K] WDX/450/19/5). Various issues of the *Westmorland Gazette* contained advertisements for farms that were for sale or to let, and an editorial comment for the issue dated 13 November 1830 drew attention to the situation thus: 'In the present state of Agriculture, when the occupier of a small farm is incapable ... to meet, in any way, his expenses' he had to resort to the Poor Laws as rents, tithes and taxes exceeded his income, and the only way to begin to redress this was to cut rents by thirty per cent.

Adverse weather conditions impacted equally on stock farmers along the South Pennine edge with the first half of 1830 having been described as 'the wettest time ever nowen by the ouldest person' in the manor of Colne (WYAS Bradford HEA/B/158). That year witnessed total failure of hay and grain harvests.

The situation, not just in the Central Pennines, was so serious that the government convened a select committee which produced its Agricultural Distress Report based on evidence-taking sessions held through 1833 (BPP Command Paper 612). This was not the first such commission, however, as others had sat in 1820, 1821 and 1822 to examine the depressed state of the industry then. The ultimate conclusion of the 1833 committee was that the situation had worsened considerably though the effects were felt more in the south than in the north or in Scotland or Ireland. One witness, on an unidentified farm somewhere in the North Riding, reported that many landowners had been forced by circumstances to reduce rents by up to a half as the only means of attracting new or renewing tenants.

The 1833 commission was not the last either. A further select committee was convened in 1836 to consider the 'extent of the distress which *still presses* upon some important branches' of farming (this writer's italics) (BPP Command Paper 464). Once again, though, southern England fared worse than the north: only the North Riding and Cumberland were northern exceptions, where agriculture was deemed to be still in decline. Here, according to Edward Cayley MP, many old yeoman families had been driven off the land by bankruptcy. In east Cumberland (and Westmorland perhaps?) many small statesmen farmers had been reduced to a 'pitiable' state, according to its MP, William Blamire. Reasons put forward for the depression in both areas were cheap grain imports from the continent after 1815, cheap imports from Ireland, and Peel's Cash Payments Act 1819. This had induced widespread deflation forcing prices of all farm commodities downwards – beef, for example, suffered a fifty per cent drop (Davis 2005, 85).

3. From 'Golden Age' to 'Great Depression'

Farming was once again adversely affected by the repeal of the Corn Laws in 1846, which acted as a catalyst for cheap imports of grain though the effects were short-lived and mainly restricted to cereal producers (Wood 1982, 125). Farming then witnessed a renaissance, lasting from the 1850s to the early 1870s, a period which has been described as a 'Golden Age' of 'High Farming', after which farming settled down into a 'state of contentment' (Perry 1973, xii; Thirsk 1997, 149). A price boom and an increased scientific emphasis in agriculture helped foster a new sense of optimism within the industry that was reflected in investment in buildings and improvements (Clark 1999, 293).

Various stimuli can be identified as relevant to the research area. Firstly, an ever growing urban population in the industrial West Riding and Lancashire provided a secure market for farm produce; and the developing rail network ensured rapid and cheap transfer from farm to market. New technology made farming more efficient, whether machinery, new stock breeds, new crop species, improved methods of draining, or artificial soil additives; and a greater effort was placed on efficient management practices, helped by the development of training farms and institutes. The extent to which farming within the Central Pennines responded to these developments will become apparent as this chapter progresses.

The good times were not to last, however, and a protracted agricultural depression set in during the late 1870s (Perry 1973, xi-xviii; Thirsk 1997, 149; Rowley 2006, 252). Effectively, it was to last well beyond the end of the century. The causes were partly economic and partly climatic. Cheap imports yet again undercut local produce, both cereal and meat, and a series of harvest failures (of cereals and hay) from 1879-82 reduced the size of home-grown supplies. Within the research area many marginal farms were abandoned as tenants found the task of making a living impossible and landowners were unable to secure tenants to take on vacant farms. Examination, by this writer, of census returns and parish records for the parishes around Ribblehead, for example, has identified a significant number of farmsteads that were given up for ever by the end of the century (Johnson 2000).

A direct result of abandonment was the further consolidation of non-viable farm holdings into larger and more efficient units as well as the increased use of upland pastures for grouse shooting. No statistical facts – other than farm censuses perhaps – can prove it but there is anecdotal evidence that stocking levels may even have increased on the more economic farms within much of the Central Pennines.

One contemporary observer was blunt in his assessment of the situation when he wrote 'A man, in 1875, could select one tenant from twenty applicants, but blessed is the man who in 1879 has no farm on his hands' (Silverstick 1879, 3). Clearly that four-year period had witnessed a deepening of the crisis and farms were becoming a liability, even allowing for any exaggeration on his part. Interestingly, that author did not blame the weather at all but focussed on economically- and politically-derived factors which imposed intolerable burdens on tax payers from increased tithes, poor-rate deductions, higher land rents, restrictive tenancy clauses and increased labour costs, plus cheap imports (1879, 4-6). Examination of sale particulars for north-east Westmorland confirms the large number of farming units that entered the market during the years of depression (CRO [K] WD/HH).

The degree to which depression was apparent within the research area will also be investigated later in this chapter.

4. The introduction of 'new' soil additives

The Agricultural Distress Report for 1833 drew attention to the introduction of bone-dust as a soil additive as early as 1797 and to its efficacy, even on limestone pastures, and its relative cheapness compared to lime (BPP Command Paper 612, 13 and 108). In fact, it reported that less than half the previous amount of lime was being applied in the North Riding, partly owing to the use of bone-dust which had been in use here since the mid 1820s (612, 112-14). Bone-dust was an early introduction and, according to Prothero (1912, 218) an accidental discovery. Apparently, a fox hunter was sweeping out the kennels and had spread the residue of the hounds' last meal on nearby ground. Someone noticed that the grass there grew more luxuriantly than elsewhere and put two and two together, deciding the dust would make a suitable soil additive.

Other additives were introduced during the middle decades of the nineteenth century, partly as a result of pioneering work at the Rothamsted Experimental Station in Hertfordshire (Carne 2006). Oilcake dust was possibly the first nineteenth-century introduction (Thompson 1968); nitrate of soda was first tried in 1835 but not widely applied until after 1850 (Prothero 1912, 369); a mix of blood, bone and sulphuric acid came in after 1850 and coprolites (naturally preserved fossilised faeces) in the same period (Prothero 1912, 370); nitrates were introduced in the 1850s and potash on a major scale in the 1860s (Thompson 1968, 71).

Sulphate of ammonia, a by-product of urban gas works, added to the plethora of inputs available to Golden Age farmers (Binns 1851, 62). Unlike most of the others mentioned, which acted in broadly similar ways to lime, ammonia could not act alone because it increased acidity levels within the soil. If liberal doses of ammonia were applied, any benefits of top-dressing with lime were negated but this was not the case if it was applied in moderation along with lime (Thurston *et al.* 1976, 1065).

A further industrial by-product was made available to farmers on the continent in 1881, and soon afterwards here, when it was realised that basic slag from Bessemer converters performed in much the same way as lime, as long as it was pulverised (Pearson and Adams 1967, 151-54). Basic slag contained thirty to forty-five per cent calcium so it was effectively a low-grade form of lime, and much cheaper.

By the 1970s lime had been entirely replaced by raw pulverised or crushed limestone but the first trials with this date from as early as the 1760s when broken rock was laid on the ground to 'slowly effervesce' (Mills 1767, 29-30). Its use did not catch on, despite further experiments, and there was no mechanism for pulverising the stone to a size small enough for effective air-slaking to occur (Dickson 1805, 235; Kames 1815, 280).[64] The first powered crushing apparatus was described by Loudon (1839, 441-42) but widespread crushing really dates from after the Great War, with Joseph Harrison of Flusco, near Penrith, having been a pioneer (Johnson 2002, 173.)

All of these products made inroads into the market for lime. Many of them were undoubtedly cheaper, an attribute of especial attraction during the years of depression when farmers were hard pressed; and, at least during the Golden Age, they were perceived as 'new' and progressive and more representative of the modern farmer than age-old lime. None of them, however, had the same impact as guano,[65] either nationally or within the research area. Estate accounts start to include purchases or applications of various alternatives from mid-century but such entries were few and far between and one gains the (purely subjective) impression that land managers were merely dabbling in something new. This was not the case for guano, however. It was adopted in a major way.

Guano was first imported through Liverpool in 1835 but not widely taken up until the next decade (Prothero 1912, 369). Within ten years it had become the prime soil input and has been dubbed the 'Wizard of the Pacific' (Goddard 1990, 179). Total imports through Liverpool had risen from 1140t in 1841 to 13,852t in 1848, a staggering increase (Garnett 1849, 42). To many progressive farmers and commentators of the time it was the 'best' manure available (Lawson 1879, 111). The Royal Agricultural Society promoted its value, as did writers of agricultural treatises such as Stephens (1844, 1235) who believed it to be 'true and valuable'. Local commentators, speaking of Lancashire, conceded that lime was still in almost universal use across the county though on high ground a mixture of guano and crushed bone had replaced lime as the first choice (Garnett 1849, 41; Binns 1851, 59).

In Westmorland a coeval commentator noted that since guano's introduction 'much less lime' was being used than hitherto (Dickinson 1850, 32), and this situation applied throughout what is now Cumbria (Bainbridge 1945, 87). Guano was probably first applied in Westmorland in 1843 (Garnett 1912, 207). One set of farm accounts, between 1852 and 1873, albeit slightly beyond the bounds of the current research area, illustrates the lime-guano position (CRO [K] WD/Hee). That period recorded a total of forty-seven purchases of lime compared to twenty-five for guano but these data should be viewed with a degree of circumspection. The average payment per purchase of lime was £27 but for guano it was £79, and over that period more than twice as much was expended on guano. These contrasts can perhaps be explained by supply logistics: lime could be purchased locally whenever and in whatever quantity it was required whereas guano had to be ordered from importers through third-party suppliers and, quite possibly, bulk orders secured a discount.

Chemically, guano is forty-three per cent phosphate of lime with a varying proportion of oxolate of lime, nitrogen and ammonia. Unlike lime, guano is not volatile and was therefore easier to manage and store (Johnston 1841, 305-06) and it was believed that its effects were longer lasting than those of lime. Unfortunately no contemporary accounts have been located to confirm why individual farmers had turned from lime to guano. Binns (1851, 110) described improvements made by Mr Nutter who took over Ireby Hall (SD647 749) in 1834. Nutter, like his neighbours in Leck and Tunstall, was making use of both lime and guano: the former was applied on 'old pasture land' at the rate of 150-180 bushels per acre and on rough grassland at 200-240, but Binns did not say why lime was the preferred option or what the guano was used for or when it was first used here.

On the other side of the Pennines Messrs Outhwaite's Bainesse Farm (SE240 971) in Catterick was described as a 'very favourable specimen' of agricultural management (Caird 1852, 323). They practised a six-year rotation applying heavy doses of manure with guano being specifically applied to turnips. In Caird's view guano worked well on crops like turnips but was of little value on, say, beans. However, he wrote nothing about the treatment of grassland.

Caird said guano was a cheap option (1852, 463) so was this the prime reason for its enthusiastic take-up? During the great depression towards the end of the century this would indeed have been a very important factor but all the accounts noted above predate it. They all refer to the period of High Farming when cost-cutting measures would not have been given the

[64] Crushed limestone breaks down more readily than burnt lime but has less than half its strength and efficacy (Ogg 1942).

[65] Guano, in the nineteenth century, was a term applied to bird dung collected from seabird colonies in Peru.

same priority. Could it be that farmers adopted guano simply because it was innovative, and because it was actively promoted by writers and by the Royal Agricultural Society? Perhaps farmers wanted to be seen to be progressive in that age of prosperity. Or was it merely that, from thrift, they chose the cheapest option?

A picture has been built up above of varying fortunes through the nineteenth century. The period under review began with post-war depression and a succession of harvest failures, continued with a series of problems through the 1820s and severe distress in the 1830s. The Repeal of the Corn Laws in 1846 once again induced temporary setbacks but this was soon followed by a time of boom that endured for more than two decades, when science and technology were introduced to farming, and 'new' additives began to replace lime as the first choice. During the late 1870s this all came to a rather sudden end as a deepening depression set in that was to last well into the next century.

This picture has been built up from contemporary accounts and from the writings of modern historians. The rest of this chapter will consider the extent to which it has parallels within the research area. Was lime knocked off its pedestal, does the documentary record mirror the trends identified, and did farming slowly sink into almost terminal decline?

5. Local improvement strategies

Draining, paring and burning

Whether or not liming by itself had the desired effect depended on the pre-existing state of the soil – on variables such as moisture content, pH, mineral and organic content. It did have a positive impact on naturally well-drained land with a pH only slightly acidic and where manuring had perhaps been practised. On ill-drained acidic ground, with high annual precipitation totals and low evapo-transpiration rates, where coarse vegetation species were dominant and no inputs had previously been added, it was vital to prepare the land prior to liming by first draining and then, in many instances, by paring and burning. Indeed, a contemporary view was that upland grassland *must* be drained before lime could be applied (Macdonald 1887, 163), as 'new' grass species such as Bent (*Agrostis*) and Fescue (*Festuca*) would not thrive on poorly drained soil (Simmons 2003, 96-97). Deep, or under-, draining could be considered fundamental to successful agricultural improvement and increased yields (Phillips and Clout 1970, 71; Williamson 2002, 164, 167; Gritt 2008, *passim*).

Seen from a different perspective, draining was perceived as fundamental by landowners who saw the potential benefit accruing to them from increased rents (Phillips 1972, 195). The disparity between drained and limed land, on the one hand, and unimproved land was demonstrated within the North Riding part of the research area: in its natural state land was valued in the middle of the nineteenth century at 2s. 6d. per acre but if drained the value exactly doubled and if drained and then limed values soared to 20s., according to a contemporary commentator whose stated belief was that the change 'almost surpasses credibility' (Milburn 1848, 202).

In the 1830s, at least, the tenant of High Burton Farm on the Constable Burton estate in Wensleydale was tied in to a nine-year rotation on nine arable fields that had been subjected to managed improvement, starting in 1829, at the behest of the landowner (NYCRO ZFW 4/2/7). During the fourth and fifth seasons each field in sequence was to be put down to 'clover or seeds', the latter referring to artificial grasses with sainfoin (*Hedysarum onobrychis*) specifically mentioned for Lime Kiln Field for 1832.

Land drainage was certainly not a new phenomenon in the nineteenth century and slow improvements in its efficacy can be traced through the early modern period; deep drains, as opposed to open shallow ditches, were introduced in 1764 (Kerridge 1967, 37) though a possibly contradictory view suggested they had only 'of late years' been introduced into the 'northern counties' (Bailey and Culley 1794, 36). Deep drains were pioneered by a Warwickshire farmer, Joseph Elkington, and his name was very quickly applied to under-draining, which practice spread across the country with amazing rapidity and became the accepted mode (Stephens 1844, 490-518). The Board of Agriculture recognised the value of his method and awarded him a prize of £1000. Charles Towneley had made use of his services and invited him to Burnley to oversee draining operations there, as did the Farrers of Ingleborough.

As late as the 1820s, open drains were still being cut on the Bolton Abbey estate in Wharfedale, possibly because this method was undoubtedly cheap and quick to implement (Devonshire BAS/3/6). For example, the steward's accounts recorded payment in September 1822 for cutting 99 roods (about 730m) at 3d. per rood.

Tile drains were first introduced around 1820, on the mosses of southern Lancashire (Beesley 1849, 38), though they were not widely used for another two decades. The exact date is uncertain. Beesley said 1840 or 1841 was the year when the first tile kiln was built, though this could have fired hand-made U-shaped tile drains, and 1842 has been given as the year when the first tile drain machine was invented, with a capacity of 500 sections per hour. These were superseded in 1843 by the introduction of a machine capable of producing up to 20,000 cylindrical sections daily at less then half the cost (Hoelscher 1963, 77-78).[66] A further improvement, the mole drain, was brought in during 1859 (Kerridge 1967, 37).

In Yorkshire a tile drain machine was first shown in 1844 at the county agricultural show in Richmond and its use mushroomed: within four years 130 such machines had been commissioned (Charnock 1848, 305). This may have been a consequence of the Improvement of Land Act 1846 which

[66] Though Thomas Scraggs won a RASE prize in 1845 for what has been described as the first such machine (Tarlow 2007, 60).

promoted drainage schemes partly financed by state loans (Gash 1979, 321).[67]

On the other hand, the Drainage Act 1838 had already set the scene for increased emphasis on drainage and its effects had an immediate impact. On the Thorp Perrow estate, near Bedale, the Milbanks took full advantage of this Act (NYCRO ZAL 4/24). A detailed book of field plans was compiled showing the network of 'pipe tiles', set just under 1m deep, in each of their fields in Snape and Well, and the book specifically noted that the work had resulted from that legislation. Tenants were required to contribute sixty per cent of the 9d. to 11d. per rood cost.

A later set of field plans shows the network of drains, set 1.2m deep, installed from 1856-57 (NYCRO ZAL 4/44). The terms were still the same as in the 1840s but by 1865 the cost had fallen below the earlier levels, no doubt reflecting manufacturing economies of scale in the Bedale area. In the more remote valley of Snaizeholme, south of Hawes, drains were being installed at a cost of 1s. 3d. per rood, again in the 1860s (NYCRO ZST/2946). It is, perhaps, worth noting that this small and marginal estate was still actively engaged in installing drains during the early 1880s, years of alleged depression.

The advantages of draining were readily understood by even the most casual observer; the dual practice of paring and burning, however, was rather more controversial. Sinclair, founder of the Board of Agriculture and Improvement, believed there was nothing better for reclaiming barren land (Sinclair 1819, 152); and a later observer deemed paring and burning to be one of the beneficial 'high farming operations' (Voelcher 1857, 342). Both writers listed a range of benefits which included destroying parasites and weeds as well as rendering potash more soluble within the soil (Sinclair 1819, 236; Voelcher 1857, 350-58). More than forty per cent of all county reports to the Board, submitted by 1794, commented that paring and burning were current and accepted, indeed sometimes 'ancient', practices (Woodward 1994, 105).

There were also contrary opinions: the dual practice was either felt to be 'controversial' (Sproule 1839, 165) or even 'destructive of the soil' (Rham 1845, 144) which latter possibly accounted for a marked reduction in usage mid-century (Johnston 1849, 194). One first-hand account of farming in the West Riding recounted how great efforts were being expended on the gritstone moors to bring them into productive use (Charnock 1848, 310). Draining was the first task followed by cutting heather and then liming but he specifically advised against paring and burning, without explaining his rationale.

Paring involved removing all vegetation growth either with a dedicated paring spade (Fig. 8.1), or breast-plough (Hitt 1761, 75-76), which effectively did what archaeologists would now

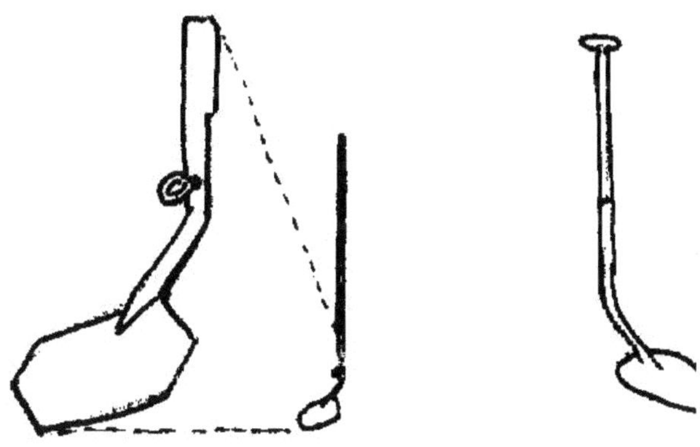

Figure 8.1 Paring spade

call de-turfing.[68] This was best undertaken in spring to give the parings, which were piled in small heaps, chance to dry out. Burning was carried out in a controlled manner to avoid the fire spreading across the entire pared area which would have destroyed any organic matter within the soil (CRO [K] WDB/35). After burning, the ash piles were raked out applying a thin veneer to the entire area.

If improvement was being performed in the time-honoured manner, the next task was liming.

Liming

There was a bewildering volume of advice available to land managers though much of it was conflicting. At least most contemporary writers were agreed that lime was of definite benefit to soil productivity; where they diverged was in methods and rates of application. Lime was described in one treatise as being so essential for reclaiming the waste that 'little good could be done without it' (Falkner 1843, 207), while to another observer no practice existed in agriculture that had more palpably obvious effects than lime (Way 1854, 491). One view, offered by a practising farmer, maintained that lime – or marl – was vital if farming was to be 'carried on in perfection' (Rothwell 1850, 89-91); and John Watson, a contemporary Kendal-based land surveyor, regarded lime 'of incalculable benefit on every kind of virgin soil in a greater or lesser degree' (CRO[K] WDB/35). He provided details of local farmers who had experimented with liming, or used it selectively, with invariably positive results.

In terms of how to apply the substance there were two main points of disagreement: firstly, whether or not it should be laid 'hot from the kiln' and thus in its most caustic state or be allowed to lie until it had been reduced to an effete state before being applied; and secondly whether lime should be spread

[67] Further legislation added to the amount of grant finance available for land drainage: the Private Money Drainage Act 1849 and the Improvement of Land Act 1864. The state loan scheme came to an end in the mid 1870s (Gash 1979, 321).

[68] A paring spade generally had a handle rather longer than 2m and a blade 300mm to 375mm long set an angle to allow it to almost glide parallel to the ground surface, held to the body with a leather strap (Donaldson 1842, 251-52; Watson 1845, 101). It had a low-angled blade and was pushed along (Fussell 1981, 67).

alone or mixed with dung. These issues were not new. Indeed they had been thrashed around for at least the previous century but now they came more to the fore. This may just have been a reflection of the increased number of agricultural writers wishing to make their mark, or it could reflect the greater level of scientific and technical thinking associated in particular with High Farming.

For Falkner (1843, 205), it was best taken straight from the kiln and piled in heaps across the field, covered immediately with soil and then spread out and ploughed in as soon as it had air-slaked. This, of course, presupposes that the kiln was very close to the given field otherwise it would have cooled, and slaked to an extent, in transit. This was at odds with advice that urged slaking prior to taking it to the field, either by deliberately adding water or allowing it to air-slake on field edges (Sproule 1839, 39-40; Briggs 1845, 7).

Both of these views were totally at odds with an earlier piece of advice that three years should elapse before the lime was spread out (Dickson 1815, 195). Perhaps nearer the target was that given in a letter, dated 1814, saying lime should be taken from the kiln and just left for three to four days to 'fall' (air-slake) prior to spreading (Lovat 1831, 68). By the latter end of the century it was being suggested that leaving lime in small heaps to be air-slaked was a 'waste of time' and that it was preferable to cover the heaps with soil to try and prevent rapid slaking (Griffiths 1889, 239).

The 'mix it with dung' issue attracted rather less polarity in thinking and most writers tended to favour mixing (see, for example, Dickson 1815, 508; Anon j 1820, 103; Donaldson 1842, 140-41; Johnston 1849, 119; Stephens 1852, 671; and Morton 1855, 534-39). There was also some convergence of thinking concerning frequency of liming: Rothwell (1850, 89-91) urged once every twelve years on light soils and twenty on heavy; while Stephens (1852, 667) concurred with twenty but did not distinguish between soil types. Falkner (1843, 217) stressed that repeated liming could have no added benefit, that once a soil had been brought into a state of lime sufficiency continued liming would do nothing more. Way (1854, 514), however, preferred to plough his own furrow by saying apply little and often. Documentary evidence, in one form or another, confirms that the practice of liming continued throughout the century under review, and it is reinforced by later observations. Russell (1920-21, 882) felt that 'every farmer' well knew how important liming was; and Turner *et al.* (2001, 82) trawled extant estate records across the country for evidence of soil additives in use (Table 8.1).

Period	Lime as a % of all soil additives	Marl as a % of all soil additives
1700-59	58	18
1760-99	74	10
1800-39	72	9
1840-79	70	6
1880-1914	40	5
Average*	63	10

Table 8.1 The position of lime as a soil conditioner, 1700 – 1914
Source: Turner et al. *2001 (82).*
** Note that the time periods in the table are not of equal length.*

Evidence for liming will be presented chronologically, and it should be stressed that the picture that emerges is necessarily based on extant evidence and may be a distortion of the actual situation.

Tenancy clauses with liming stipulations were still being applied after the end of the French wars as, for example, on Bank House Farm in Barbon (SD635 821) (CRO[K] WD/Whelp/T53). The new tenants, William and John Langstroth, were required to apply 1000 bushels of lime within the first three years with all but 100 paid for by the tenant, and it was to be mixed with 'slutch that arises out of the two puddles in the lane', which seems an odd stipulation to make. One can only assume that this mud was (deemed to be) rich in nutrients.

Also just after the end of those wars tenant Jonathan Umpleby, at Eastby near Skipton, had apparently volunteered to 'set 300 loads of lime upon the Roods' on condition that his landlords met the cost of coal to burn the lime in either of the kilns that he had chosen (Keighley BK 79/150). This is a rare instance of a tenant leading the way and, in effect, telling his landowner what to do.

George Smith, on the Hornby Castle estate in Lunesdale, kept a detailed day book and there are frequent entries of relevance to this account (Hornby, Smith, diaries, unpub.). On 8 November 1819 he had noted that labourers were employed 'mixing up lime and earth in the Pointer Field' prior to spreading it; on 3 December John Walling, of Storrs at Ingleton, had agreed to lime part of Hornby Park at 13½d. per load though the total was not stated; and on 11 December Walling further committed himself to liming Foxholes Field, this time at 13d. In a later entry Walling was described as a limeburner and it is quite possible that he produced the lime at one of two known lime kilns on the edge of Storrs Common and carted it to Hornby. The direct route between Storrs and Hornby is 16km and what Smith's diary does not make clear is whether the 13d. was the cost of just spreading the lime or inclusive of purchase cost and transportation. The fact that Walling was paid £40 in January 1820 'for lime' might suggest the latter as that sum would otherwise represent about 800 loads. However, liming entries continued on a regular basis with one agreement, dated

21 January 1823, for 4000 loads of lime, a staggering quantity by any standard. Tenancy agreements here, too, stipulated liming.
In the 1820s Smith started sourcing agricultural lime from a kiln at Wennington, presumably to reduce costs.[69] The last entry of relevance to this narrative was dated 26 May 1842: liming had continued on the estate continuously from the post-1815 downturn and through the troubled 1830s.

Ashes Farm in Waldendale, a tributary valley of Wensleydale, was put up for auction in August 1823 (*Leeds Mercury* 2 August 1823). The farm was advertised as possessing, *inter alia*, a lime kiln (SE0109 8176) and nearby coal reserves and it was described as having pastures that were 'very improveable by draining and lime'. We cannot know if the previous occupant had succumbed to economic pressures but the sale notice does suggest that even in those straightened times improvement was still in mind.

This was the case in Ribblesdale, too, as in the same year Dub Cote Farm in Horton (SD820 715) was relet with identical conditions to a 1727 lease of nearby Top Farm, namely that 1000 loads of 'well-burnt stone lime' were to be applied during the twelve-year tenancy (Horton History Group 1984, 43). In 1832 Richard Garrel, tenant at Birk Knott Farm (SD721 634) in Keasden in the Wenning valley, was ordered to 'lay upon the aforesaid farm 233 loads of lime before the 26th day of April next': he was covenanted to do this but had disputed the agreement (YAS DD 107). The key point here may be that this dispute came to a head in 1832 and Garrel may also have been suffering the same hardships as so many other farmers at this time (see Section 2).

A lease for Bleaflatt farm (NY736 037) in Ravenstonedale may suggest that this area was immune from the problems of the 1830s, or that the lessor was taking measures to mitigate an adverse situation (CRO[K] WD/MG). Thomas Guy agreed to renew his lease of the farm in 1836 for a further three years, at £33 per year, and he accepted the requirement to 'lay and spread ... sixty bushels of lime mixed up with soil so as to make a good compost in each and every year'. If the situation in the 1830s was as serious as the Select Committee on agricultural distress maintained, it is questionable that he would have been in a position to extend his lease and equally unlikely that the landowner, Mary Scarborough of Kirkby Stephen, could have held out for the stringent terms he accepted.

Meanwhile, in Stainforth in Ribblesdale, a case of trespass was heard before the assizes in 1843 between two prominent landowners in the township (YAS MD335/14/39 and MD335/2/1/4/21). James Foster owned meadows near the village while Thomas and Marmaduke Redmayne owned pastures beyond his. The only legal means of access to their lands was from a road on a steep ascent but they claimed a (shorter and less steep) private right of way 'from time immemorial' through Foster's fields. Foster maintained that the right of way was restricted to driving stock; the Redmaynes, however, had engaged the services of Thomas Tatham, limeburner, to lead lime to their lands through Foster's 'on pony's backs'. This enraged Foster who thus took them to court for breaching his rights.

There are several points of note here. Most significantly, it shows the Redmaynes were intent on improving their fields – which still bear the clear signs of improvement – and, according to one witness, this was the first time in forty to forty-five years that lime had been laid in this particular area. Once again, we cannot know the rationale as the court would not be concerned with why they were doing it. Two possibilities remain: either the land had not been fully improved in the past or its lime requirement had fallen below the critical level and the Redmaynes wanted to increase its value and profitability.

To the south of Settle William Johnson renewed his lease of an unnamed farm at Wham in 1849 for three years (NA, unref.). He was to apply manure 'in a husbandlike manner' on meadow lands annually; to cut 100 roods of 'good substantial & suitable stone drains'; and to spread 200 loads of lime each year, the purchase costs to be met by the lessors.

That lease is virtually identical to one agreed for Bank House Farm (SD630 818) in Barbon in 1851 (CRO[K]WD/Whelp/T57). Henry Carr was to lead and spread 100 bushels of lime each year at his own expense for the five years of the agreement though the lessor was to pay for the purchase of 150 bushels, and the lessee was to be allowed £10 against his rent for cutting drains.

The evidence so far has covered the decades between the post-war slump and the onset of the High Farming era, and encompassed the downturn of the 1830s. For the High Farming period itself – the 1850s to early 1870s – there is rather less extant material on liming. In fact, in 1871, one tenant on the Kiplin Hall estate, near Richmond, was even told to apply a 'liberal top dressing (of manure) instead of lime', though no reasons were given (NYCRO ZBL IV/5). However, liming still seems to have been an established practice across the research area, if available data can be extrapolated more widely.

For example, in 1854, Thomas Simpson, of West Witton in Wensleydale, leased a portion of ground near Spigot Lodge for an unspecified period with the intention of working a seam of coal to feed a lime kiln in the vicinity, at SE080 870 (NYCRO ZBO V Y/37). The rationale behind the lease was for Simpson to supply Lord Bolton's tenants thereabouts with the lime they needed: he was forbidden from selling any elsewhere.

The Trust Lords of Kettlewell, in Upper Wharfedale, found it to their financial advantage to lease out a large lime kiln with a quarry at Knipe Wood, starting in 1866 and continuing beyond the end of the century, suggesting that the market for lime there was still buoyant (Brit. Speleological Soc.; NYCRO ZKK II).

[69] This kiln (SD6162 7007) still survives and was very unusual for this part of the research area in that it probably burned limestone cobbles extracted from within the River Wenning on whose bank it stands.

There are very comprehensive accounts for the Marrick and Hurst estates in Swaledale, with regular purchases of lime in virtually every year between 1861 and 1881 (NYCRO ZPL 10). A few purchases are tied into building work but the vast majority leave the researcher guessing; only two refer specifically to liming and they were both in tenancy agreements from 1866, based on generic covenants. Each required the tenant to apply a minimum amount, stated in chaldrons, of 'well-burnt lime' each year.

Elsewhere in that area similar agreements existed. Ravensworth Farm, for instance, was re-let in 1863, on a fifteen-year tenancy, with the new tenant being bound to drain and spread manure at his own expense with any unused manure or lime to be compensated for at the expiry of the period, should the tenant not renew (NYCRO ZQX 4/13). Interestingly, later leases on the Marrick estate gave tenants the option of draining, at the landowner's charge, with compensation being granted for unused lime or crushed bone at double the rate for bone dust but with superphosphate of lime calculated at twenty per cent of cost (NYCRO ZQX 4/11). This lease dates from 1885 by which time agricultural depression was well established and landowners were no longer in a position to dictate unreasonable terms to potential tenants.

The Milbanks' Thorp Perrow estate near Masham in lower Wensleydale was still being actively managed and improved through the middle years of the century (NYCRO ZAL 3/5). Estate rental accounts for the period 1855-67 include frequent entries for allowances against lime procured by tenants, though some of this was intended for building repairs, and for a major programme inserting field drains (NYCRO ZAL 4/24; ZAL 4/44).

On a much more extensive scale, Lord Wharncliffe owned vast areas at the head of Wensleydale, based at Simonstone Hall (SD872 915) during the latter half of the century and his agent went to extraordinary lengths to try and develop the property's economic potential, including opening up coal pits and exploiting limestone reserves. This included building a new kiln, in 1880, at Sedbusk, using local low quality coal reserves that were deemed adequate for producing lime for his lordship's own agricultural use thereabouts (Sheffield WhM235/36).

Six years later the agent, J H Scott, was writing to his employer bemoaning the realities of life in the area. One tenant, in Cotterdale, had given notice and could not be persuaded to stay complaining, along with his neighbours, that rents were set too high for them to be able to afford to buy lime despite the fact that they knew the land was 'in very poor condition and much requires lime' and many of the existing drains were no longer functional (WhM241). In the agent's own words, it was proving 'difficult to know what is the best thing to do, for if the old inhabitants *cant* live it will be most difficult to find any one else who *will* live in Cotterdale' (original emphasis). Matters were desperate indeed.

Scott must have possessed persuasive powers as he wrote to his master less than a month later to the effect that the tenants had finally seen sense and had realised that 'agitation' would get them nowhere (WhM 242). They had, mostly, agreed to sign up to liming again and had collectively admitted the land was suffering from a lack of it. The carrot was that Scott committed the estate to sourcing and paying for lime-coal and to renovating the long-idle kilns in the valley as long as the tenants agreed to burn, lead and spread the lime 'free of charge'. They were reportedly delighted with the new arrangements and the estate put pressure on the ring leader to move on.

If ever there was an instance of pragmatism, this was it. Times were harsh, tenancies were hard to fill and tenants were undoubtedly suffering, but common sense initially prevailed, here helped by a healthy dose of mutual compromise. However, Scott bemoaned again, in November 1890, that 'very little attention had been paid to the Liming Clause in the New Agreement' despite the 'great expense' incurred by Lord Wharncliffe in repairing kilns on the estate and in providing his tenants with a new local source of lime (WhM 245). The outcome was not recorded but the implication is clear: tenants *must* comply with the terms agreed.

The Wyvills, on the Constable Burton estate lower down Wensleydale, also adopted a pragmatic approach through the years of depression, by granting rent allowances to tenants for liming and drainage carried out, at least from 1879 when extant accounts began (NYCRO ZFW4/2/13; 4/2/20). They contain relevant entries continuously up to 1890, again in 1893 and from 1898 to the 1900s when basic slag first entered the scene: it was considerably cheaper than lime. The estate had its own limestone reserves and kilns so production costs were not a significant factor in its decision-making process.

6. Garth of Summer Lodge

It has been said that 'ordinary' farmers rarely recorded their daily routine and that, even if they had, such records are unlikely to have survived (Turner *et al.* 1996, 27). Successive generations of the Garth family, however, were a notable exception to this rule. Francis started his day book in 1795, his son Richard continued the tradition after 1810 and his grandson, also Francis, from 1848 to 1911 (Barker MSS 251, private collection). They lived and farmed at Summer Lodge and Haverdale, in a tributary valley of Upper Swaledale, and were just ordinary farmers though they maintained a series of sidelines to boost their income and standing. Entries in the books are very brief, often one-liners, and the purpose was evidently to record what tasks they performed on the farm and in their wider activities. There is an entry for every single day from 1795 to 1911 and, as such, the books provide a valuable, and unparalleled, insight into farm life in Swaledale. As far as this research is concerned, entries connected with liming encompass the period from 1795 to 1879, from the early years of the French wars to the early years of the great depression.

In keeping with the general tone of the day books, the liming records provide only minimal detail. In June 1795, for example, one entry states 'burning lime', the next the same, and the following two days 'at lime kiln'. On 22 June Francis wrote 'about lime kiln all week'. What can such entries add about their involvement with liming? In isolation they say very little as no end use of the lime is recorded but by comparing entries and reading between the lines it becomes apparent that they were indeed undertaking liming as well as using some for building purposes. In March 1797 two entries noted that some lime was led 'without ashes' whereas other loads did not have that notation. Lime destined for the land could be polluted with lime ash, indeed the ash would be of benefit to the soil in its own right, but lime for mortar or other building uses needed to be purer and ash-free. The records do provide some clarification here. In November 1798, a quiet period on the farm, Francis was busy 'plastering at Nettlebed', now just a barn (SD967 970); and in April 1802 he was plastering and whitewashing his own home.

Some entries indicate where the lime was led to: twelve days, between 17 September and 30 October 1812, were spent leading lime by cart with seven of them giving Whitaside as the destination. This is a steep-sided area divided into many small and irregular enclosures lying between Summer Lodge Beck and the unenclosed Whitaside Moor: it is a purely pastoral area, so all these loads were unequivocally intended for liming.

Richard Garth employed others to help him in his many activities. In 1813 William Mare was engaged to repair Garth's own lime kiln while William Birkbeck appears in the books in 1812 'kiln burning', and in 1814 'set on kiln'. He was clearly a limeburner who was assisted by someone merely recorded as 'Math' who led lime – on 26 October 1814 he managed '133 barrows', though how far he had to push them was not recorded. Given that they also purchased lime from a kiln above Low Row, north of the Swale, with seventeen loads recorded from there on 10 May 1802 alone, it would not have been feasible to do everything without hired help.

The hillside west of Summer Lodge is known as Crackpot Side and it was a 190ha-stinted pasture bounded by a curvilinear headwall, sub-divided in 1809 when Garth was allocated 9ha. This area became known as Sunside (it faces south-east). On 15 September 1814 Richard Garth noted that he had 'begun limekiln at Sunside', and it was already burning its first charge before the end of that month. This kiln (SD9602 9605) has now been demolished but its site has been determined from field survey. Birkbeck was set to work at this new kiln.

They were busy for the next twelve months leading lime, some of it 'fallen' (air-slaked), and initially piled in heaps, and also draining. The post-1815 problems seem not to have had much effect, at least in the short term, as similar work continued through 1816, with many entries noting that the lime applied was either fallen or part-fallen, and through 1818 to 1820. For an unknown reason no lime was led to the fields in 1817 or from 1821 to 1824, though work did resume in a small way in 1825 and picked up the following year. Liming of pastures, and on one occasion of potatoes, was again undertaken in 1829, from 1831 to 1833, in 1836 and 1838, and from 1841 to 1847. By this time lime was being purchased from Bouldershaw Kiln (NY9946 0185), over 6km to the north in Arkengarthdale. The distance is not exceptional but the outward journey entailed a total ascent of 220m with a particularly steep climb out of, and laden descent into, Low Row in the valley bottom. These journeys were taxing enough but to fire their own kiln they normally purchased lime-coal from Tan Hill colliery, and this journey was more than four times longer, again with very steep hills to negotiate.

The distances clearly impacted on the costs of leading lime and coal, and further expenditure was incurred in paying limeburners' wages, but the Garths would not have done it had they not perceived it to be of long-term economic value to their farm. Unfortunately the day books do not permit comparison of total farm profit through the period covered but one must assume liming did boost productivity.

Detailed analysis of seasonal patterns in liming has been demonstrated for the Colsterdale area from lime-coal sales on the Swinton Park estate (see Chapter 6.6). It was shown that sales occurred in every month of the year, and the hypothesis was put forward that sales of lime-coal could be correlated with lime purchases and therefore with liming. April was the busiest single month with late spring and early summer accounting for over half of all sales, whereas winter months were relatively quiet. A similar analysis is possible from the Garth day books, as shown in Table 8.2. The overall pattern is rather different from Swinton's. In Summer Lodge most leading was done in summer and autumn, rather than spring and early summer, with October displaying the greatest contrast. At Swinton October only accounted for 7.4 per cent of the total whereas in Summer Lodge it was twenty-nine per cent, four times greater, with November showing a broadly similar difference. In spring the position was reversed with April accounting for twenty-five per cent at Swinton compared to only 3.4 per cent in Summer Lodge. No logical reasons can be suggested for such marked contrasts: they may have been due to personal preferences.

In 1851 Francis junior started burning lime on Riggate, north of Crackpot Side, where field surveying has located two kilns, and he extracted limestone from within a nearby gill rather than from a quarry. He also recorded, on 4 June, that he was 'making sod limekiln in Low Pasture' and, on 29 October, he 'kindled a lime heap' in the same field. These would have been similar to the seventeenth-century sod kilns in Craven discussed earlier (see Chapter 5.4). Francis had no need to say why he resorted to clamp kilns but it is easy to explain his rationale. Low Pasture (the place-name is now lost) was being improved around that time and a sod kiln was the quickest and by far the cheapest structure for burning lime. If the amount of lime required was relatively small, and if none was needed for adjacent enclosures, the expense of building a masonry kiln could not have been justified. In that October, after the 'lime heap' was fired, only two days were spent leading lime confirming the temporary and small-scale nature of this particular liming operation.

Month	Total number of days	Percentage
January	0	
February	3	1.7
March	4	2.3
April	6	3.4
May	8	4.6
June	36	20.4
July	18	10.2
August	11	6.3
September	26	14.8
October	51	29
November	11	6.3
December	2	1

Table 8.2 Monthly patterns of lime leading for Summer Lodge, 1795 – 1874

There is a further entry in April 1853, 'begun building lime kiln', and this must also have been a sod kiln because three days later lime was being led from it. A masonry kiln could not possibly have been built, let alone fired up, in three days. The location was not given. As late as 12 July 1879 there is an entry 'men leading coals and stone to sod kiln' – they were still improving, albeit using the cheapest method available.

Within the entire period of the day books there is only one entry for an alternative soil additive: on 17 May 1855 'John' was despatched to Richmond railway station, some 25km to the east, to collect guano. They bought no more. Did Francis find it unsatisfactory, perhaps, or was it too expensive given the length of the journey? What his day book does confirm, however, is that he was still liming up to July 1879.

7. Farrer of Ingleborough

From 1833 onwards the Farrrers of Ingleborough Hall at Clapham kept detailed records and accounts which contain a wealth of detail on improvements to both land and buildings across their ever-expanding estate. The first relevant entry is contained under the heading 'Draining and Limeing' within the first of many estate ledgers (NYCRO ZTW) and account books (NYCRO ZTW III.7). The last pertinent entry, before the turn of the century, was dated 1895.

In September 1833 Henry Wilson was paid £3 6s. 8d. for supplying 100 loads of lime to be applied on Thin Oaks Farm (now Oaklands, SD707 685) and throughout that year tenants and estate employees were engaged in fencing, draining and liming across the estate. The Farrers had their own lime kilns, and leased out at terms favourable to themselves the large commercial Mealbank Quarry at Ingleton, from which tenants were permitted to go and collect lime as and when they wished, much to the chagrin of the quarry operators (NYCRO ZTW III/12/6/2164). The estate purchased extensive lands around Ribblehead and immediately started improving the quality of the pastures, again by draining and liming. They worked a kiln (SD779 794) at the hamlet of Thorns and had it repaired in 1833 to produce the lime needed to transform acidic drumlin-based rough grassland into productive pasture (NYCRO ZTW). The kiln's small size and crude build suggest that it was not intended for anything more than initial improvement of this area (Fig. 8.2).

In that same September William Thistlewood was paid £4 7s. 6d. for having burnt 1608 loads through the summer for application around Thorns, in addition to 17s. 6d. for having repaired the kiln, while £46 1s. was paid to Robert Fothergill for coal to fuel the kiln. At current values total expenditure totalled £3890, a considerable sum even now, and a testament to the Farrers' determination to improve their new property.

November 1835 saw another kiln (SD747 678), at Bowsber Farm south of Clapham, being rebuilt at a cost of £7. 10s. and hundreds of loads were soon being produced there. Liming continued on a regular basis for the rest of the decade and only spasmodically were there purchases of alternative inputs such as manure and, in November 1837, 'black mould' for mixing with lime: in all, 557 loads of mould were brought in, though what it was is unclear. At times lime was spread having first been mixed with earth (for example, in 1839 on Crina Bottom farm, SD722 682, and Waters farm, SD752 673); while at others kiln waste – lime ashes – was applied, presumably to reduce costs (as at Waters in the same year); and the estate purchased a 'liquid manure cart' in November 1848 at a cost of £18. 10s. plus 5s. 6d. delivery charge from Preston.

Similar work continued each year with tenants either being allowed to collect lime at a reduced rate from Mealbank or being granted rental allowances against lime spread. Allowances were also made for draining carried out by tenants. In 1861 the first use of guano was recorded on the estate, as well as of bone manure, but not at the expense of lime. Regrettably the accounts do not always state quantities bought or unit costs so it is not possible to compare the relative costs of lime and guano.

One significant change in the accounts occurred at the end of 1864 when 'Limeing' costs were no longer accorded a separate heading: henceforth it was subsumed within general

Figure 8.2 Lime kiln at Thorns, repaired in 1833.

improvements. Nevertheless, liming and draining continued apace every year up to 1895. The importance still attached to liming is perhaps illustrated in a memorandum from the company secretary of Mealbank Limeworks, dated July 1895 but referring back to 1874, which noted that the estate had collected 250t at 6s. per ton rather than the normal 10s. and that 'it appears to be the practice on the Ingleburg Estates to apply this lime to the drained grasslands and arable fields' (NYCRO ZTW III/12/6/2164). This is certainly borne out in a letter from the estate's agent, Robert Lipscombe, to Oliver Farrer in April 1861 in which he reported that 'The whole of the meadow round Ingleborough is now either guanoed, limed or night soiled' (NYCRO ZTW III/11/155). This was, in essence, a phenomenal achievement.

The Farrers were not alone in pushing liming during the High Farming era, and they were instrumental in assisting a neighbouring landowner to do likewise. In May 1863 the estate received a letter from the agent of Mr Pilkington, who owned estates in Widdale and Snaizeholme, between Ribblehead and Hawes (NYCRO ZTW III/11/168). Pilkington sourced the lime he needed for his intended improvements (Fig. 8.3) from his own kiln high above Redshaw Moss but had to import coal by packhorse by way of Cam Fell and Gavel Gap. This necessitated crossing a short section of land ('say 200 or 300 yards') owned by the Farrers and the agent requested permission for this, with the offer of compensation for any damage caused. The response has not been located but the improvements were carried out.

Signs of change appeared in 1880 (NYCRO ZTW). Liming was still being undertaken but new soil additives were slowly eroding its dominance. Bones came more to the fore, first dissolved in sulphuric acid and presumably spread in liquid form, but still alongside lime. A comment in a letter from James Whinray at Reeby's Farm, Keasden (SD722 666) to the agent, dated December 1881, stressed that he had drained most of the farm and 'covered every rood with lime and bones', all within a five-year period at a total cost of £263 14s. 3½d., over £13,000 at current values (NYCRO ZTWIII/II/168) – also an impressive undertaking.

That year saw the first purchase of a large quantity of soot for unspecified meadows, and basic slag appeared in the accounts from 1887 to add to the mix of treatments which also included shoddy (shredded rags), and phosphate powder in 1893. From 1895 lime was completely replaced by basic slag, bones and phosphate with the latter dominant.

8. Conclusion

Chapter 8 has considered the role of liming within land improvement during most of the nineteenth century. During the eighty-five years reviewed, farming across the country endured periods of regression interspersed with prosperity. Immediate post-war slump coincided with a climatic downturn to bring farming to its knees, and further slumps occurred in the 1830s and, briefly, after the Repeal of the Corn Laws in 1846. This was followed by three decades of progress and economic boom during the era of High Farming which came to a rather abrupt end in the 1870s when severe depression set in, lasting beyond the end of the century. The years 1902-14 were seen as a time of 'convalescence after the buffeting' of the previous three decades but after the Great War long-lasting recession set in again (Harwood Long 1969, 5).

Figure 8.3 Meadows in Widdale reclaimed using lime from Redshaw kiln

This rather gloomy picture has been emphasised by modern writers. Fieldhouse (1980, 192-93) expressed the view that evidence for widespread improvement in the 1840s in Carperby, Wensley and Coverham did not exist, according to tithe valuations, either by 'high or low farming'. Perry (1973, xi-xii) took a rather different view preferring to use the term crisis rather than depression for the closing decades. On the other hand, in his view, many farmers responded to crisis positively by innovating and modifying inefficient practices. Fletcher (1961, 37) believed that depression largely passed Lancashire by, as did Perry (1972) who also showed that the North and West Ridings fared better than southern counties.

Aristocratic landowners, such as the earl of Zetland at Aske Hall near Richmond, were instrumental in furthering progress and change in agriculture from the 1840s to the 1870s (Hall 1987, 39); and the role of external processes has been pointed up in recent literature (Wilmot 1990, 8-30). Greater scientific understanding, training for land managers, a more professional approach to farming, and more targeted agricultural writings, have all been flagged up for turning farming round in the 1840s.

Apart from Perry, few modern writers have seriously interrogated archival sources to determine regional contrasts in the state of farming, and whether or not the various recessions extended nationwide. Neither has any specific published analysis concentrated on chronological patterns of liming after 1815, or sought to identify correlations between the rate of liming and the state of the economy.

Contemporary writers provide some guidance here. Falkner (1843, 204) concluded that lime was still being extensively applied, especially so in Scotland and the North; and liming was still dominant in Westmorland prior to the High Farming era (Mannex 1849, 47) in contrast to a rather subjective view that lime was 'far more used by the farmers than at present' in the immediate post-war years (Rothwell 1850, 89-91).

The direct link between external forces and liming locally was picked up, for example by John Watson, a land surveyor in Westmorland, who maintained in 1845 that Prime Minister Peel's siren call to produce more 'in the shortest time' while still managing land in a sound manner had led to a surge in improvement, including liming (CRO [K] WDB/35). Another Kendal-based surveyor concurred with Rothwell's rather than Watson's conclusion, describing some parts of Westmorland as a 'monument of folly' because marginal land dragged into production during the war years was by mid-century 'left to Dame Nature' (Webster 1868, 7). Active management had ceased and natural vegetation was beginning to reassert itself, though he did make the point that land that had been properly drained and limed still maintained its old value

What of archival evidence for the period under review? No decade is unrepresented in the documentary record and most of the research area is encompassed with the exception of north-east Lancashire: after Charles Towneley's death in 1805 few extant estate records are available here. The correlation between decade and area is summarised in Table 8.3. It should be noted that gaps shown in the table could be due to gaps in the documentary record or to actual gaps in usage during those decades.

	1815- 19	1820s	1830s	1840s	1850s	1860s	1870s	1880s	1890s
Westmorland	Y		Y	Y	Y				
Swaledale	Y	Y	Y	Y	Y	Y	Y	Y	
Wensleydale			Y	Y	Y	Y	Y	Y	Y
Lunesdale	Y	Y	Y	Y					
Craven	Y	Y	Y	Y	Y	Y	Y	Y	Y

Table 8.3 Documentary references to liming, 1815 – 1900

Tenancy clauses and liming allowances still played an important part in ensuring that tenants carried out the wishes of their landowner in husbandry. This has been shown to have applied to draining, continuously from the 1820s to the 1880s, on large landed estates (*eg* Bolton Abbey and Thorp Perrow) and smaller estates (*eg* Constable Burton and Snaizeholme). It has also been demonstrated for liming, again on estates large (*eg* Lord Wharncliffe's) and small (*eg* Hornby Castle, Constable Burton and Ingleborough), and on individual farms (*eg* Barbon, Eastby, Horton in Ribblesdale and Summer Lodge) throughout the century. The idea of improvement did not wither during times of stress and there is sufficient evidence to confirm that liming remained a key element during the so-called Great Depression in farming that started so suddenly during the 1870s. The Garths, the Farrers, the Wyvills and Lord Wharncliffe all maintained their improving zeal well into the depression, and these four examples include the individual farmer, the small (gentry) and the large (aristocratic) estate.

Within the research area drainage regimes were maintained and extended throughout the century, and new (cheaper) soil inputs were introduced, yet lime still continued to play a prominent role alongside them, into the following century, rather than being eclipsed by them.

Despite clear evidence that marginal farms were abandoned late-century, what has been presented above belies the oft-portrayed picture of widespread gloom and despondency in farming, at least within the Central Pennines and in Lancashire.[70] Though few records specifically mention the process equivalent to today's rationalisation, reorganisation and consolidation were going on. For example, the Farrers bought up a number of discrete small, marginal farming units around Ribblehead in the 1850s and very soon they were merged and the remoter farmsteads abandoned (Johnson 2000, 7-9). The consolidated units were subjected to a determined effort to improve productivity and profitability, continuing into the following century. This estate's owners and stewards accepted that the pressures were profound but they adapted and innovated rather than succumb to the challenges they faced. Unfortunately, the primary evidence does not exist to confirm if this was the response elsewhere in the general area.

[70] Two branches of this writer's family gave up arable farming in Nottinghamshire in the mid-1890s and moved to farms in Lancashire, suggesting perhaps that they saw more of a future in the latter county.

Chapter 9
Synthesis and Conclusion

1. Introduction

This study began with the contention that agricultural lime burning within the Central Pennines and contiguous areas has been misunderstood and occasionally misrepresented by previous writers. It has raised a number of pertinent issues, each of which formed the core of a different research strand:

1. Hitherto it has generally been accepted that agricultural lime burning was largely a phenomenon of the parliamentary enclosure era, roughly between 1750 and 1850, and that very little liming had been done in the research area prior to about 1700, or later than 1850. This research set out to extend significantly the length of time over which land was limed.
2. Similarly, received knowledge has suggested that liming was mainly associated with winning productive land from uplands, especially given that the majority of enclosure awards embraced such 'wastes'. The corollary of this must be that correspondingly little lime was applied in dale bottoms and in the lowland fringes of the Central Pennines. This research set out to investigate the possibility that such areas had indeed been improved by liming, but at an earlier date, thereby extending the spatial as well as the chronological extent of liming.
3. Furthermore, the notion that most lime was burned on or close to outcrops of limestone bedrock, from which it was quarried, was also challenged. If that hypothesis were to be accepted at face value, it would presumably follow that most of the lime was applied on or adjacent to limestone pastures. In this respect it has been shown that, within the Central Pennines and along the northern edge of the South Pennines, lime was frequently burned far from limestone bedrock either by importing raw stone to where it was to be burned or by sourcing the raw material by rather arcane, and formerly little known, localised methods.
4. Many contemporary, and some modern, observers were inclined to dismiss the role of the small tenant or freehold farmer in soil improvement, believing it was the aristocracy and gentry who possessed the necessary intelligence, drive and financial wherewithal. This research took the view that this was too simplistic a picture and sought evidence to indicate that all tiers in the agricultural hierarchy had a view on improvement.

This chapter draws together the various research strands and summarises the different forms of evidence to support the four contentions put forward. The evidence has been culled from archival searches of manorial, estate and farm records, from contemporary written accounts (treatises, journal papers and diaries), by extensive field surveying across the entire extent of the research area and, very importantly, by targeted archaeological excavation of early post-medieval lime kilns under the direction of this writer.

It then goes on to relate the key findings to pre-existing research, published or otherwise, with the aim of either confirming or questioning the validity of such interpretations, and identifies areas where further archival, field and archaeological research might corroborate, or contradict, the conclusions drawn here.

The chosen time frame encompassed six centuries, from before 1300 to the end of the nineteenth century. It was felt appropriate to begin the process in the medieval period because that is when documentary sources indicate the first real attempts, dominantly by monastic houses, to improve existing farmland and to extend the cultivated area by piecemeal assarting and enclosure of common land. It was felt important to determine if evidence of liming could be isolated within the medieval era. The end point was set after more than two decades of severe depression in the agricultural sector when upland farming, at least, was thought to have been reduced to a parlous state with abandonment of marginal land and farmsteads, consolidation of smaller unviable units, and a drying up of investment in improvement. The research sought evidence of continued improvement involving liming throughout the period.

In the same vein the spatial extent was determined, firstly, because it is not valid to separate upland pastures within the Central Pennines from the lowland fringes as land ownership extended across both and, secondly, because partial geographical coverage would have presented an incomplete picture. Agricultural liming was practised across the entire area but there were significant differences in the detail of how stone was sourced, who was behind liming, when it was at the forefront of improvement, and how significant the practice was locally.

It has been posited elsewhere that much published landscape history has lacked both spatial breadth and temporal depth (Whyte 2002, 17). This research programme has maintained breadth and depth and the choice of both time frame and spatial extent has been justified by the results, with each research aim having been achieved.

2. Time frame

It would be axiomatic to say that the quantity of available written records increases through time: the further back one looks, in broad terms fewer records were created, and the chance of their having survived is much reduced. It therefore follows that this summary may well be partial and biased as it can only be based on extant and accessible sources.

There is sufficient archival evidence to state with a degree of confidence that agricultural change and improvement were underway, across much of the research area, during the thirteenth and fourteenth centuries on monastic and aristocratic estates. Assarting, piecemeal enclosure of common land, gradual conversion from arable to pastoral farming, and the rise of a more progressive and aware tenantry, were all phenomena of the high Middle Ages (Fryde 1996; M Johnson 1996; Higham 2004). Circumstantial evidence, such as references to kilns from contemporary written sources and monastic documents, suggests that these improvements within the area almost certainly included a growing realisation that soil potential could be increased by draining and by the addition of dung, marl or lime. In areas where marl was a common resource and limestone scarce or absent, such as lowland Lancashire and Cheshire (and Norfolk), marl ranked as highly as a soil input as lime did within or accessible to limestone-rich areas such as the Central Pennines.

There are also specific references within monastic cartularies and *compoti* to lime kilns – in Otley during the reign of Edward II, in Caton in the mid-thirteenth century, in Kilnsey in the mid-fifteenth, and in Threshfield in 1517. The references are brief, often just as boundary markers, but their isolated and field-based location does give credence to the view that they *may* have been producing agricultural lime, even if some of the lime was destined for other purposes.

Extant sources suggest that during the later sixteenth and through the early seventeenth century liming came more to the fore, confirming processes recognised elsewhere in the country by Kerridge (1967, 248). Writers such as Outhwaite (1986) and Smith (1970) noted sixteenth-century improvements and encroachments within the research area but neither was able to make any reference to liming. It is this research that has highlighted lime production north of Richmond, well documented from the 1570s; while manor court proceedings in Newby, Middleham and Cononley, the Giggleswick bylaws of 1602, and entries in late-sixteenth-century wills for Tosside, Wharfedale and Newby testify to the probable application of agricultural lime before 1600. Liming is documented in adjacent Bowland from the mid-sixteenth century and much of the lime (or limestone) used there could have been brought in from Craven, as well as from Chatburn near Clitheroe, or sourced from localised outcrops as with the excavated medieval clamp kiln at Halsteads.

The earliest categorical evidence of liming within the research area, however, is from a manor court ruling in Lothersdale about 1600 which formalised the pre-existing use of limestone for 'manuering' the land. This confirms that liming was an established practice in Craven prior to 1600 and it is surely inconceivable that liming was unknown elsewhere in the region. Indeed, documentary sources have pointed up the practice of liming on the Shuttleworth estates in north-east Lancashire from at least 1605.

Two reliable contemporary writers corroborate this hypothesis. Camden described the large-scale exploitation of limestone and its transportation to Airedale for application on the land. Though first published in 1586, he started working on his opus, *Britannia*, in 1577 having returned from his extensive travels two years earlier: Camden wrote of a practice that was well established long before the Lothersdale court made its ruling. John Speed, Camden's close contemporary, first published his map of Yorkshire in 1610 and the accompanying text has a telling comment in the margin, in Latin, of direct relevance to the research area, that 'lime gladdens the fields'. This, too, is unequivocal evidence of liming before 1610 on the eastern margins of the Central Pennines, as there is no indication in the text or margin that the practice was new.

After the Restoration in 1660 there was a new-found emphasis on scientific and aesthetic considerations in the nation; the Royal Society was granted its royal charter two years later and it spawned the Georgic movement which had at its heart the ideal of improvement (Low 1985); and landowners large and small felt able to undertake large-scale investment in land and property, with an aesthetic, moral and economic basis. One of the consequences of this 'new age' of improvement was a surge in the use of lime as a soil input, within the Central Pennines and more widely.

There is detailed and widespread archival testimony to liming within the research area after 1660. Indeed, by then it was the accepted method of improving soil, here as elsewhere (Dilley 1991; Wrigley 1993). Landowners from the gentry and aristocracy ensured that their tenants limed their fields by inserting clauses into new agreements, or by granting allowances against rents for lime purchased and applied. This was the established practice on the estates of the earl of Burlington at Bolton Abbey, Sir William Chaytor at Stainton, Smales father and son at Gilling, Christopher Crowe (the acknowledged Improver) at Kipling, William Danby at Swinton Park, the Scropes at Danby Hall, and the Improver Charles Towneley at Towneley Hall, with estate documentation variously covering the period from the 1690s to the end of the eighteenth century. In addition, Sir John Lowther was actively enclosing and liming in Ravenstonedale around 1669.

The likelihood is that neighbouring small landowners and freehold farmers gradually came to see the economic benefits brought about by liming and followed suit. The distance-decay model can arguably be applied with new ideas permeating outwards from demesne land to peripheral areas (Walton 1978, 258), from large estates to smaller farms. Certainly Joseph Watkinson's submission to the Royal Society confirms the take-up of liming in Lower Wharfedale, as do post-mortem inventories for that area after 1690, and manor court rulings in North Craven (from 1664 to 1748) and Burley in Wharfedale (1667).

Two generations of Middletons were leasing rights of stonary in the Wharfe in 1620 and 1678, while the Towneleys had been central to the possibly unique ways of hushing for limestone cobbles in north-east Lancashire from at least 1548 right through into the later eighteenth century. Such arcane methods, along with the digging of limestone boulder pits, are testament to the emphasis placed on liming in these areas from a relatively early date.

This writer's archaeological excavations of early clamp, or sod, kilns within Craven add weight to the hypothesis being put forward here that liming was in widespread use across the research area well before the era of parliamentary enclosure. The excavated sod kiln above Newby Cote, near Clapham, proved a calibrated radiocarbon date range between 1440 and 1640, while the one at Feizor, west of Settle, provided contemporary calibrated dates between 1470 and 1660. The date range of 1205-80 for the remote site just across the Craven border in the Forest of Bowland suggests that liming was practised here at that time.

A further strand of evidence lends credence to this view. Estate and mining maps that pre-dated the enclosure period frequently included mention of lime kiln field names. In addition, Smith (1961) recorded similar place-names. Many of these fields were not marked with kilns on First Edition six inch mapping, or else were depicted as 'Old Limekilns', and they have left no trace on the ground. It has to be assumed that such fields did once have kilns and that they had disappeared, or gone out of use, before six inch mapping began. To have been totally expunged from the landscape, without even leaving the characteristic earthwork profile, suggests these kilns were considerably older than the ones of the same type that still stand today.

There can be no argument against the contention that liming rates increased exponentially during the main period of enclosure (Havinden 1974; Dent 1983; Whyte 1999). A note of caution should be sounded, however, as the greater volume of evidence suggesting such an increase could be reflecting a greater survival of documentary confirmation rather than actual increased usage of lime.

Attention was drawn earlier to a range of external stimuli that contributed to this situation, despite the economic downturn following peace in 1815. It has been shown that there was a strong correlation between enclosure and liming, with almost seventy per cent of all areas enclosed within the research area having possibly been associated with the practice. Fourteen per cent of awards made direct provision for lime burning for agricultural use, and a further quarter made no specific provision but lime kilns have been identified on the enclosed ground from field surveying. The balance is made up of awards which provided for the burning of lime without actually stipulating agricultural usage.

Three peaks of liming were isolated from examination of enclosure awards: 1760 to 1780 with twenty-four per cent of all lime-specific awards, 1801 to 1820 with an identical percentage, and 1841 to 1860 with thirty-three per cent. The first grouping can be equated with a period when there was a concerted drive nationwide to improve and to enclose land from the waste; the second largely corresponded to political unrest in Europe and constraints (and opportunities) generated by the French wars; while the third grouping encompassed the mid-century High Farming era, with just temporary turmoil caused by Repeal of the Corn Laws in 1846.

Documentary references to liming have been located for each sub-region within the overall research area for the entire nineteenth century: in Swaledale it continued into the 1880s, in Wensleydale to the 1890s, and in Craven beyond the end of the century. In each case liming was still being undertaken long after the process of parliamentary enclosure had come to an end and, more significantly perhaps, through the decades after 1870 when farming was mired in deep depression. In addition, in every sub-region liming was promoted during the 1830s, another time of rural and agricultural depression and distress.

Landowners large and small had maintained policies during the eighteenth and nineteenth centuries to ensure that land was treated with lime, namely tenancy clauses and rental allowances, and these have also been identified for every decade of the nineteenth century. Detailed records have demonstrated the widespread and continuous use of, and faith in, agricultural lime by successive generations of Garths in Upper Swaledale and Farrers on the Ingleborough Estate. In spite of the introduction of cheaper and more 'modern' soil additives, like guano, crushed limestone and chemical substitutes, from mid-century, lime still figured large in the farmer's panoply of strategies to wrest a livelihood from the Central Pennines, and remained so beyond 1900. This is contrary to conclusions drawn by other writers (see, for example, Havinden 1974; Fieldhouse 1980).

3. Spatial extent

Because parliamentary enclosure mainly affected upland commons and stinted pastures within the Central Pennines, and because enclosure had such a profound impact on the scale of liming, it is only a short step to assume that liming as a phenomenon of improvement was largely confined to the uplands (Dent 1983). As has become patently evident during this research programme, improvements were made in the valley bottoms within the Central Pennines and in the lowland fringes bordering the uplands. The Swale-Ure Washlands downstream of Richmond and Middleham respectively, the broad Wenning Valley in Craven, Lunesdale south of Sedbergh, Mallerstang and lower Ravenstonedale in Westmorland, Lower Airedale and Wharfedale, and the Brun valley in Lancashire, all show signs of improvement having been carried out to the full. Contemporary comments by travellers and diarists indicate that this process had largely been completed – or was at least well underway – by the late sixteenth century.

The evidence to confirm that these areas were indeed improved, with lime having been given a major role, derives from several sources. Estate accounts and records provide limited evidence prior to the eighteenth century simply because they have either not survived or were perfunctory in nature, so documentary sources cannot provide the answer. What has been of vital importance has been gleaned through fieldwork.

This writer's field surveying of lime kiln sites across the entire area – kilns extant and those now long since gone – has demonstrated that lime was burned in the lowlands, in field situations, and that it has been possible in many cases to (tentatively) link particular kiln sites to discrete areas of improvement. Of the 1500 or so kiln sites surveyed on the ground, less than a half (forty-two per cent) are situated above 300m, with only 7.2 per cent above 460m; one in five sites are located in valley bottoms or on the lowlands. The full picture of kiln distribution, the sheer number that once existed and the part they played in shaping the modern landscape have only become apparent as a result of this field surveying.

In addition, the series of dated archaeological excavations of early clamp kilns across Craven, and the concomitant plotting of over 100 clamp kiln earthworks, linked with examination of manor court records and post-mortem inventories, has added to the strength of the argument being proffered. It has already been noted that the findings of these excavations, in terms of dating, have been of exceptional importance. Liming was a feature of the lowlands and was in no way confined to, or concentrated in, the uplands.

4. Liming and geology

There has been a common, and understandable, misconception that lime in the Central Pennines was normally burned at or close to the source of raw material, in other words where limestone is the bedrock (Davies-Shiel 1972, 97; Raistrick 1972; Wright 1986; Williamson 2002, 122).[71] One could assume that the lime so produced was applied not too far from the kiln, quicklime being a caustic and unstable substance difficult to transport. Given that limestone tends to outcrop in the uplands, this would bring us back to the conception that liming was concentrated in those areas.

Field surveying by this writer has pointed up a very different reality, at least across the North and West Riding and north-east Lancashire sections of the research area. Whereas the majority of kilns were sited where limestone was accessible, either by quarrying into hillsides or scars, by working quarry holes or by picking stone from natural scree slopes, a significant proportion (twenty-one per cent) of Yorkshire kilns were sited far from any source of limestone. These kilns burned stone transported from elsewhere, sometimes over considerable distances, or sourced it locally in other ways. The negative correlation between limestone geology and the incidence of lime burning/agricultural liming across large parts of the research area is strong.

Ongoing fieldwork in Westmorland, beyond the area encompassed by the current study, is so far showing a rather different picture though. Thus far, forty-nine parishes have been surveyed on the ground, yielding 511 kiln sites, and only around ten per cent were *not* sited on limestone. This can be readily explained, however, as those parts of Westmorland already surveyed have a proportionately greater surface area where limestone is the bedrock than the corresponding areas within Yorkshire where much of the limestone is buried by later deposits of sandstones, gritstone or glacial till.

Within parts of Yorkshire, especially but not exclusively in the West Riding, the stone was obtained from a relatively early date by what can only be considered innovative methods. Cobbles were picked from gravel spreads within river beds, as on the Wharfe, Cowling Beck, and tributaries of the Brun in north-east Lancashire; and from boulder pits dug into layers of till on moorland interfluves between the Lower Aire, the Lower Wharfe and the Washburn Valley. Perhaps much more enterprising was the development of hushing, a process normally associated with North Pennine lead mining, along the fringe of the Pennine moors in north-east Lancashire and adjacent parts of the two Yorkshires. Heads of water were built up, or an intricate system of goits fed water to hushing grounds – the many 'scars' – to flush out and loosen limestone cobbles from glacial deposits.

Each of these methods is supported by archival evidence, and together they emphasise the importance attached to liming as a means of transforming highly acidic pastures into (marginally) productive farmland in areas where geological logic might suggest otherwise.

In a different vein the existence of coal resources, admittedly of generally low quality but perfectly adequate for burning in lime kilns, was also a significant geological factor impacting on the take up of liming. The process was delayed in south-east Westmorland where coal supplies were hard to come by, and thus very expensive, whereas it is possible to recognise a symbiotic relationship between early liming and coal resources across much of the research area, particularly in Craven and north-east Lancashire.

5. Prime movers

If one takes at face value what contemporary commentators like Arthur Young or John Tuke wrote, one would assume that liming, and improvement in general, were the preserve of the landowning classes, and that the tenantry were by and large incapable of initiating it. They were dismissed as lazy, slovenly, or incapable. They were said to lack the education, to be hidebound by conservative ideas, and to be so often too old to accept that new did not necessarily equate to bad. It has been assumed, perhaps unfairly, that tenants merely did as they were bid by their masters.

The hapless William Chaytor, languishing in the Fleet Prison, did have endless trouble from two of his tenants for over twenty years from the 1690s; the steward of Colonel Wood of Middleham went on record as describing one of his tenants (Ryder) as 'slovenly'; and Lord Wharncliffe's agent met some resistance in Cotterdale, but such instances are rare in the documentary record.

There is no doubt that landowners, and stewards acting on their behalf, were instrumental in furthering the theory and

[71] Even limestone-based soils (rendzinas) in the uplands need liming as prevailing cool and wet climatic conditions render them slightly acidic.

practice of improvement and in ensuring that land was limed at specified intervals and in specified quantities (Mingay 1975; Whyte 2000; Wade Martins 2009, xi). As we have seen, several landowners within the research area have gone down in agricultural history as 'Improvers', those who made what was perceived in their own time to be outstanding contributions to the aesthetic and practical renewal of their estates, to the benefit of themselves, their tenants and the nation. Charles Towneley of Towneley and Christopher Crowe of Kiplin were accorded this epithet though others could well have sat alongside them: Willam Danby of Swinton Park, Sir John Lowther of Lowther, Simon Scrope of Danby Hall, Thomas Elliot of Fremington, were all intent on improvement. All put their faith in liming – along with draining and, very often, paring and burning.

Opposition to liming from tenants may have derived, not really from their misconceptions about its value but because stewards often required liming to be carried out too frequently and at too high a dosage. The chemistry of lime was not understood until well into the nineteenth century, and even today lime is perceived by some to be a fertiliser in its own right rather than a catalyst. Excessive applications did lead to crops being 'burned' by lime's caustic properties: tenants blamed the lime *per se* yet it was the rate of application that was at fault. As determined by exhaustive experimentation, rates above 4-7 tonnes per hectare on pastures based on peaty soils, and 2-6 tonnes on clays and loams, depending on pH, are now considered counter-productive and harmful to plant growth.[72]

Set against this notion, of landowners and stewards dictating to tenants, is an alternative scenario, one supported by archival evidence. It has been shown that many minor estate owners and 'ordinary' farmers across the region were aware of the benefits of liming and were instrumental in using it, correlating with conclusions reached by Macdonald (1975). The Garths of Upper Swaledale are arguably the most visible in this respect because they maintained a day book for 116 years but they were far from alone, as has been shown. Among the minor estates promoting liming can be counted those at Middleham and Braithwaite, Ingmire Hall near Sedbergh, Marrrick and Hurst, Hornby Castle, and the Ingleborough Estate.

What of the tenants, though? It is unlikely that any tenant had the wherewithal to introduce liming for the first time in his locality and the chances are that the distance-decay model can indeed be profitably applied here (Walton 1978). Supposing that improvement had not been imposed from above, new ideas would inevitably have slowly filtered down through the farming hierarchy, being disseminated by word of mouth, by example from neighbours, by regional and local agricultural societies, and through discussion at stock markets. Change in tenants' attitudes was also induced by example of their peers (Horn 1982, 316). Several instances, admittedly mostly from the eastern fringes of the Central Pennines, will illustrate the point in chronological order.

The heroic efforts of tenant James Croft in Cotterdale in the 1750s and 1760s have been discussed at depth and it is encouraging to read that both the Swinton Park steward and Arthur Young gave due recognition to his improving zeal and, significantly, to the influence he had on his neighbours large and small who saw what a marked difference had been achieved on his new enclosures. It was Young who was quick to slate tenants for lacking drive and commitment, so this was praise indeed.

Colonel Wood, of Middleham, also enjoyed the benefits of committed tenants, if the record of 1795 was in any way typical of their attitude. Several of them had expressed a 'strong desire' to lime and one had even offered to do all he could to encourage his fellow tenants to do likewise. Though Wood had had problems with Ryder, a sitting tenant, he was probably an exception. Similar behaviour has been identified on the Aske Hall estate in 1814 when one progressive tenant, Richard Pearson, proved willing to take the liming message to his peers (NYCRO ZNK X 2/1/2007); and, just after the end of war in 1815, tenant Jonathan Umpleby voluntarily took up liming on his farm in Eastby and even felt himself able to 'impose' conditions on his landlord.

In the Cotterdale properties of the earl of Wharncliffe, in the depressed 1880s, the tenants faced seemingly insuperable problems. Initially reluctant to lime, through financial distress, they were soon brought round and said to be 'delighted' with the situation as they understood perfectly well that their pastures desperately needed lime. Once a single and vociferous troublemaker had been pushed out of harm's way, the rest did what they knew was right – with a little prodding from the land agent.

There is rather a dearth of incontrovertible evidence to support this statement but the notion that tenants were merely there to do their landlords' bidding, that they were unthinking and non-rational receptors of innovation, must be discarded. In some cases, as shown by this research, tenants and smaller players were innovators and proponents of improvement.

6. Context

Having stressed the scale and nature (and dominance) of liming in the Central Pennines, over many centuries, as a soil additive and as a means of agricultural improvement, this might be the juncture at which to stress that lime was but one facet in the panoply of methods utilised by land managers and farmers. Marling was dominant long before the value of liming was fully appreciated and in areas such as north Cheshire and south Lancashire marl remained paramount into the modern era. Liming alone was often not the answer: draining, followed by paring and burning, often preceeded the application of lime. Indeed, it would be an exaggeration to suggest that every improved tract of ground had been limed. In many instances draining followed by paring and burning were deemed sufficient by the farmers concerned. Within the last three years, one large upland field above Settle has been transformed from a soggy and unproductive carpet of soft rush into a

[72] Searle (1935, 535, 553) considered optimum applications of lime to be 2.5t/ha on slightly acidic soils and 5t/ha on very acidic soils; Gardner and Garner (1953) advised 0.6-5.6 and 7.4-11 respectively.

pasture that is now able to support a flock of Texel sheep. This was achieved through the repeated application of chemical inputs rather than lime. However, in the days before chemical inputs came on stream, it has to be doubted whether ground such as this could have been brought into intensive use without the addition of calcium carbonate. The Ribble floodplain, to the south of Settle, was systematically brought into more productive use during the nineteenth century starting with an enclosure award in 1799 which had drainage central to its enactments. It is highly doubtful if any of this particular land was ever limed.

Drainage was far from a relatively modern strategy. Examination of medieval and post-medieval manor court rolls across the research area would point up the high frequency with which tenants were arraigned for failure to 'scour their ditches'. The need to keep farmland dry was as important then as it is now.

Nevertheless, in the light of the findings of this research, it follows that much pre-existing analysis (such as it is) of liming as a land improvement strategy within the Central Pennines needs modification and re-interpretation. This research suggests that liming had a much greater spatial and temporal extent than previously appreciated. The former understanding was of liming having been largely a phenomenon of parliamentary enclosure and of upland tracts, and on land where limestone was readily available, with the clear caveat though that some writers on agricultural change did recognise that liming was well established before 1600 in the areas they had focussed their research on. Kerridge (1967) and Porter (1977) are two such commentators. Within and around the Central Pennines, however, the view that liming was most prominent during the era of parliamentary enclosure is not really the case, as has been demonstrated. The notions that lime kilns were low status features, and that liming made little contribution to the national economy (Taylor 2000; Leach 2002), are also in need of revision. Exactly at what point in history agricultural lime entered the market economy is difficult to quantify: extensification of production could be ascribed to the coming of canals, and perhaps more so of railways, as distribution costs were significantly reduced, but this would be to ignore the network of packhorse routes that enabled lime to be led from production centres in the limestone uplands to those parts of the gritstone Pennines where agricultural and building lime were in ever-increasing demand. Lime has been an important and integral part of the Pennine (and wider) economy since at least the very beginning of the fourteenth century. Commercial lime burning has clear origins in the medieval period.

To extend knowledge and understanding of liming historically, and to appreciate more fully past improvement policies and practices, further research elsewhere is required. This would show whether the situation within this research area was unique or typical of improvement across upland Britain. Detailed archival research into estate and farm records in other regions would highlight who were the prime movers and what motivated landlord, steward and tenant to invest in improvement. It would also illustrate the causal links between improvement, tenancy clauses and rents on the one hand and between improvement and profit levels on the other.

Across Britain, in broad terms, lime was burned wherever limestone or chalk occurred as for example in the High Peak, on the Gower, in the Lothians, on the Downs, and widely across Westmorland. No similar research has been carried out in these areas, save this writer's current work in Westmorland. This study has re-evaluated our knowledge of agricultural improvement through liming in the Central Pennines: there is a need to extend the research methods employed here to other parts of the country, whether limestone-based or otherwise.

Like any body of research, this study established clear spatial and temporal boundaries and made use of a specific but broad range of methods. It is not claimed that its results are necessarily applicable elsewhere. For instance, field research by this writer across Westmorland is beginning to show that there is a clear and direct relationship between lime kilns and quarries, much more so than in the Central Pennines. In Westmorland landed estates were not as common as in the North Riding, the Lowthers and earl of Thanet excepted, so the same patterns and processes are unlikely to be replicated. The specific geological and geomorphological conditions that provided opportunities for hushing and river stonary may not be found elsewhere, so the fact that lime could be applied within the non-limestone parts of the current research area may not be found in other areas. On the other hand, the discovery and excavation of the clamp kiln in Bowland in August 2009 lends support to this writer's belief that such early kilns have a very broad spatial distribution.

At times the source material used has not proved sufficient to enable firm conclusions to be drawn, particularly for the earlier centuries reviewed. In some instances the evidence has been circumstantial and partial and the links between documentary kiln references and liming somewhat tenuous. Nevertheless, given the added benefit of extensive field surveying and targeted archaeological excavation, it has been deemed appropriate to suggest such conclusions with the proviso that their consideration must be nuanced. This applies, for example, to the links made between references to valued lime and limestone in post-mortem inventories and liming, or to those between mention of isolated lime kilns in monastic cartularies and liming.

It may well be that a firmer picture of early improvement and liming could have been gleaned by greater recourse to manorial and monastic records in their original Latin form: the number and range consulted was limited, and the number of valid references found was small. A further, unused documentary source that may have offered an insight into the relationship between size of land holdings and potential for carrying out improvements is the Land Tax registers, first imposed in 1693.

It is also probable that the excavation element could usefully have been extended to clamp kiln earthworks beyond Craven as this would either have confirmed or contradicted the date range of such kilns, thereby strengthening the conclusions drawn from excavation. However, excavation is by definition invasive, it can be destructive, and should be limited in extent.

7. Revisiting landscape theory

It would be axiomatic to state that landscapes contain elements from earlier periods of occupation, that Muir's (1999) 'historical text' or Lake's (2007) 'complex tapestry' are realities, at least for rural areas such as the Central Pennines. Anyone with an attuned eye – and maybe a touch of imagination – can recognise within the Dales landscape successive layers of past activity and even a superficial viewing would draw out the straight lines and right-angles of parliamentary enclosure fields or the irregular and curvilinear walls of 'ancient' enclosures within the valleys and lowland fringes. The more attuned eye would note the differences in the style of walls: narrow, slightly battered and neatly coursed in the former; broad, irregular and founded on massive orthostats in the latter.

However, it could be argued that only the approach of the landscape historian/archaeologist can truly isolate the individual facets that were inherent in forming any given landscape, be it 'lost' field kilns, sub-surface clamp kilns or formerly improved fields now 'going back' to their former state, or indeed the locational factors behind the siting of individual kilns away from sources of limestone. If it is accepted that these skills are a pre-requisite for meaningful interpretation of landscape, then it must also be considered that Historic Landscape Characterisation (HLC) has little to offer: it is too one-dimensional, it is desk-based (so arguably ignores landscape) and does not provide a mechanism for exploring past landscapes despite the claims made by its proponents that it has time-depth. It does not.

HLC cannot pick out those facets of a landscape that give it a measure of real meaning from the perspective of those who created and modified it over successive generations. It can pick out dominant themes – landscapes of parliamentary enclosure, perhaps, or medieval cultivation terraces – but it cannot deal with facets that are only recognisable on the ground. In short, it eschews the empirical approach.

The debate within landscape studies, between the processualists and the post-processualists, the theoreticians and empiricists, was visited in Chapter 2.2. Certainly, much analysis of landscape is subjective and empirical but one has to question whether the precise details of liming and land improvement within the Central Pennines could have been explained by theoretical modelling or objective hypothesising. Could such an approach identify the many variables that came into play, be they physical, economic, social, aesthetic or purely personal? It is strongly argued here that it could not. This research has been firmly grounded in the muddy boots approach.

8. Revisiting the chronology of agricultural improvement

Chapter 2.6 offered a review of agricultural improvement from the monastic era to the nineteenth century. It began by reiterating the oft-spoken view that improvement was largely a phenomenon of the late eighteenth and early nineteenth centuries but went on to offer an alternative analysis, namely that the seeds of improvement can be identified within the medieval period. Economic specialisation, commodification, soil improvement and an increase in the area farmed were all features that the medieval husbandman was aware of, regardless of whether or not he was in a position to benefit from such changes. Even the climatic, demographic and economic calamities of the fourteenth century acted as the catalyst for far looking farmers to seize the opportunities presented, for example by engrossing and by adapting from arable-based cultivation to stock rearing. In these pioneering entrepreneurial farmers we can perhaps detect the embryonic beginnings of the sixteenth century's rise of the yeoman class.

After the Restoration, in 1660, farming reached new heights with a renewed vigour, a new scientific approach and a focus (at least among larger landowners) on aesthetics. Over the following century and a half farming and improvement became ever more inextricably linked, and were perceived as a moral imperative and a patriotic duty. The techniques of improvement – at first marling and dunging, later on liming tied in with draining, paring and burning – were widely taken up until the late nineteenth-century depression brought about a hiatus within the industry.

All these trends and processes can be recognised within the research area: so, too, can the hypothesis that changes through the period under review represent a gradual phased agricultural *evolution* rather than a time-specific agricultural *revolution*.

9. Final thoughts

This has been a broad-based study which has sought to identify common aspects of agricultural land improvement through liming, by empirical means, across the Central Pennines and across six centuries. At the same time, it has focussed on regional diversity in approaches to liming to highlight contrasts between the large landed estates of the North Riding with the small-scale efforts of those involved in hushing or those using lime from clamp kilns. In short, it has identified definite local diversity within common methods and approaches.

Much of the region has been improved at some point and lime played a central role in this transformation of the landscape, at least from the late sixteenth to the late nineteenth century. It drew in all levels of the agricultural hierarchy from the humblest uneducated tenant to the most aristocratic landowner. It extended across limestone areas and into those parts where limestone does not outcrop. Indeed, even as late as the 1960s more lime was applied in the Yorkshire Dales than in any other part of the county (Harwood Long 1969, 148).

Signs of this improvement are to be seen in every nook of the Central Pennines, either as clearly-improved fields contrasting starkly with neighbouring heather-clad or soft rush-dominated commons (Fig. 9.1), or by the ubiquity of lime kiln remains. For much of the twentieth century upland farming was in a state of crisis, and perhaps still is. All too common are the formerly improved enclosures, reverting to their natural state as rushes and coarse grasses spread (Fig. 9.2), and the decaying

Figure 9.1 Contrasts between limed and unlimed land in Arkengarthdale. There are remains of four lime kilns in this view, at Eskeleth

Figure 9.2 Formerly limed meadows reverting to their former state at Widdale head. A kiln used to stand in the left-hand field

Figure 9.3 A telling advertisement from 1929
Source: Craven Herald *26 July 1929*

ruins of neglected field kilns. Old drains have failed and the soil is depleted of calcium. As upland farming is marginal, the will and the wherewithal to invest have evaporated.

Landscapes such as those depicted in Figures 9.1 and 9.2 bear the clear hallmarks of past improvement and illustrate how very different agricultural fortunes were when the farmers concerned felt able to invest in their holdings. Both views include unenclosed land that was never improved (probably never could be without inordinate investment) as well as walled-in improved pasture or hay meadow. We can only surmise that the latter satisfactorily repaid the financial and labour outlay, and in Eskeleth those meadows are still productive, though for silage or haylage rather than hay. In upper Widdale, in contrast, very few meadows are still managed as such and certainly not the ones in Figure 9.2: one has reverted to damp soft rush and the other three are now down to pasture. Current and recent economic returns from marginal land such as this cannot begin to justify expenditure on liming, or on any other form of soil treatment. Instead, there has been a change to extensification in stocking.

The ageless couplet 'lime without manure makes both farm and farmer poor' has long been redundant because, in the poignant words of Corrie (1926, foreword),

> This old and oft repeated rhyme,
> Hath little point at present time,
> When farms are poor for want of lime.

Corrie's statement is as valid now as it was eighty years ago. So, too, is the advertisement from P W Spencer Ltd, quarry owners and limeburners at Lothersdale and Giggleswick Quarries in Craven, which could not have been clearer in stating the then obvious: grasslands were 'going back' (reverting) to their unimproved state with a commensurate decline in stock quality (Fig. 9.3). By the 1920s it was generally accepted that soils across the county were in desperate need of calcium input and considerable pressure was put on the government to do something (Johnson 2002, 172-73).[73]

In September 1937 the Land Fertility Scheme was introduced with a flat rate subsidy on agricultural lime of fifty per cent: applications flooded in at over 100 per day with an eventual ninety per cent take-up rate among Craven farmers (*Craven Herald* 12 November 1937). It was replaced with the very similar Agricultural Lime Scheme in 1947 which was given a life of five years. However, it was only finally scrapped in 1976, such was the need and demand for lime (MAFF 1981, 6). That effectively brought the use of agricultural lime in this country to an end. In spite of the application of chemical inputs over many decades before and since 1976, much of the Central Pennines is once again lime-deficient. In short, it is no longer as productive as it could be.

Despite the inevitability of a rising global population leading to increased food output wherever food can be produced, it

[73] In 1932 PW Spencer issued an advertisement with a cartoon showing a desperate farmer and a very skinny cow, with the caption 'On Craven Moor bah't LIME!!!'.

is highly unlikely that the once-improved marginal farmland of the Central Pennines will ever be brought back into fully productive use. European Union-inspired and funded agri-environment schemes have substantially reduced stocking levels (which have not recovered their pre-foot and mouth levels anyway) with an emphasis now on encouraging farmers to manage their land in a much less intensive way, without recourse to artificial soil inputs, and with wildlife in mind. It was also clear during the foot and mouth crisis of 2000-01 that tourism was on a par with farming in terms of income generation and employment within the Yorkshire Dales, at least. Perhaps the uplands have seen their best agriculturally and perhaps it is now proper to regard them as places of recreation and as wildlife havens. What the farmer regarded as 'improved' grassland, having a uniform sward of rye grass or similar species, is perceived very differently by environmentalists. Since 1990 English Nature, and from 2008 Natural England, have been actively turning rye grass fields back into flower-rich meadows at Colt Park in Ribblesdale in long-running field trials: these bodies did not classify the flower-rich plots as 'improved' and the aim was to rid this estate of what was until very recently deemed to be improved, as well as to garner experiential knowledge of how to restore 'improved' grassland to species-rich hay meadows (Smith *et al.* 2008).

In the lowland fringes of the Central Pennines, on the other hand, there is a very different situation. Here, climate, topography and soils favour arable and intensive pastoral farming as they have always done. Agri-business is dominant, certainly east of the uplands, and farming concerns are able to afford the inputs necessary to maintain soil fertility. Does lime still have a role here? Pulverised limestone dust is still spread, alongside scientifically-determined chemical additives. Indeed, liming is still perceived by some farmers to be a necessary soil treatment, albeit using limestone dust. A case in point is Daggerstones Farm, at Healaugh in Swaledale. A change of ownership in 2004, and entry into an environmental stewardship scheme, saw a determined effort to renovate the farm with particular emphasis on restoring traditional features and hay meadows. The current farmer's first action was to 'lime the whole farm in moderation' as it had not been limed for at least four decades (Morrogh-Ryan, 2010, 241).

Liming in the form of limestone dust still has a part to play; quicklime, however, has had its day as a soil input.

Appendix 1

References to Lime Kilns and Lime Kiln Field Names, not referred to elsewhere in the Text

Township	Detail	Date	Reference
Arkengarthdale	boundary survey	1764	NYCRO ZQA I. 6/1/1
Aske	township plan	1761	NYCRO ZNK. M1/3/90
Carperby	Bolton Estate plans	1810	NYCRO ZBO (M)
Danby	Danby Estate survey	1785	NYCRO ZPT 4/1/1
East Witton	Plan of E. Witton Bank	n.d.	NYCRO ZS
Feldom	field book	1786	NYCRO ZQH 11/14
Garsdale	Danny Bridge	n.d.	CRO (K) WD/Big/1211
Gilling	Dundas Estate map	1813	NYCRO ZNK. M1/6
Grinton	Map of Grinton Manor	1774	NYCRO ZRT.2/3
Keld	Clarkson's maps	1824	NYCRO ZCL
Langcliffe	Preston of Mearbeck	1801	NYCRO ZXC
Leyburn	Dundas Estate map	1822	NYCRO ZBO (M)
Middleham	estate surveys	1765	NYCRO ZS
Muker	Clarkson's maps	c. 1824	NYCRO ZCL
Preston-u-Scar	Dundas Estate map estate plan	1824 1869	NYCRO ZBO (M) NYCRO ZTS (M) 1/112
Richmond	Dundas Estate map estate plan, West Side	1813 1798	NYCRO ZNK. M1/6 NYCRO ZXF (M) 1/9/1
Snaizeholme	Estate plan	1877	NYCRO DC/RMB
Snape	Thorp Perrow estate	1727	NYCRO CRONT 1798
Spennithorne	estate plan estate plan	1821 1838	NYCRO ZQH 6/22 NYCRO ZQH 6/34
Thoralby	estate plan	1866	NYCRO TD159
Well	'Maps of Well' estate map	1845 n.d.	NYCRO ZM NYCRO ZMI 140

Appendix 2
Estates, not discussed in the Text, with Archival Evidence of a Direct Link between Lime-coal Production and Agricultural Lime Usage

Estate	Location and detail	Source
Danby Hall	Stainton 1775 farm lease with right to mine coal for an estate lime kiln	NYCRO ZPT 4
East Scrafton	West Witton 1718 lease of coal & lime 1853 sale of coal and lime rights as part of estate sale	NYCRO ZBO 5 NYCRO ZBO 1
Bolton Hall	West Witton 1718-19 sales of lime-coal 1854 lease to mine coal & produce agricultural lime	NYCRO ZBO C NYCRO ZBO 5
Wyvill	Coverdale 1832-37 lime-coal sales	NYCRO ZFW 4
Morley	Marrick 1870 lime-coal lease	NYCRO TD 79
Spittle farm	Stainmore 1726, 1754 & 1762 leases of farm with rights to coal & lime	NYCRO ZAW 138
Raikes farm	Threshfield 19th c farm with coal & lime rights	WYAS (B) 68D82
Trust Lords	Kettlewell 1866 lease of coal mine & kiln 1868 lease of coal mine & kiln	Brit. Spel. Soc. NYCRO ZKK II
Ribblesdale	Fountains Fell 1814-15 lime-coal sales	YAS MD335/1/4/5
Top farm	Horton in Ribb 1727 farm lease of coal rights and kiln	Horton Hist. Group
Dubcote	Horton in Ribb 1823 farm lease of coal rights and kiln	Horton Hist. Group
Langstone Fell	Garsdale 1797-1896 lease of coal mines & kilns	Sed. & Dist. Hist. Society
Hall Intack	Ravenstonedale 1779 lime-coal allowance	CRO (K) WDX/94
Hornby Castle	Hornby 1829-34 coal and lime kilns	Diary of Geo. Smith
Hornby Castle	Farleton 1582 estate coal mine	Chippindall 1939
Towneley Burnley	1716 coal mines	NA E 317/Lancs/8

Glossary of Technical Terms

boll a measure of capacity, used in Scotland and northern England, equal to 2-6 bushels (*q.v.*) or 2 cwt.

breasting paring off turf using a breast-plough which was a specialised spade with a long haft and a long cross-handle against which the chest was pressed.

Bushel a measure of capacity varying across the country. The standard Winchester bushel contained 8 gallons. In the 1860s in Westmorland 10 bushels equated to 1 tonne but with considerable variation, and the Carlisle, Chester and Kirkby Stephen bushels were different again. Tyson (1998, 81) suggested that one bushel could vary from 8 to 24 gallons even though the standard measure was 8, as set by an Act in 1543 (Statutes of the Realm III, 34-35 Henry VIII). An earlier statute set a Winchester bushel at 1072 cubic inches (7 Henry VII 1491).

calcination the process of driving carbon dioxide out of the limestone at temperatures in excess of 900 degrees, in a lime kiln.

carbonaceous containing large amounts of carbon.

chalder a measure of capacity generally equating to 92-96 bushels (*q.v.*).

chaldron a measure of weight varying across the country: 1321 or 1372kg in London, 2692kg in Newcastle-upon-Tyne after 1694 but 914kg prior to then. Often equated to 36 bushels (*q.v.*).

chowcher an archaic word referring to a person who packed stone and fuel into a kiln.

clamp an early form of lime kiln, consisting of a shallow bowl cut into a bank or slope, piled up with limestone and fuel, then rounded off like a flattened igloo, and sealed with turf or peat (*q.v.* sod and sow kiln).

compotus a regular set of accounts kept by a religious or manorial body.

coop a small cart with solid sides used for leading lime and dung.

corf/corve originally a basket used to haul coal up mine shafts. Later also a sled used to drag coal along adits. No fixed weight or volume, but 1cwt (50.8kg) approx.

delf a small and shallow quarry working.

fallen lime slaked lime.

fire-coal coal whose main end use was domestic.

harl roughcast plaster thrown onto external walls.

hushing the process of channelling water to flush out and loosen limestone cobbles from deposits of glacial till.

hydrated lime lime reduced to powder that has been slaked and then dried.

leading a term used in the Central Pennines, and elsewhere, for cartage of bulk products.

lime deficiency the extent to which a soil lacks the optimal proportion of calcium carbonate in its mineral content.

lime requirement the amount of lime required to raise soil pH from an acidic state to a level of that can sustain optimal growth.

lime status the state at which a soil has the requisite amount of lime.

lime-coal coal whose (almost) sole end use was in lime kilns.

limewash a milky mixture of slaked lime and water.

load a horse load equated to 100-150kg; a cart load to 200-650kg, depending on size of cart.

paring stripping off the turf layer on land being reclaimed so it can be burned with the resultant potash being worked into the soil.

quarter a measure of capacity equal to 8 bushels (*q.v.*) or 6 cwt (304.8kg).

quicklime lime resulting from calcination (*q.v.*). Calcium oxide. The same as burnt lime:
$CaCO_3$ (calcium carbonate) less H_2O (water) = CaO (calcium oxide).

slaked lime quicklime (*q.v.*) with water added either by hand in a trough or naturally by rain; transforms calcium oxide into calcium hydroxide:
CaO (calcium oxide) + water (H_2O) = $Ca(OH)_2$ (calcium hydroxide).

sea-coal a generic term for coal, originally used for coal mined on the coast of north-east England.

sod kiln a common vernacular term for a clamp kiln (*q.v.* sow kiln).

sow kiln term noted in Scotland and northern England for a clamp kiln (*q.v.* sod kiln).

unfallen lime unslaked lime.

waste a term used through much of the period under review, not referring to actual unused waste land but to land not deemed to be in productive use, thus described as a 'weasel word' (Tarlow 2007, 45).

whitewash a very liquid solution of lime and water. Also called whitening.

Bibliography

Abbreviations

CRO (C)	Cumbria Record Office, Carlisle.
CRO (K)	Cumbria Record Office, Kendal.
ChRO	Cheshire Record Office.
LRO	Lancashire Record Office.
NA	National Archives.
NMRS	Northern Mine Research Society.
NRO	Norfolk Record Office.
NYCRO	North Yorkshire County Record Office.
SDHS	Sedbergh and District History Society.
WYAS (B)	West Yorkshire Archive Service, Bradford.
WYAS (W)	West Yorkshire Archive Service, Wakefield.
YAS	Yorkshire Archaeological Society.

Primary Sources

State papers

National Archives

Calendar of the Charter Rolls preserved in the Public Record Office. Vol. 1, 14 Henry III (20 April 1230), p. 121; vol. 5, 31 Edward III (10 May 1357 and 28 December 1383), membrane 4 and 16; vol. 6, 22 Edward IV (10 February 1483), membrane 2.

Calendar of Close Rolls, 14 Henry III (April 1230); Edward I (1302-07), vol. 5, 35 Edward I (12 June and 28 June 1307), pp. 537 and 539.

Calendar of the *Liberate* Rolls. Vol. 1, 12 Henry III (24 April 1228), p. 78; vol.1, 13 Henry III (25 January 1229) p. 116; vol. 5, 45 Henry III (4 July 1261), p. 50.

Statutes at Large, vol. 1, *From Magna Carta to the 14th year of King Edward III, inclusive.* 9 Henry III, Cap. XII, 1224-25. Pickering, D (ed.), (Cambridge: Joseph Bentham), p.20.

Statutes of the Realm, 9 Henry III 1224-25, 20 Henry III 1235-36, vol.1, pp. 2-3; III 6 Henry VIII, 1514-15, 'An Act concerning artificers and labourers', vol. 4, chapter. 3, p. 124; III 34-35 Henry VIII, 1543, 'An Act for the assize of wood and coal', vol. 4, chapter 3, p. 899; 2 & 3 Edward VI, 1548, 'An Act towchinge victuallers and handycrafte men', vol. 4, chapter 15, p. 59; 3 & 4 Edward VI, 1549, 'An Acte concerninge the ymprovement of Comons and waste Groundes, vol. 4, part 1, p. 102-03; 5 Elizabeth, 1562-63, 'An Act touching divers orders for artificers, labourers, servants of husbandry and apprentises', vol. 4, chapter 4, p. 420.

D.L. Min. Acc. 18-19 Henry VI. Bdle. 89, No. 1634 (Duchy of Lancaster), Reparations of manor house at Ightenhill, 1440 and 1443.

British Parliamentary Papers

British Sessional Papers, House of Commons, vol. XL 1794/5, No. 796., Board of Agriculture's Committee on Draining, 5 June 1795.

Command Paper 464. Reports from Committees: 1837. Agriculture: 464. Report from the Select Committee of the House of Lords on the state of agriculture in *England* and *Wales*; with minutes of evidence, appendix and index, vol. 5. (Shannon: Irish University Press, 1968).

Command Paper 612. Agricultural Distress Report, 29 January – 29 August 1833. Report from the Select Committee on Agriculture; with the minutes of evidence, appendix and index, vol. 5 (Shannon: Irish University Press, 1968).

Archival Sources

Borthwick Institute for Archives

Borthwick wills, vol. 19, fol. 340. Thomas Bullock, 1570; vol. 21, fol. 5. William Browne, 1576; Bundle August 1647-48, William Hall, 1647.

Borthwick, Cause papers CP.G 2812, 2 February 1594.

Borthwick, Burley bills, CC.Ab.12. Archbishopric estates. Otley Manorial Papers 1662-95.

Borthwick, Prog. Exch. Probate inventories, Ainstie, original wills.

Bradford Museums, Galleries and Heritage (Cliffe Castle Museum)

Bradford, Glusburn Box 1, K.1.1, Glusburn charters, assignment 76, 25 April 1300.

British Museum Library

BL MS Add.40011A, *Liber propter seruos of Fountains Abbey, Memorandum Book of Thomas Swynton 1446 to 1458.*

Brit. Museum TY 1/7, Towneley Archive, Charles Towneley, Diaries, 1796 – 1804.

Brit. Museum TY 1/21, Towneley Archive, Accounts, 1770.

Brit. Museum TY 7/140, Towneley Archive, Agricultural Surveys, 1793.

Brit. Museum TY 7/1064, letter to Charles Towneley from Andrew Fountaine in London, 27 May 1796.

British Speleological Society

Yorkshire Records, Lease of Cam Head, 3 October 1866.

Cheshire Record Office

ChRO DCH/X/15/1, Scrope of Bolton cartulary.

Cumbria Record Office, Carlisle

CRO (C) D/Lons/L3/1/5, Sir John Lowther's Long Vellum Book

CRO (C) D/Lons/L12/1/39, A valuation of the estates in Westmorland, n.d. (1808).

Cumbria Record Office, Kendal

CRO (K) WDB/35, Box 16, An essay on the best method of reclaiming heath land by John Watson Junr, surveyor in Kendal, n.d. (1844).

CRO (K) WD/Hee, Heelis of Appleby Archive, farm account books, 1852-73, purchases.

CRO (K) WD/HH, Heelis and Heelis, solicitors of Appleby, Box 2, uncatalogued.

CRO (K) WD/Hoth/Box 26, uncatalogued, miscellaneous Westmorland leases.

CRO (K) WD/KILV/3/7, Indenture, 12 February 1579.

CRO (K) WD/MG, Metcalfe-Gibson archive, Ravenstonedale, lease, 16 November 1835.

CRO (K) WD/PP, Box 8, Underley estates: draft depositions, Deposition of James Windle of Barbon, 27 April 1843.

CRO (K) WD/Whelp/T53, lease, 19 September 1817; WD/Whelp/T57, lease, 6 September 1851.

CRO (K) WDX/94. acc. 463/63, Hall Intack Farm, 11 February 1779.

CRO (K) WDX/450/19/5, pocket and note books, Riggs of Applegarth, Middleton, Richard Rigg's book.

CRO (K) WDX/668, Waller family of Winton, Kirkby Stephen, lease for nine years, 9 June 1784.

CRO (K) WPR/9/Z11, Orton enclosure award map, 1779.

CRO (K) WQR/I/14, Casterton Moor enclosure award, 1816.

CRO (K) WQR/I/73, Orton enclosure award, 1779.

Devonshire Collection

Carr, steward to His Grace the Duke of Devonshire, by Henry Pickup, 1815-30. Bolton Abbey Archive, BAS/3/6, Ledger. Miscellaneous Book, kept at Bolton Abbey.

Chatsworth, Bolton Abbey, L/45/7, Cononley and Cowling Court Roll 1590; L/45/19, Cononley Court Roll 1606, 1617-18.

Chatsworth, Bolton Abbey, Clifford household accounts, Book 2, 1515-16; Miscellaneous cash disbursements: Bolton Abbey, 1754 – 1765.

Londesborough MSS, Box J, Craven Estates, x, tenancy agreements, 1730s; xiv, petition, 21 January 1725.

Hornby Castle Muniments

Burton manor court roll 1742-1806, 9 November 1748.

The diary of George Smith, vol. 1, January 1819 to December 1826; vol. 2, 1827-35.

The Honor and Manor of Hornby Lancashire, court roll, vol .2 (1), 1578 – 1588.

The manor of Tatham, Lancashire. Court book, vol. 1 1584 – 1664; vol. 2 1665 – 1711.

Horton Parish Council

Horton enclosure award, 1771.

Ironbridge Gorge Museum and Library

The Raistrick Collection, Box 8, RS 796.10, Petition to Sir Rauffe Sadler, 1576.

The Raistrick Collection, Box 13, RS 1026.11, Yoredale coals, n.d.; Box 14, RS 1057.25, n.d.

The Raistrick Collection, Box 13, RS 1039.1, hand-written notes.

The Raistrick Collection, Box 14, RS 1057.20, Inventory of William Hill of Cottingley, 13 July 1627.

The Raistrick Collection. Box 14, RS 1057.25, Agreement, 28 August 1622.

Keighley Local Studies Library

Keighley BK 79/157, Kenneth Wilson family papers, letter, 18 October 1819.

Keighley BK 79/205, Kenneth Wilson family papers, Raygill lime kiln, typescript notes.

Lancashire Record Office

LRO DDHCl 3/51, Court rolls, Colne and Trawden, Colne 9 December 1568.

LRO DDPa, Box 1, Parkinson of Hornby (uncat.), pamphlet: 'Thoughts on the design of making a navigable canal, from the vicinity of Kendal, to join some of the canals in the south parts of Lancashire, by way of Lancaster', 1791.

LRO DD Pa/Box 3 (uncatalogued), pamphlet: 'To the committee on ther intended canal from Kendal by way of Lancaster to the southern part of Lancashire', Lancaster: Samuel Gregson, 1792.

LRO DDTO, Box CC, uncatalogued, estate accounts, rentals 1781-87; cash book 1781-87.

LRO DDTO, Box II, uncatalogued, Charles Towneley's estate correspondence c. 1770 – 1800.

LRO DDTO, Box J, uncatalogued, 'A short account of Mr Elkington's drainage at Towneley made in the course of the last 14 months', September 1795.

LRO DDTO, Box OO, uncatalogued, Towneley correspondence 1792 – 1804.

LRO DDTO, Box UU, uncatalogued, drainage papers, Towneley rental 1712, letter; Extracts from the minutes of the proceedings of the Board of Agriculture concerning Mr Elkington's mode of draining, December 1794 (it should be dated July 1795); 'Produce of the meadows at Towneley in the years 1793 and 1796', 25 August 1796.

LRO DDX 70/34, Hornby Verdict Book 1732-53, 1 November 1748.

LRO DL 29/86/1603, Duchy of Lancaster, Ministers' Accounts 7 and 8 James I (1609-10).

LRO RCHY 2/2/128, Hornby Catholic Mission Collection, Indenture 28 March 1628.

LRO WCW, Inventory James Hartley of Wynewalle, yeoman, will proved 12 December 1615; Inventory John Ambler of Wycoller, husbandman, 14 July 1674.

Lancaster University, Rare Book Archive

Stan Lawrence Collection, Burton manor court roll, 1742 – 1806.

The National Archives

NA DL 5/27, Folio 1314, Duchy of Lancaster, Manor of Colne, terms and conditions, 20 June 1618.

NA DL29/86/1603, Duchy of Lancaster, Ministers' Accounts, 7 & 8 James I (1609-10).

NA E 317/Lancs/8/Parliamentary Surveys, Towneley coal mines, 1648.

NA E 320/J5, Duchy of Lancaster, Records of the Exchequer, Trustees for Crown lands and fee farm rents, Particulars for sale of estates of Charles I, Grant to Thomas Smith of London, clothier, 7 November 1632.

NA PRO IR30.18/349, Tithe apportionment map, Wray with Botton, 1849.

NA unreferenced, Settle, indenture 11 March 1850.

Norfolk Record Office

NRO WLS LXVI/210/4, Walsingham (Merton) Collection, Minutes of agreement with Blubberhouses tenants, n.d. (1790s).

North Craven Historical Research Group

Settle Banks, High Scarr, and Scaleber enclosure award, 1759.

North Yorkshire County Record Office, Northallerton

NYCRO Bolton MSS 437 MK/2.

NYCRO CRONT 1840, Richmond Coucher Book, 1589 – 1833.

NYCRO DC/RMB 2/1/1, Richmond Borough Council Coucher Book, 1590 – 1833.

NYCRO NRRD no.3, Masham enclosure award, 1797.

NYCRO NRRD AX. No. 54, Gayles enclosure award, 1774.

NYCRO OUTFAC, 'Account book of Thomas Smailes, Henry and William Chaytor', n.d.

NYCRO PC/BEL, Bellerby enclosure award, 1772.

NYCRO QDD/I no. 9, Bainbridge enclosure award, 1849.

NYCRO TD 31, Greenfield estate, Chamley family, 1803-92, bond of indemnity 12 May 1803; release and surrender, 13 February 1811, of Green Haw House and closes formerly occupied by Thomas Elliott (*sic*).

NYCRO TD 53, Muker enclosure award, 1843.

NYCRO TD 124, Low Abbotside, Yorescott farm accounts (Willis family), 1822-50.

NYCRO TD 125, Storey family of Castle Bolton: mason's quarrying accounts and papers, c.1818-58.

NYCRO Z 2, Covenant for building a lime kiln in Green Bank Pasture, Burtersett, 3 December 1857.

NYCRO Z 760, Muker papers relating to draining, liming, coals, 1834.

NYCRO Z 775, Middleham manor, transcripts of letters and papers, 1677 – 1856.

NYCRO ZAG 279, Swale Navigation papers, 1767-72.

NYCRO ZAL 3, Mark Milbank Esq. Rentals, 1876-83.

NYCRO ZAL 4/24, Snape and Well field drainage plans book, 1847-69.

NYCRO ZAL 4/44, Snape and Well field drainage plans (Milbank), 1881.

NYCRO ZAL 3/5, Snape and Well Rentals of Milbank Esq. of Thorp Perrow, 1855-67.

NYCRO ZAL 6/6/8, Milbank of Thorp Perrow and Barningham, Mary and John Milbank defendants v the Hon. Charles Cecil plaintiff.

NYCRO ZAW 31, Title deeds High and Low Burton, 1 March 1793.

NYCRO ZAZ (M) 1, Stainton Moor, estate map, 1694.

NYCRO ZBL IV 2/1/5, Kiplin Hall archive, leases with conditions, 22 January 1778; ZBL IV 2/1/82, 1836.

NYCRO ZBL IV/3 Kiplin Hall archive: Robert Hutton's steward's accounts to Christopher Crowe the Elder and C. Crowe the Younger, 1737 – 1761.

NYCRO ZBL IV/5, Kiplin Hall archive – estate surveys and valuations: early 17 century to 1915.

NYCRO ZBO. MD 1, Bolton MSS, Deeds and leases, indenture 14 September 1378. (In Norman French).

NYCRO ZBO 1, West Witton manor, 1853.

NYCRO ZBO IV.8, Bolton Estate Archive: estate papers, 29 July 1760; Observations concerning Lady Bolton's properties in 1809, 15 July drawn up by Wm Sadler.

NYCR0 ZBO 5 Y37, Bolton Archive, letters etc, Lord Bolton's mining ground etc, Memorandum of Agreement, 15 September 1854.

NYCRO ZFW, Wyvill of Constable Burton, title deeds, 1832-37; estate leases, Denton.

NYCRO ZFW 4.1, Wyvill of Constable Burton, Denton survey 1596, March.

NYCRO ZFW 4/2/7, Wyvill of Constable Burton, A scheme or course in husbandry, 20 October 1829.

NYCRO ZFW 4/2/13, Wyvill of Constable Burton, valuation of farms; ZFW 4/2/20, Denton estate survey.

NYCRO ZHG, Box 19, Documents relating to Richmond and Swaledale, 'A map of lands within East Witton in the North Riding of Yorkshire belonging to G. Elsley Esq. 1772'.

NYCRO, ZKK II, Trust Lords of Kettlewell, Agreement for use of lime kiln, 1868.

NYCRO ZNK X 2/1/2007, Letter to Lord Dundas from Richard Smithson, steward, 15 January 1814.

NYCRO ZPL 5, Purchas family archive of Braithwaite and West Burton, lords of the manor, 1750s, vouchers and accounts, 22 November 1759.

NYCRO ZPL 10, Marrick and Hurst estate papers, conditions for letting a farm, 8 February 1866.

NYCRO ZPT, Scrope family farming account book, 1730-87; ZPT 4/1/1, 13 September 1731; ZPT 4/1/5/5, 3 June 1732; ZPT 4/1/5/6, 8 May 1779; ZPT 4/1/14/10, 17 November 1771; ZPT 4/2/10, 14 December 1775; ZPT 4/4/3, Scrope of Danby, farm and household accounts.

NYCRO ZPT 4/1/4, Thornton Steward enclosure award, 1694.

NYCRO ZQA I.4.1, Barningham enclosure award, 1807.

NYCRO ZQH 2/2/4, Yorke MSS, conveyance, 11 June 1652.

NYCRO ZQH 2/2/124, Yorke MSS, conveyance, 7 September 1774.

NYCRO ZQH 2/2/165, Yorke MSS, conveyance, 11 May 1790.

NYCRO ZQH 7/46, Smales family, Swaledale estate rentals etc, 1742.

NYCRO ZQH 7/17, Smales papers, letter from Matthew Smales to Jack Smales, 6 June 1727.

NYCRO ZQH 7/17/6, Smales family papers relating to Wharton Estates, letter from Matthew Smales to Jack Smales, 23 May 1727.

NYCRO ZQH 7/53/1, Yorke MSS, Articles agreed upon at Richmonde the xth daie of Marche 1562 betwixt Sir George Bowes knighte (and) the Bailiffes and Burgesses of the towne of Richmonde.

NYCRO ZQH 7/53/3, Yorke MSS, Richmond, 5 March XV Eliz I (1573).

NYCRO ZQH 9/14/24, The papers of Sir William Chaytor of Croft (1639 – 1721), letter from Lady Chaytor at Croft to Sir William in London, 10 February 1700 or 1701.

NYCRO ZQH 9/19/84, The papers of Sir William Chaytor of Croft, letter from C. Pinckney of Dalton to Sir William, 11 January 1719.

NYCRO ZQH 9/22/4, The papers of Sir William Chaytor of Croft, lease to John Croft, 24 February 1690.

NYCRO H 9/22/10, The papers of Sir William Chaytor of Croft, covenant to Memorandum of Agreement to John Croft, 22 June 1703.

NYCRO ZQX 4/11, Arkengarthdale Estate papers, Marrick and Hurst leases, agreement 10 March 1885;

NYCRO ZQX 4/13, lease of Ravensworth Farm, 1 June 1863.

NYCRO ZRL 11/3, Letter book of Richard Wright, 1843-75.

NYCRO ZRL 1/23, 'Building Catterick Bridge 1421'.

NYCRO ZS*, Middleham Estate, surveys, rentals etc 1617 – 1866, Box 1, Bundle 11, A survey of Middleham and Braithwaite belonging to Thomas Wood, Esquire, 1765 – 75, Mr Head's remarks May 1775; particulars and valuation of the Estate of Thomas Wood Esq., October and November 1795;Letter from Mr Head at Gayles, 1 March 1776.

NYCRO ZS uncatalogued, Swinton Archive, Colsterdale deeds, 1334-1567.

NYCRO ZS* uncatalogued, Swinton Archive.

NYCRO ZS* uncatalogued, Swinton Lime Kiln, August 1696.

NYCRO ZS* uncatalogued, Swinton Archive, Box 88, 'Counterpart of lease from Abstrupus Danby Esq. to Robt Rucroft and Jas Rucroft of South Leighton Farm for 21 years commencing Lady Day 1733.

NYCRO ZST/2946, Willan family deeds and papers, n.d.

NYCRO ZTO 2, indenture, 10 June 1751.

NYCRO ZTO 5/2, Preston of Flasby, list of land in Flasby, 1763.

NYCRO ZTS (M) 1/92, Specification of works to be done in building a hay barn and cowhouse at Melmerby, 1878.

NYCRO ZTW, Farrer of Ingleborough, estate ledgers, 1833 – 1900, 'Draining and limeing'.

NYCRO ZTW III/11/155, Ingleborough Estate, correspondence files, letter 13 April 1861; ZTW III/11/168, letter 4 May 1863.

NYCRO ZTW III/7, Ingleborough Estate, accounts, c. 1806 – 1947.

NYCRO ZTW III/12/6/2164, Ingleborough Estate, memorandum, 6 July 1895.

NYCRO ZUC, Newby court books and papers.

NYCRO ZXF 4/5/1, Armitstead: John Foster's estate 1793 – 1825, conditions of agreement, 10 October 1793.

NYCRO ZXF 4/6/1 (i), Lease for 21 years of East Pallener in Horton in Craven, 9 November 1790; ZXF 4/6/1 (ii); lease Stockdale to Dearlove, n.d. (c. 1790).

Otley Museum

Otley O/O/dc/7/13 The Extent of Otley 1307.

Otley Parish Registers O/pb/1 1640.

Private collections

Account book of Ursula Towneley, 1712-16.

Barker MSS 251, Garth day books, 1795 – 1911.

Cowside, Stainforth, indenture, 30 January 1682.

Diary of George Smith, January 1819 to May 1856, 4 vols.

Fox Commonplace Diaries.

Letter book of William Stackhouse of Giggleswick, 1808-41, letter 23 September 1816; letter 20 March 1828.

William Foster, Stainforth, Indenture 31 December 1677, SHG/ind/1677.

Royal Society

Cl. P/10 iii, Classified papers 1660-1740, vol. 10, Agriculture, vol. 3, 22 and 24.

Sedbergh and District Historical Society

Ingmire account book, rentals, 1 January 1787.

Minute and account book for the manor of Garsdale, 1797-1896.

MF 80, Robert Foster 1782-83, Hebblethwaite Hall, Cautley.

TCC Archive, Box 17, Sedbergh A1, 13 August 1568.

Sheffield Archives

Wharncliffe Muniments, WhM 235/36, letter to Lord Wharncliffe, 15 June 1880; WhM 241, letter from J H Scott, 18 September 1886; WhM 242, letter from J H Scott to Lord Wharncliffe, 9 October 1886; WhM 245, letter from J H Scott, 15 November 1890.

Skipton Reference Library

Raistrick MSS 557, 15 July 1806.
Raistrick MSS 601, conveyances.

West Yorkshire Archive Service, Bradford

WYAS (B) BCC, Bradford Canal Company, Records 1661-1923.
WYAS (B), pamphlet, 'Summary view of proposed canal from Lds to Liverpool and of its importance to the public', Leeds: Griffith Wright, 1768.
WYAS (B) DBS2.C, Bradford Limekiln Company.
Ferrand C/218, Letter Richard Metcalfe to J A Busfield, 22 December 1847.
WYAS (B), Heaton Papers, HEA/A/346, Feoffment, 15 May 1623; HEA/A/462, Conveyance, 30 March 1739; HEA/B/158, Commonplace Book of John Halstead of Nave Hill, Colne.
WYAS (B) 10D 76/3/24, Horsfall Family of Burley-in-Wharfedale, Records, 1589 – 1924.
WYAS (B) 68D 82/5/628, Letter, William Dean, n.d (19th c.).

West Yorkshire Archive Service, Wakefield

Hammond Papers, uncatalogued, indenture 31 December 1617.
Indenture of lease and release for lands in Rathmell, 1783.
WYAS (W) WYL 162. GA/2/61.4, 'Plan of the intended navigable canal from Settle proposed to communicate with the canal now making, between Leeds and Liverpool'
WYAS (W), vol. 202, folio 119, Will of William Procter of Close House, Giggleswick, July 1840.

Yorkshire Archaeological Society

YAS DD 107, deed of award, 22 August 1832.
YAS DD 121/Add/6, Field book of the estate in the manor and lordship of Silsden ... belonging to Earl of Thanet, 1823.
YAS DD 146/13/9, Farnley Hall, estate and house building accounts, 1706-45.
YAS MD 59/12/78, indenture, 30 October 1620.
YAS MD 59/14/103, indenture 25 March 1678.
YAS MD 247/6, conveyance ,28 January 1572.
YAS MD 335/1/4/5/30, An account of John Brotherton's expenses, 1815.
YAS MD 335/1/4/5/29, Lime vouchers, 1814.
YAS MD 335/1/10/6/15, Paper, 'Case in favour of the Settle Canal', 1768 or 1769.
YAS MD 335/2/1/4/21, Foster v Redmayne, 26 July 1843; MD 335/14/39, Map drawn 1843 to support a court case of trespass by Foster v Redmayne.
YAS MS 1186, 'Plan of a proposed canal from Parkfoot Bridge to Settle', 1780.

Published Primary Sources

Advertisement, 'To be sold by auction', *Blackburn Mail*, 6 May 1795.
Anon a, 'A farm in Flanders' *Annals of Agriculture* 20 (1793), pp. 463-67.
Anon b, 'Rules and regulations of the Georgic Society, for the promotion of agriculture and husbandry' *Annals of Agriculture* 20 (1793), pp. 339-49.
Anon c, *The farmer's compleat guide*. (London: G Kearsly, 1760).
Anon d, 'Instructions. The methods of improving the whole arable lands within the said bounds' *Transactions of the Honourable Society for the Encouragement of Agriculture*. Gilbey Pamphlet no. 4 (1776).
Anon e, 'Letters from a Scots farmer, during a tour through England, to his friend in Edinburgh' *The Farmer's Magazine* Old series 1 (1800), pp. 147-53.
Anon f, *The farmer's letters to the landlords of Great-Britain* vol.2. (London: W Strahan and W Nichol, 1771).
Anon g, 'Letters from a Scots farmer, during a tour through England, to his friend in Edinburgh' *The Farmer's Magazine* 1 (1800), pp. 147-53.
Anon h, 'On the application of lime' *The Farmer's Magazine* 2 New Series (1839), p. 386.
Anon j, 'An essay on lime and marl' Part 2 *The Lonsdale Magazine* 1 (1820), pp. 102-04.
Bailey J and Culley G, *General view of the agriculture of Northumberland, Cumberland and Westmorland*. (Newcastle: Frank Graham, 1972, facsimile edition, first published 1794).
Banister, J, *A synopsis of husbandry*. (London: G G and J Robinson, 1799).
Beesley, G, *A report of the state of agriculture in Lancashire*. (Preston: Dobson & Son, 1849).
Binns, J, *Notes on the agriculture of Lancashire, with suggestions for its improvement*. (Preston: Dobson and Son, 1851).
Blith, W, *The English Improver improved or the survey of husbandry*. (London: John Wright, 1652).
Board of Agriculture and Fisheries, *The uses of lime*. Leaflet no. 170. (London: HMSO, 1907).
Bradley, R, *A general treatise of agriculture, both philosophical and practical*. (London: W Johnston, 1757).
Briggs, H, *On lime and its uses in agriculture*. Gilbey Pamphlet 41. (London: Simpkin, Marshall, & Co.; Wakefield: Bowland Hurst, 1845).
Brown, R, *General view of the agriculture of the West Riding of Yorkshire, Appendix No.1 Extracts from the journal kept during the survey*. (Edinburgh: James Watson, 1799).

Brown, W, *Yorkshire inquisitions*. vol. IV, no. XLII, 18 July 1304. Yorkshire Archaeological Society Record Series 57. (Leeds, 1906).

Byng, Hon. J, *The Torrington Diaries* vol. 3, ed. C Bruyn Andrews. (New York: Barnes and Noble; London: Methuen, 1970 edition).

Caird, J, *English agriculture in 1850-51*. (London: Longman, Brown, Green and Longmans, 1852).

Camden, W, *Britannia: or a chorographical description of Great Britain and Ireland*. Second edition revised by E Gibson 1722. (London: William Taylor, 1610).

Charnock, J, 'On the farming of the West Riding of Yorkshire' *Journal of the Royal Agricultural Society of England* 9 (1848), pp. 284-311.

Clarke, C, *The true theory and practice of husbandry*. (London: G Robinson; Newcastle: T Slack, 1777).

Clarke, S, *Clitheroe in its railway days*. (Clitheroe: J Robinson, 1900).

Clarkson, C, *The history of Richmond in the County of York*. (Richmond: T Bowman, 1814).

Cochrane, A, earl of Dundonald, *A treatise showing the intimate connection that subsists between agriculture and chemistry*. (London: J Murray and S Highley, 1795).

Cochrane, J, *A letter addressed to the members of the Society of Agriculture at Darlington, and the farmers of the County of Durham and northern part of the County of York*. (Richmond: W King, 1795).

Davy, H, *Elements of agricultural chemistry in a course of lectures to the Board of Agriculture*. (London: Longman; Edinburgh: Constable, 1814).

Defoe, D, *Tour through the whole island of Great Britain*. (London: Dent, 1962 edition). First published 1726.

Dibdin, C, *Observations on a tour through almost the whole of England, and a considerable part of Scotland, in a series of letters* vol. 2. (London: G Goulding, 1802).

Dickinson, W, *Essay on the agriculture of West Cumberland*. (London: Whittaker and Whitehaven: R Gibson, 1850).

Dickson, A, *A treatise of agriculture*. (Edinburgh: A Kincaid and J Bell, 1765).

Dickson, R W, *Practical agriculture; or, a complete system of modern husbandry* vol 1. (London: Richard Phillips, 1805).

Dickson, R W, *General view of the agriculture of Lancashire*. (London: Sherwood, Neely and Jones, 1815).

Dodgson, J, *Communications to the Board of Agriculture; on subjects relative to the husbandry and internal improvement of the country* IV. (London: W Bulmer, 1805).

Donaldson, J, 'On the husbandry of the Carse of Gowrie' *Annals of Agriculture* 10 (1788), pp. 422-25.

Donaldson, J, *Modern agriculture, or, the present state of husbandry in Great Britain* vol. 2. (Edinburgh: Adam Neill, 1796).

Donaldson, J, *A treatise on manures*. (London: Robert Baldwin, 1842).

Dossie, R, *Memoirs of agriculture and other oeconomical arts* vol. 3. (London: C Nourse, 1782).

Dundonald, earl of, *A treatise shewing the intimate connection that subsists between agriculture and chemistry*. (London: J Murray and S Highley, 1795).

Falkner, F, *The muck manual*. (London: John Murray, 1843).

Farrer, W, *The court rolls of the Honor of Clitheroe in the county of Lancaster* 1. (Manchester: Emmott, and Burnley: Lupton Bros, 1897).

Farrer, W, *The chartulary of Cockersand Abbey of the Premonstratensian Order*. (Manchester: Chetham Society New Series III, 1898).

Farrer, W, *Lancashire inquests, extents, and feudal aids. Part II AD1310 - AD1333*. (Manchester: Lancashire Record Society, 1907).

Farrer, W, *The court rolls of the Honor of Clitheroe* vol. 2. (Edinburgh: Ballantyne Press, 1912).

Fisher, J, *The history and antiquities of Masham and Mashamshire*. (London: Simpkin, Marshall & Co., 1865), pp. 66-67.

Fitzherbert, Master (John), *The boke of husbandrie*. (London: Richard Pynson, 1523).

Fitzherbert, Master (John), *The book of husbandry*. (London: English Dialect Society, 1882), transcript of 1534 edition.

Fitzherbert, Master (John), *The boke of surveying and improvements, newly corrected and amended, very necessary for all men*. (London: Thomas Marshe, 1539).

Fitzherbert, (John), *The boke of surveying and improvements, London 1523*. (Amsterdam: Theatrum Orbis Terrarum, 1974 reprint).

Fuller, Thomas *The Holy State*. (Cambridge: John Williams, 1642).

Garnett, F W, *Westmorland agriculture 1800 - 1900*. (Kendal: Titus Wilson, 1912).

Garnett, W J, 'Farming of Lancashire' *Journal of the Royal Agricultural Society of England* 10 (1849), pp. 1-51.

Googe, Barnaby, *The whole art and trade of husbandry*. (London: Richard More, 1614). First published in 1577.

Grantham, R B, 'A description of the works for reclaiming and marling parts of the late Forest of Delamere, in the County of Cheshire' *Journal of the Royal Agricultural Society of England* Second Series 25 (1864), pp. 369-80.

Griffiths, A B, *A treatise on manures*. (London: Whittaker & Co., 1889).

Hale, Thomas, *A compleat body of husbandry*. (London: T Osborne and J Shipton , 1756).

Harland, J, *The house and farm accounts of the Shuttleworths of Gawthorpe Hall, in the County of Lancaster, September 1582 to October 1621*. (Manchester: Chetham Society, 1856).

Hartlib, Samuel, *His legacy of husbandry*. (London: Richard Wodnothe, 1655).

Hawkins, E (ed.), *Travels in Holland, the United Provinces, England, Scotland and Ireland 1634-35 by Sir William Brereton, Bart*. (Manchester: Chetham Society, 1844).

Headrick, J, *Essays on manures*. Gilbey Pamphlet no. 166. (London: W Bulmer, 1796).

Henry, D, *The complete English farmer, or a practical system of husbandry*. (London: F Newberry, 1771).

Heresbach, Conrad, *Foure bookes of husbandry*. (London: Thomas Wright, 1601). First published in German in 1567, and translated into English in 1577.

Hitt, T, *A treatise of husbandry on the improvement of dry and barren lands*. (Dublin: R Watts and W Whitestone, 1761).

Hodgson, J, *A topographical and historical description of the County of Westmoreland* (sic). (London: Sherwood, Neely and Jones, 1820).

Holt, J, *General view of the agriculture of the county of Lancaster*. (London: G Nicol, 1795).

Home, F, *The principles of agriculture and vegetation*. (Dublin: G & A Ewing, 1759).

Home, F, *The principles of agriculture and vegetation*. Gilbey Pamphlet no. 67. (London: A Millar; Edinburgh: A Kincaid and J Bell, 1762).

Houghton, J, *A collection for the improvement of husbandry and trade* vol. 1. (London: Woodman and Lyon, 1727).

Housman, J, *A topographical description of Cumberland, Westmorland, Lancashire and a part of the West Riding of Yorkshire*. (London: C Law, 1800).

Hunter, A, *Georgical essays*. (York: A Ward, 1777).

Hustler, J, *An explanation of the plan of the canal from Leeds to Liverpool*. (Bradford: George Nicholson, 1788).

Hutton, J, *A tour to the caves in the environs of Ingleborough and Settle in the West Riding of Yorkshire*. (London: Richardson and Urqhuart, and J Robson; Kendal: Pennington, 1781).

Johnson, C W, *British husbandry; exhibiting the farming practice in various parts of the United Kingdom* vol. 1. (London: Robert Baldwin, 1834).

Johnston, J F W, 'On guano' *Journal of the Royal Agricultural Society of England* 1st Series, 2 (1841), pp. 310-21.

Johnston, J F W, *On the use of lime in agriculture*. (Edinburgh: W Blackwood, 1849).

Kames, Lord (Home, H), *The gentleman farmer*. (Edinburgh: Bell and Bradfute; London: Longmans, 1815).

Kent, N, *Hints to gentlemen of landed property*. (London: J Dodsley, 1775).

Killick, H F, *Notes on the early history of the Leeds and Liverpool Canal*. (Bradford: H Gaskarth, 1897).

Lamond, E, (translator) *Walter of Henley's husbandry*. Includes *Seneschaucie: a book of the office of estate seneschal*. (London: Longmans, Green & Co, 1890).

Lawson, T, 'On results of experiments with lime and various artificial manures' *Transactions of the Highland and Agricultural Society of Scotland* 4th series 11 (1879), pp. 106-21.

Leadbetter, C. *The royal gauger, or gauging made easy, as practised by the officers of His Majesty's Revenue of Excise*. (London: E Wicksteed, 1739).

Loudon, J C, *An encyclopaedia of agriculture*. (London: Longman, Orme, Brown, Green, and Longmans, 1839).

Lovat, E, *A sketch of the rural economy of the County of Lancaster*. (Burnley: T Rickard, 1831).

Lyons, P A, *Two "Compoti" of the Lancashire and Cheshire manors of Henry de Lacy, Earl of Lincoln, XXIV and XXXIII Edward I*. (Manchester: Chetham Society, 112, 1884).

Macdonald, A, 'The improvement of hill pasture without breaking it up' *Transactions of the Highland and Agricultural Society of Scotland* 4th Series, 19 (1887), pp. 158-74.

Mannex, P J, *History, topography and directory of Westmorland*. (London: Simpkin, Marshall & Co, 1849).

Markham, Gervase, *Markham's farewell to husbandry or the enriching of all sorts of barren and sterile grounds in our Kingdome*. (London: John Harison, 1631).

Marshall, W, *The rural economy of Yorkshire* vol. 1 (London: T. Cadell, 1788); vol. 2 (London: G Nicol and G G and J Robinson, 1796).

Marshall, W, *The review and abstract of the county reports to the Board of Agriculture. Vol. 1 Northern Department*. (York: Thomas Wilson, 1808).

Mavor, W (ed.), *Five hundred points of good husbandry by Thomas Tusser*. (London: Lackington, Allen, and Co, 1812).

Maxwell, R, *The practical husbandman*. (Edinburgh: C Wright, 1757).

Milburn, M M, 'On the farming of the North Riding of Yorkshire' *Transactions of the Yorkshire Agricultural Society* 11 (1848), pp. 182-207.

Mills, J, *A new system of practical husbandry* vol. 1. (London: F Johnson and B Davenport, 1767).

Moore, J, 'Remarks (chiefly agricultural) made during a short excursion into Westmoreland and Cumberland in August 1815' *Transactions of the Manchester Literary and Philosophical Society* 3 (1817), pp. 179-203.

Mortimer, J, *The whole art of husbandry or, the way of managing and improving of land*. (London: Mortlock and Robinson, 1712).

Morton, J C, *A cyclopedia of agriculture, practical and scientific*. (Glasgow: Blackie, 1855).

Norden, John, *The surveyors dialogue*. (London: Thomas Snodham, 1618).

Overman, F W, 'On claying or marling land' *Journal of the Royal Agricultural Society of England* Third Series 3 (1842), pp. 234-36.

Page, W, *The Victoria County History of the County of York North Riding* vol. 1. (London: Constable, 1914).

Parkinson, R, *The experienced farmer* vol. 2. (London: GG and J Robinson, 1798).

Peters, M, *Winter riches, or a miscellany of rudiments, directions and observations*. Gilbey Pamphlet no. 72. (London: W Flexney, 1771).

Peyrouse, Baron Picot de la, *Agriculture of a district in the south of France*. (London: John Harding, 1819).

Phillips, J, *A report on the probability of the occurrence of coal and other minerals in the vicinity of Lancaster*. (Lancaster: W Barwick, 1837).

Plattes, Gabriel, *Practicall husbandry improved: or, the discovery of infinite treasure*. (London: Edward Thomas, 1656).

Potts, T, *The British farmer's cyclopaedia*. (London: Scatcherd and Letterman and M Jones, 1807).

Pringle, A, *General view of the agriculture of the County of Westmoreland* (sic) in J Bailey and G Culley *General view of the agriculture of Northumberland, Cumberland and Westmorland*. (facsimile edition, Newcastle upon Tyne: Frank Graham, 1972. Originally published 1794).

Prothero, R E (Lord Ernle), *English farming past and present*. (London: Heinemann and Frank Cass, 1912).

Radcliff, J, 'On burning lime with peat' *Prize Essays and Transactions of the Highland Society of Scotland* II (1803), pp. 160-63.

Radcliff, T, *A report on the agriculture of Eastern and Western Flanders; drawn up at the desire of the Farming Society of Ireland.* (London: John Harding; and Dublin: W Porter, 1819).

Raines, F R, 'The rent roll of Sir John Towneley of Towneley 1535-6' *Chetham Miscellanies* vol. 6, Chetham Society 103 (1878).

Reeve, G, *Directions left by a gentleman to his sons: for the improvement of barren and heathy land, in England and Wales.* (London: R Royston, 1670).

Rees, A, *Rees's Manufacturing Industry (1819-20)* vol.3 (facsimile edition, edited N Cossons, Newton Abbot: David & Charles, 1972 edition).

Reid, C, *Submerged forests.* (Cambridge: Cambridge University Press, 1913).

Rennie, G, Brown R and Shirreff, J, *General view of the agriculture of the West Riding of Yorkshire.* (London: W Bulmer, 1794).

Rham, W L, *Dictionary of the farm.* (London: C Knight, 1845).

Rigby, E (translator) *Italy, its agriculture etc, from the French of Mons Chateauvieux, being letters written by him in Italy, in the years 1812 and 1813.* Written by J F Lullin de Chateauvieux. (Norwich: Burks and Kinnebrook, 1819).

Robinson, Mr, 'Statistical account of the Parish of Orton, in the county of Westmoreland (*sic*)' *Monthly Magazine* 98 (1803), pp. 109-13.

Robinson, Mr, 'Account of Kirkby Stephen in Westmoreland' (*sic*) *Monthly Magazine* 127 (1805), pp. 213-15.

Rothwell, W, *Report of the agriculture of the County of Lancaster.* (London: Groombridge , 1850).

Silverstick, *What shall we do? A treatise on the present agricultural depression, its causes, effects and remedies.* (Bishop's Stortford: Arthur Boardman, 1879).

Sinclair, Sir John, *The code of agriculture.* (London: Sherwood, Neely, and Jones, 1819).

Sinclair, Sir John, *Memoirs of the life and works of the late Right Hon. Sir John Sinclair, Bart.* vol.2 (Edinburgh, 1837).

Smith, J, *England's improvement reviv'd: in a treatise of all manner of husbandry and trade by land and sea.* (London: Benjamin Southwood, 1673).

Society of Gentlemen, *The complete family-piece and country gentlemen, and farmer's best guide.* (London: J Roberts, 1736).

Somerville, R, 'On the use of lime as a manure' *Outlines of the fifteenth chapter of the proposed general report from the Board of Agriculture on the subject of manures.* (London: W Bulmer, 1795).

Speed, J, *Eboracensis comitatus descriptio.* (1610). Private collection.

Speight, H, *Chronicles and stories of Old Bingley.* (London: Elliot Stock, 1899).

Sproule, J, *A treatise on agriculture, suited to the soil and climate of Ireland.* (Dublin: W Curry; London: Longman; Belfast: Simms and McIntyre, 1839).

Stephens, H, *The book of the farm.* (Edinburgh and London: William Blackwood, 1844).

Stephens, H, *The book of the farm.* (Edinburgh: William Blackwood & Sons, 1852).

Taylor, Silvanus, *Common-good: or, the improvement of commons, forests, and chases, by inclosure.* (London: Francis Tyton, 1652).

Trimmer, J, 'On the geology of Norfolk as illustrating the laws of the distribution of soils' *Journal of the Royal Agricultural Society* 7 (1847), pp. 444-85.

Tuke, J, *General view of the agriculture of the North Riding of Yorkshire.* (London: W Bulmer, 1794).

Turner, N, *An essay on draining and improving peat bogs.* Gilbey Pamphlet no. 83. (London: R Baldwin; and J Bew, 1784).

Ure, D, 'Lime burnt with peat' *Annals of Agriculture* XXIV 134 (1795), pp. 8-9.

Verral, H, 'On the use and misuse of lime' *Annals of Agriculture* 32 (1799), pp. 99-136.

Voelcher, A, 'On paring and burning' *Journal of the Royal Agricultural Society of England* 2nd series, 18 (1857), pp. 342-68.

Walker, A, *Remarks made in a tour from London to the Lakes of Westmoreland* (sic) *and Cumberland in the summer of M, DCC, XCI.* (London: G Nicol, 1792).

Walter of Henley, *Boke of husbandry.* (London: Wynkyn de Worde, 1508).

Watson, J, 'On reclaiming heath land' *Journal of the Royal Agricultural Society of England* 6 (1845), pp. 79-102.

Way, J T, 'On the influence of lime on the "absorptive properties" of soils' *Journal of the Royal Agricultural Society of England* 1st Series, 15 (1854), pp. 491-514.

Webster, C, 'Farming of Westmorland' *Journal of the Royal Agricultural Society of England* 2nd Series, 4 (1868), pp. 1-37.

Weston, Sir R, *Discours of husbandrie used in Brabant and Flanders.* (London: Wm Du-Gard, 1645).

Whitaker, T D, *An history of the original parish of Whalley and Honor of Clitheroe.* (Blackburn: Hemingway and Crook, 1800-01).

Williamson, W, *The trial at large of Nicholas Wilkinson, Doctor Herd, and Henry Warswick for the wilful murder of George Battersby, at Clithero* (sic), *in the night of the 25th of March, 1773.* (York: W. Blanchard; Lancaster: Mr Ashburner, n.d.).

Wilson, J M, *The rural cyclopedia, or a general dictionary of agriculture* vol. 3. (Edinburgh: A Fullarton, 1849).

Winter, G, *A new and compendious system of husbandry.* (Bristol: W Routh, 1787).

Worlidge, J, *Systema agriculturae. The mystery of husbandry discovered.* (London: John Worlidge, 1675).

Worlidge, J, *Systema agriculturae being the mystery of husbandry discovered and laid open.* (London: Thomas Dring, 1687).

Young, A, *The farmer's letters to the people of England.* (London; W Nicoll, 1768).

Young, A, *Letters concerning the present state of the French nation.* (London: W Nichol, 1769).

Young, A, *A six months tour through the North of England.* Four volumes. (London: W Strahan, 1770).

Young, A, *The farmer's letters to the landlords of Great-Britain* vol. 2. (London: W Strahan; Edinburgh: J Balfour, 1771).

Young, A, *Observations on the present state of the waste lands of Great Britain.* (London: W Nicoll, 1773).

Young, A, *A tour in Ireland.* (London: H Goldney, 1780).
Young, A, *Travels during the years 1787, 1788 and 1789 in France,* vol.2. (Bury St. Edmund's: J Rackham, 1794).
Young, A, 'Excursion to Yorkshire including the account of a waste land farm' *Annals of Agriculture* 27 (1796), pp. 287-308.
Young, A, *The farmer's calendar.* (London: Richard Phillips, 1804).
Young, A, 'An essay on manures. Letters and papers on agriculture, planting etc' *Bath Society Papers* 10 (1805), pp. 95-198.

Secondary Sources

Allen, R C, 'Tracking the agricultural revolution in England' *Economic History Review* 52 (1999), pp. 209-35.
Allen, R C and Ó Gráda, C, 'On the road again with Arthur Young: English, Irish, and French agriculture during the Industrial Revolution' *Journal of Economic History* 48 (1988), pp. 93-125.
Appleby, A B, 'Disease or famine? Mortality in Cumberland and Westmorland 1580 - 1640' *Economic History Review* 26 (1973), pp. 403-32.
Appleby, A B, *Famine in Tudor and Stuart England.* (Liverpool: Liverpool University Press, 1978).
Arthurton, R S, Johnson, E W and Mundy, D J C, *Geology of the country around Settle.* (London: HMSO, 1988).
Ashmore, O, *Industrial archaeology in Lancashire.* (Newton Abbot: David & Charles, 1969).
Aston, M and Rowley, T, *Landscape archaeology.* (Newton Abbot: David & Charles, 1974).
Austin, D, 'Character or caricature? Concluding discussion' *Landscapes* 8 (2007), pp. 92-105.
Bagot, A and Munby, J (eds.), *'All things is well here'. Letters from Hugh James of Levens to James Grahme, 1692-95.* Cumberland and Westmorland Antiquarian and Archaeological Society Record Series X. (Kendal: Titus Wilson, 1988).
Bainbridge, T H, 'Some factors in the development of Cumbrian agriculture, especially during the nineteenth century' *Transactions of the Cumberland and Westmorland Antiquarian and Archaeological Society* New Series 44 (1945), pp. 81-92.
Barker, D and Cranstone, D (eds.), *The archaeology of industrialization.* (Leeds: Maney, 2004).
Barnatt, J and Dickson A, 'Survey and interpretation of a limekiln complex at Peak Forest, Derbyshire: and a review of early limeburning in the North-West Peak' *Derbyshire Archaeological Journal* 124 (2004), pp. 141-215.
Barnes, B, *Man and the changing landscape. A study of occupation and palaeo environment in the Central Pennines.* (Liverpool: Merseyside County Council, Merseyside County Museums, University of Liverpool Department of Prehistoric Archaeology, 1982).
Bayley, J and Crossley, D, 'Archaeological science as an aid to the study of post-medieval industrialization' in D Barker and D Cranstone (eds.) *The archaeology of industrialization.* (Leeds: Maney, 2004), pp. 15-23.
Beckett, J V, 'The decline of the small landowner in eighteenth- and nineteenth-century England: some regional considerations' *Agricultural History Review* 30 (1982), pp. 97-111.
Beckett, J V, *The agricultural revolution.* (Oxford: Basil Blackwell, 1990).
Bentley, J, Bond, B R and Gill, M C, *Ingleton coalfield.* British Mining No. 76. (Sheffield: NMRS, 2005).
Beresford, M, *History on the ground,* (London: Lutterworth Press, 1957).
Beresford, M W and St. Joseph, J K S, *Medieval England. An aerial survey.* (Cambridge: Cambridge University Press, 1958).
Berry, E K, (ed.), *Swaledale wills and inventories 1522 – 1600.* Yorkshire Archaeological Society Record Series CLII (Leeds, 1998).
Blackett-Ord, M, 'Lord Wharton's deer park walls' *Transactions of the Cumberland and Westmorland Antiquarian and Archaeological Society* 86 (1986), pp. 133-39.
Blake, J B, 'The medieval coal trade of North East England: some fourteenth-century evidence' *Northern History* 2 (1967), pp. 2-26.
Bond, J, 'Landscapes of monasticism' in D. Hooke (ed.) *Landscape the richest historical record.* Supplementary Series 1 (Westbury: Society for Landscape Studies, 2000), pp. 63-74.
Bond, J, *Monastic landscapes.* (Stroud: Tempus, 2004).
Born, A, 'Limestone, limekilns and the limeburning industry north and west of Dartmoor' *Report and Transactions of the Devonshire Association for the Advancement of Science, Literature and Art* 123 (1991), pp. 213-40.
Bouch, C M L and Jones, G P, *A short economic and social history of the Lake Counties, 1500 – 1830.* (Manchester: Manchester University Press, 1961).
Bourne, S, *An introduction to the architecture of Towneley Hall.* (Burnley: Burnley Borough Council, 1979).
Bowden, P, 'Agricultural prices, farm profits and rents' in J Thirsk (ed.), *The agrarian history of England and Wales, IV, 1500 - 1640.* (Cambridge: Cambridge University Press, 1967).
Boynton, R S, *Chemistry and technology of lime and limestone.* (New York: John Wiley, 1980).
Braudel, F, *L'identité de la France.* (Paris: Flammarion, 1986).
Brayshaw, T and Robinson, R M, *A history of the ancient parish of Giggleswick.* (London: Halton, 1932).
Brenner, R, 'Agrarian class structure and economic development in pre-industrial Europe' *Past and Present* 70 (1976), pp. 30-75.
Brimblecombe, P, 'Industrial air pollution in thirteenth-century Britain' *Weather* 30 (1975), pp. 388-95.
Bristow, J, *The local historian's glossary of words and terms.* (Newbury: Countryside Books, 2001).
Brock, J, 'Lime kilns around Richmond' *The Richmond Review* (1987), pp. 6-21.
Brown, I, *Discovering a Welsh landscape. Archaeology in the Clwydian range.* (Macclesfield: Windgather Press, 2004).
Buchanan, R A, *Industrial archaeology in Britain.* (Harmondsworth: Penguin, 1972).

Campbell, B, 'Commodified origins – the regional dimension' in T Rowley (ed.), *The origins of open-field agriculture* (London: Croom Helm, 1981).

Cantor, L, *The changing English countryside 1400 - 1700.* (London: Routledge and Kegan Paul, 1987).

Carne, R, '100 years of fertilizers' *Journal of the Royal Agricultural Society of England* 167 (2006), pp. 32-39.

Catling, C, 'Drapers' Gardens. Unearthing the lives of Roman Londoners' *Current Archaeology* XXI (2010), pp. 20-27.

Challands, A, 'A Roman limekiln at Helpston' *Durobrivae* 4 (1976), pp. 22-24.

Chambers, J D and Mingay, G E, *The Agricultural Revolution 1750-1880.* (London: Batsford, 1966).

Chandler, J (ed.), *John Leland's itinerary.* (Stroud: Alan Sutton, 1993).

Chandler, J, 'The discovery of landscape' in D Hooke (ed.) *Landscape the richest historical record.* Supplementary Series 1 (Westbury: Society for Landscape Studies, 2000), pp. 133-41.

Chapman, J, 'Parliamentary enclosure in the uplands' in I D Whyte and A J L Winchester (eds.) *Society, landscape and environment in upland Britain.* Society for Landscape Studies Supplementary Series 2 (Westbury: Society for Landscape Studies 2004).

Clark, K, 'The workshop of the world. The industrial revolution' in J Hunter and I Ralston (eds.) *The archaeology of Britain.* (London: Routledge, 1999), pp. 280-96.

Cleasby, I, 'Limekilns and limeburning in Sedbergh, Garsdale and Dent' Part 1, *The Sedbergh Historian* 3 (1994), pp. 2-9.

Cleasby, I, 'Limekilns and limeburning in Sedbergh, Garsdale and Dent' Part 2, *The Sedbergh Historian* 3 (1995a), pp. 2-8.

Cleasby, I, 'Limekilns in Sedbergh, Garsdale and Dent' *Current Archaeology* 145 (1995b), pp. 16-20.

Cook, J, Huckerby, E, Quatermaine, J and Wearing, N, 'The Ribble valley: a methodological study in landscape development' *Landscapes* 9 (2008), pp. 28-49.

Corrie, F E, *Lime in agriculture.* (London: Chapman and Hall, 1926).

Cossons, N, *The BP book of industrial archaeology.* (Newton Abbot: David & Charles, 1978).

Cossons, N, (ed), *Perspectives on industrial archaeology.* (London: Science Museum, 2000).

Cossons, Sir N, 'New directions in industrial archaeology' in E C Casella and J Symonds (eds.) *Industrial archaeology. Future directions.* (New York: Springer, 2005), pp. ix-x.

Cottam, A, 'The granges of Furness Abbey, with special reference to Winterburn-in-Craven' *Transactions of the Historical Society of Lancashire and Cheshire* 80 (1928), pp. 58-85.

Cowper, A D, *Lime and lime mortars.* (London: HMSO, 1927).

Craig, W J, 'James Ryther of Harewood and his letters to William Cecil, Lord Burghley' *Yorkshire Archaeological Journal* 56 (1984), pp. 95-118.

Cranstone, D, 'The archaeology of industrialization – new directions' in D Barker and D Cranstone (eds.) *The archaeology of industrialization* (Leeds: Maney, 2004), pp. 313-20.

Crawford, O G S, *Archaeology in the field.* (London: Phoenix House, 1953).

Crossley, D, 'English woodlands and the supply of fuel for industry' *Industrial Archaeology Review* 27 (2005), pp. 105-12.

Crossley, D W, '*Industrial Archaeology.* By Michael Rix. Review' *Post-Medieval Archaeology* 1 (1967), pp. 137-38.

Crow, J, *The English Heritage book of Housesteads.* (London: Batsford, 1995).

Cunliffe, B, *Facing the ocean. The Atlantic and its people.* (Oxford: Oxford University Press, 2001).

Cunliffe Shaw, R, *The Royal Forest of Lancashire.* (Preston: Guardian Press, 1956).

Curwen, J F, 'Records relating to the Barony of Kendale', vol. 3 *Cumberland and Westmorland Antiquarian and Archaeological Society Extra Series IV* (1926).

Curwen, J F, *The later records relating to North Westmorland or the Barony of Appleby.* Cumberland and Westmorland Antiquarian and Archaeological Society Record Series VIII. (Kendal; Titus Wilson, 1932).

Dalglish, C, 'Understanding landscape: inter-disciplinary dialogue and the post-medieval countryside' in A Horning and M Palmer (eds.) *Crossing paths or sharing tracks?* Society for Post-Medieval Archaeology Monograph 5. (Woodbridge: Boydell and Brewer, 2009), pp. 233-54.

Dancaster, E A, *Limes and cements.* (London: Crosby Lockwood, 1916).

Darby, H C, *A new historical geography of England after 1600.* (Cambridge: Cambridge University Press, 1976).

Davies, C S, *The agricultural history of Cheshire 1750 - 1850.* New Series 10 (Manchester: Chetham Society, 1960).

Davies, P B S, *Pembrokeshire limekilns* (St. David's: Merrivale, 1997).

Davies-Shiel, M, 'A little known Late Medieval industry, part 1. The making of potash for soap in Lakeland' *Transactions of the Cumberland and Westmorland Antiquarian and Archaeological Society* 72 (1972), pp. 85-111.

Davis, T, *Ricardo's macroeconomics. Money, trade cycles & growth.* (Cambridge: Cambridge University Press, 2005).

Day, M, 'The karstlands of Antigua, their land use and conservation' *Geographical Journal* 173 (2007), pp. 170-85.

Dent, J G, 'Mechanics and effects of parliamentary enclosure of common grazings' *Folk Life* 21 (1983), pp. 83-99.

Derry, J W, *Reaction and reform.* (London: Blandford, 1970).

Dilley, R S, 'Agricultural change and common land in Cumberland 1700 -1850', unpubl. PhD. thesis, McMaster University, 1991.

Dix, B, 'The manufacture of lime and its uses in the western Roman provinces' *Oxford Journal of Archaeology* 1 (1982), pp. 331-45.

Dodgshon, R A, 'Land improvement in Scottish farming: marl and lime in Roxburghshire and Berwickshire in the eighteenth century' *Agricultural History Review* 26 (1978), pp. 1-14.

Dodgshon, R A, 'Preface' in D Hooke (ed.), *Landscape the richest historical record.* Supplementary Series 1 (Westbury: Society for Landscape Studies, 2000), unpaginated.

Donkin, R A, 'Changes in the early Middle Ages' in H C Darby (ed.), *A new historical geography of England.* (Cambridge: Cambridge University Press, 1973), pp. 75-135.

Drake, M, 'An elementary exercise in parish register demography' *Economic History Review* 14 (1961), pp. 427-45.

Dunham, K C and Wilson, A A, *Geology of the North Pennine Orefield.* Vol. 2 Stainmore to Craven. (London: HMSO, 1985).

Dyer, C, *Standards of living in the later Middle Ages. Social change in England c.1200 - 1520.* (Cambridge: Cambridge University Press, 1993).

Dyer, C, 'Peasants and farmers: rural settlements in an age of transition' in D Gaimster and P Stamper (eds.) *The age of transition: the archaeology of English culture 1400 – 1600.* Society for Medieval Archaeology Monograph 15 (Oxbow Monograph 98, 1997), pp. 61-76.

Dyer, C, *An age of transition? Economy and society in England in the later Middle Ages.* (Oxford: Clarendon Press, 2005).

Eckel, E C, *Cements, limes and plasters.* (London: Chapman and Hall, 1922).

Edwards, W and Trotter, F M, *The Pennines and adjacent areas.* British Regional Geology. (London: HMSO, 1978).

Emery, F, 'The farming regions of Wales' in J Thirsk (ed.), *The agrarian history of England and Wales* vol. IV 1500 - 1640. (Cambridge: Cambridge University Press, 1967), pp. 113-60.

English Nature, Joint Character Areas map, 1996. Available at http://www.countryside.gov.uk/LAR/Landscape/CC/jca.asp. Accessed 19 June 2007.

Everitt, A, 'River and wold: reflections on the historical origin of regions and *pays*' *Journal of Historical Geography* 3 (1977), pp. 1-19.

Everitt, A, *Landscape and community in England.* (London and Ronceverte: The Hambledon Press, 1985).

Everson, P and Williamson, T, *The archaeology of landscape.* (Manchester Manchester University Press, 1998).

Fairclough, G, *Historic Landscape Characterisation.* Available at www.pcl-eu.de/project/agenda/hlc.php. Accessed 18 October 2007.

Fieldhouse, R T, 'Agriculture in Wensleydale from 1600 to the present day' *Northern History* 16 (1980), pp. 169-95.

Fieldhouse, R, 'Some evidence of surviving open fields in the seventeenth-century Pennine Dales and the gradual elimination of communal agriculture' *Yorkshire Archaeological Journal* 54 (1982), pp. 111-18.

Fieldhouse, R and Jennings, B, *A history of Richmond and Swaledale.* (London: Phillimore, 1978).

Finch, J, 'The estate: recognising people and place in the modern landscape' in P S Barnwell and M Palmer (eds.) *Post-medieval landscapes. Landscape history after Hoskins* 3. (Macclesfield: Windgather Press, 2007), pp. 39-54.

Fleming, A, *Swaledale. Valley of the wild river.* (Edinburgh: Edinburgh University Press, 1998).

Fleming, A, 'Post-processual landscape archaeology: a critique' *Cambridge Archaeological Journal* 16 (2006), pp. 267-80.

Fleming, A, 'Don't bin your boots' *Landscapes* 8 (2007), pp. 85-99.

Fleming, A, 'Debating landscape archaeology' *Landscapes* 9 (2008), pp. 74-76.

Fletcher, T W, 'Lancashire livestock farming during the Great Depression' *Agricultural History Review* 9 (1961), pp. 17-42.

Forbes, R J, *Studies in ancient technology,* 1. (Leiden: E J Brill, 1964), pp. 58-59; 3 (1965).

Fox, H S A, 'Local farmers' associations and the circulation of agricultural information in nineteenth-century England' in H S A Fox and R A Butlin, *Change in the countryside: essays on rural England 1500 - 1900.* Special Publication, No. 10. (London: Institute of British Geographers, 1979), pp. 43-63.

Frost, R, *A Lancashire township. The history of Briercliffe-with-Extwistle.* (Briercliffe: Rieve Edge Press, 1982).

Fryde, E B, *Peasants and landlords in later medieval England c. 1380 - c. 1525.* (Stroud: Alan Sutton, 1996).

Fussell, G E, 'Crop nutrition in Tudor and early Stuart England' *Agricultural History Review* 3 (1955), pp. 95-106.

Fussell, G E, *The farmer's tools.* (London: Orbis Publishing, 1981).

Gardiner, M and Rippon, S, 'Introduction: the medieval landscapes of Britain' in M Gardiner and S Rippon, *Medieval landscapes. Landscape history after Hoskins* vol. 2. (Macclesfield: Windgather Press, 2007), pp. 1-8.

Gardner, H W and Garner H V, *The use of lime in British agriculture.* (London: Farmer and Stock-Breeder Publications, 1953).

Garnett, M E, 'The Great Rebuilding and economic change in South Lonsdale 1600 - 1730' *Transactions of the Historical Society of Lancashire and Cheshire* 137 (1987), pp. 55-75.

Gash, N, *Aristocracy and people.* (London: Edward Arnold, 1979).

Gauninga, R, Karlsons, G, Uzars, D and Ojaveer, H, 'Gulf of Riga (Baltic Sea) fisheries in the late 17th century' *Fisheries Research* 87 (1988), pp. 120-25.

Gazley, J G, *The life of Arthur Young 1741 - 1820.* (Philadelphia: American Philosophical Society, 1973).

Gilchrist, R and Sloane, B, *Requiem. The medieval monastic cemetery in Britain.* (London: MoLAS, 2005).

Gill, M C, *Keighley coal.* British Mining No. 74. (Sheffield: NMRS, 2004).

Gill, M C, 'The Great Dales coalfield, eastern areas' *Memoirs British Mining* 86 (2008), pp. 68-108.

Goddard, N, 'The development and influence of agricultural periodicals and newspapers, 1780 – 1880' *Agricultural History Review* 31 (1983), pp. 116-31.

Goddard, N, 'Information and innovation in early-Victorian farming systems' in B A Holderness and M Turner (eds.) *Land, labour and agriculture, 1700 - 1920. Essays for Gordon Mingay.* (London: Hambledon 1990), pp. 165-90.

Gourdin, W H and Kingery, W D, 'The beginnings of pyrotechnology: Neolithic and Egyptian lime plaster' *Journal of Field Archaeology* 2 (1975), pp. 133-50.

Gritt, A, 'Making good land from bad: the drainage of West Lancashire, c. 1650 - 1850' *Rural History* 19 (2008), pp. 1-27.

Grove, J M, *The Little Ice Age.* (London: Routledge, 1988).

Gwyn, D, *Gwynedd. Inheriting a revolution.* (Chichester: Phillimore, 2006).

Gwyn, D, 'The narrow-gauge nations: industrial archaeology beyond the leading sectors', The Rolt Memorial Lecture 2006. *Industrial Archaeology Review* 29 (2007), pp. 3-12.

Gwyn, D, 'An amorphous farrago? The contribution of Industrial Archaeology' in A Horning and M Palmer (eds.) *Crossing paths or sharing tracks?* Society for Post-Medieval Archaeology Monograph 5. (Woodbridge: Boydell and Brewer, 2009), pp. 19-30.

Hadfield, C, *The canals of Yorkshire and north east England.* (Newton Abbot: David & Charles, vol. 1 1972, vol.2 1973).

Hadfield C and Biddle, G, *The canals of north west England.* (Newton Abbot: David & Charles, vol. 1 and 2, 1970).

Hainsworth, D R, *Stewards, lords and people.* (Cambridge: Cambridge University Press, 1992).

Hall, V, *A history of the Yorkshire Agricultural Society 1837 - 1987.* (London: Batsford, 1987).

Handley, J E, *The agricultural revolution in Scotland.* (Glasgow: Burns, 1963).

Harris, H, *Industrial archaeology of the Peak District.* (Newton Abbot: David & Charles, 1971).

Hartley, M and Ingilby, J, *Life in the moorlands of North-east Yorkshire.* (London: J M Dent, 1975).

Harwood Long, W, *A survey of the agriculture of Yorkshire.* County Agricultural Surveys No. 6. (London: Royal Agricultural Society of England, 1969).

Havinden, M A, 'Lime as a means of agricultural improvement: the Devon example' in C W Chalklin and M A Havinden (eds.) *Rural change and urban growth 1500 - 1800.* (London: Longman, 1974), pp. 104-34.

Hewitt, W, 'Marl and marling in Cheshire' *Proceedings of the Liverpool Geological Society* 13 (1923), pp. 24-28.

Higham, N J, *A frontier landscape. The North West in the Middle Ages.* (Maclesfield: Windgather Press, 2004).

Highwood, E J and Stevenson, D S, 'Atmospheric impact of the 1783-1784 Laki eruption: Part 1 chemistry modelling' *Atmospheric Chemistry and Physics* 3 (2003), pp. 1177-89.

Hoelscher, L, 'Improvements in fencing and drainage in mid-19th century England' *Agricultural History* 37 (1963), pp. 75-79.

Horn, P, 'The contribution of the propagandist to eighteenth-century agricultural improvement' *The Historical Journal* 25 (1982), pp. 313-29.

Horton History Group, *Horton-in-Ribblesdale, the story of an upland parish.* (Settle, North Craven Heritage Trust, 1984).

Hoskins, W G, 'The rebuilding of rural England, 1570 - 1640' *Past and Present* 4 (1953), pp. 44-59.

Hoskins, W G, *The making of the English landscape.* (London: Hodder and Stoughton, 1955a).

Hoskins, W G, Preface to R Millward, *Lancashire. An illustrated essay on the history of the landscape* (London: Hodder and Stoughton, 1955b).

Howells, B E and K A (eds.), *The Extent of Cemaes, 1594* (Haverfordwest: Pembrokeshire Record Society, 1977).

Howson, W G, 'Plague, poverty and population in parts of North-West England, 1580 - 1720' *Transactions of the Historical Society of Lancashire and Cheshire* 112 (1961), pp. 29-55.

Hoyle, R W, 'Land and landed relations in Craven, Yorkshire, c. 1520 - 1600', unpubl. PhD thesis, Oxford University, 1986.

Hoyle, R W, 'New markets and fairs in the Yorkshire dales, 1550 - 1750' in P S Barnwell and M Palmer (eds.) *Post-medieval landscapes. Landscape history after Hoskins*, vol. 3 (Macclesfield: Windgather Press, 2007), pp. 93-106.

Hudson, K, *Patriotism with profit.* (London: Hugh Evelyn, 1972).

Hudson, P J, *Coal mining in Lunesdale.* (Settle: Hudson History, 1998).

Hueckel, G, 'Agriculture during industrialisation' in R Flood and D McCloskey (eds.) *The economic history of Britain since 1700* vol. 1 1700 - 1860. (Cambridge: Cambridge University Press, 1981), pp. 182-203.

Isham, K, *Lime kilns and limeburners in Cornwall* (St. Austell: Cornish Hillside Publications, 2000).

Jackson, D A, Biek, L and Dix, B F, 'A Roman lime kiln at Weekley, Northants' *Britannia* 4 (1973), pp. 128-40.

Jennings, B, (ed.), *A history of Nidderdale.* (Pateley Bridge: Nidderdale History Group, 1992).

Jobey, G, 'A note on "sow" kilns' *Journal of the Newcastle upon Tyne Agricultural Society* 20 (1966), pp. 37-38.

Johnson, D S, 'Nether Lodge, a nineteenth century upland estate' *North Craven Heritage Trust Journal* (2000), pp. 7-9.

Johnson, D S, *Limestone industries of the Yorkshire Dales* (Stroud: Tempus, 2002).

Johnson, D S, *Excavation of Broadwood enclosure, Thornton in Lonsdale, North Yorkshire.* (Ingleton: Ingleborough Archaeology Group, 2004).

Johnson, D S, *The Sow Kiln Project. Excavation of clamp kilns in the Yorkshire Dales.* (Ingleton: Ingleborough Archaeology Group, 2006).

Johnson, D S, *Report on the excavation of a 17th century lime kiln at Town's Piece, Kilnsey, North Yorkshire.* (Ingleton: Ingleborough Archaeology Group, 2007a).

Johnson, D S, 'Lime-kilns, "improvement", and superstition' *Current Archaeology* 213 (2007b), pp. 40-44.

Johnson, D S, 'The archaeology and technology of early-modern lime burning in the Yorkshire Dales: developing a clamp kiln model' *Industrial Archaeology Review* 30 (2008a), pp. 127-43.

Johnson, D S, 'Excavation of a seventeenth-century lime kiln at Kilnsey, North Yorkshire' *British Mining* 86 (2008b), pp. 31-46.

Johnson, D S, 'Hushes, delfs and river stonary: alternative methods of obtaining lime in the gritstone Pennines in the early modern period' *Landscape History* 31 (2010a), pp. 37-52.

Johnson, D S, 'Lime kilns in the Central Pennines: results of a field survey in the Yorkshire Dales and contiguous areas of North and West Yorkshire' *Yorkshire Archaeological Journal* 82 (2010b), pp. 231-62.

Johnson, M, *An archaeology of capitalism.* (Oxford: Blackwell, 1996).

Johnson, M, *Ideas of landscape.* (Oxford: Blackwell, 2007a).

Johnson, M, 'Don't bin your brain' *Landscapes* 8 (2007b), pp. 126-28.

Johnstone, M, *Agrarian change in two Scottish Lowland counties: East Lothian and Lanarkshire 1700 – 1820s.* Unpublished PhD thesis, Lancaster University, 2007.

Jones, R and Page, M, *Medieval villages in an English landscape. Beginnings and ends.* (Macclesfield: Windgather Press, 2006).

Jordan, T E (ed.), *Gerard Boate's Natural History of Ireland.* (New York, Ontario and Lampeter: Edwin Mellon Press, 2006).

Kain, R J P, Chapman, J and Oliver, R R, *Enclosure maps of England and Wales 1595 - 1918.* (Cambridge: Cambridge University Press, 2004).

Kelly, M, *Geology of the Lune valley and Upper Ribblesdale coalfields.* British Mining 85. (Nelson: NMRS, 2008).

Kerridge, E, *The Agricultural Revolution.* (London: George Allen & Unwin, 1967).

Kerridge, E, 'The Agricultural Revolution reconsidered' *Agricultural History* 43 (1969), pp. 463-75.

Kershaw, I, *Bolton Priory rentals and ministers' accounts 1473 - 1539.* Yorkshire Archaeological Society, Record Series 132 (Leeds, 1970).

Kershaw, I, 'The great famine and agrarian crisis in England 1315-1322' *Past and Present* 59 (1973a), pp. 3-50.

Kershaw, I, *Bolton Priory. The economy of a northern monastery 1286 - 1325* (London: Oxford University Press, 1973b).

Kershaw, I, 'A note on the Scots in the West Riding, 1318 - 1319' *Northern History* 17 (1981), pp. 231-39.

Kershaw, I and Smith, D M, *The Bolton Priory Compotus 1286 – 1325.* Yorkshire Archaeological Society, Record Series 154 (Leeds, 2000).

Kidd, A, Cooper, A and Brayson, J, *Your Dales Rocks. Local Geodiversity Action Plan.* (Grassington: North Yorkshire Geodiversity Partnership, n.d. [2007]).

La Rochefoucauld, F de, *A Frenchman's year in Suffolk.* Trans. and ed. by N Scarfe, Suffolk Records Society XXX. (Woodbridge: The Boydell Press, 1988).

Lake, J, 'The English *pays*: approaches to understanding and characterising landscapes and places' *Landscapes* 8 (2007), pp. 28-39.

Lamb, H H, 'The early medieval warm epoch and its sequel' *Palaeography, Palaeoclimatology, Palaeoecology* 1 (1965), pp. 13-37.

LancashireCountyCouncil. Available at **www.lancashire.gov.uk/environment/archaeologyandheritage/lhlcpcover** Accessed 8 March 2008.

Law, S, 'Archaeology report on Towneley Hall, Lancashire, up to 1902'. Unpublished MA dissertation, University of Bristol, 2006.

Leach, J, 'Limeburning in Tiverton' *The Devon Historian* 60 (2000), pp. 8-14.

Leach, J, 'The Devon lime industry' *Southern History* 24 (2002), pp. 1-26.

Legg, K L, *The lost Cartulary of Bolton Priory. An edition of the Coucher Book and Charters.* Yorkshire Archaeological Society Record Series 160. (Woodbridge: The Boydell Press, 2009).

Lennard, R, 'English agriculture under Charles II: the evidence of the Royal Society's "Enquiries"' *Economic History Review* 4 (1932), pp. 23-45.

Lewis, C, Mitchell-Fox P and Dyer C, *Village, hamlet and field.* (Manchester: Manchester University Press, 1997).

Lewis, M E, Roberts, C A and Manchester, K, 'Comparative study of the prevalence of maxillary sinusitis in later medieval urban and rural populations in northern England' *American Journal of Physical Anthropology* 98 (1995), pp. 497-506.

Long, P R, *The wealth of the magisterial class in Lancashire, c 1590 - 1640.* Unpublished MA dissertation, University of Manchester, 1968.

Low, A, *The Georgic Revolution.* (Princeton: Princeton University Press, 1985).

Low, A, 'Agricultural reform and the love poems of Thomas Carew; with an instance from Lovelace' in M Leslie and T Raylor (eds.), *Culture and cultivation in early modern England.* (Leicester and London: Leicester University Press, 1992), pp. 63-80.

Lucas, A T, *Furze. A survey and history of its uses in Ireland.* (Dublin: Natural Museum of Ireland, 1960).

Macdonald, S, 'The role of George Culley of Fenton in the development of Northumberland agriculture' *Archaeologia Aeliana* 5th Series, 3 (1975), pp. 131-41.

Macdonald, S, 'The diffusion of knowledge among Northumberland farmers, 1780 – 1815' *Agricultural History Review* 27 (1979), pp. 30-39.

Mackenzie, B R, Bager, M, Ojaveer, K, Awebo, K, Heino, V, Holm, P and Must, A, 'Multi-decadal scale variability in the eastern Baltic cod fishery 1550 - 1860 – evidence and causes' *Fisheries Research* 87 (2007), pp. 106-19.

MAFF, *Lime and liming.* Reference Book 35. (London: HMSO, 1981).

Marshall,, T H, 'Jethro Tull and the "New husbandry" of the eighteenth century' *Economic History Review* 2 (1929-30), pp. 41-60.

Mathew, W M, 'Marling in British agriculture: a case of partial identity' *Agricultural History Review* 41 (1993), pp. 97-110.

McDonnell, J, 'Upland Pennine hamlets' *Northern History* 26 (1990), pp. 20-39.

McNeil, R and Newman, R, 'The industrial and modern period resource assessment' in R Newman *The archaeology of North West England. An archaeological resource framework for North West England: volume 1 resource assessment.* (Manchester: CBA North West, 2006), pp. 165-94.

McRae, A, 'Husbandry manuals and the language of agrarian improvement' in M Leslie and T Raylor (eds.), *Culture and cultivation in early modern England. Writing and the land.* (Leicester: Leicester University Press, 1992), pp. 35-62.

McRae, A, *God speed the plough. The representation of agrarian England, 1500 - 1660.* (Cambridge: Cambridge University Press, 1996).

Manley, B, *The Penguin historical atlas of Ancient Egypt.* (Harmondsworth: Penguin, 1996).

Manley, G, 'Central England temperatures: monthly means 1659 to 1973' *Quarterly Journal of the Royal Meteorological Society* 100 (1974), pp. 389-405.

Marshall, J D, ' "Statesmen" in Cumbria: the vicissitudes of an expression' *Transactions of the Cumberland and Westmorland Antiquarian and Archaeological Society* 72 (1972), pp. 248-73

Marshall, J D, 'Agrarian wealth and social structure in pre-industrial Cumbria' *Economic History Review* 33 (1980), pp. 503-21.

Michelmore, D J H, *The Fountains Abbey Lease Book* Yorkshire Archaeological Society Record Series 140 (Leeds, 1981).

Miles, D (ed.), *The Description of Pembrokeshire.* (Llandysul: Gomer Press, 1994). Facsimile edition, first published 1603 by George Owen of Henllys.

Millward, R, *Lancashire.* An illustrated essay on the history of the landscape. (London: Hodder and Stoughton, 1955).

Millward, R and Robinson, A, *Cumbria.* Landscapes of Britain. (Basingstoke and London: Macmillan, 1972).

Mingay, G E (ed.), *Arthur Young and his times.* (London and Basingstoke: Macmillan, 1975).

Moore-Colyer, R, 'Of lime and men: aspects of the coastal trade in lime in south-west Wales in the 18th and 19th centuries' *Welsh History Review* 14 (1988), pp. 54-77.

Morrogh-Ryan, J, 'Swaledale' in Gamble, D and St. Pierre, T (eds.) *Hay time. The natural, cultural and land management history of hay meadows in the Yorkshire Dales.* Appendix 2 (Lancaster: Scotforth Books for the Yorkshire Dales Millennium Trust, 2010), pp. 241-42.

Muir, R, *Approaches to landscape.* (Basingstoke: Macmillan,1999).

Muir, R, *The NEW reading the landscape.* (Exeter: Exeter University Press, 2000).

Muir, R, *Landscape encyclopaedia.* (Macclesfield: Windgather Press, 2004).

Nef, J U, *The rise of the British coal industry.* (London: Frank Cass, 1966).

Nevell, M, 'Industrial archaeology or the archaeology of the industrial period? Models, methodology and the future of industrial archaeology', The 2005 Rolt Memorial Lecture *Industrial Archaeology Review* 28 (2006), pp. 3-15.

Nevell, M, 'People versus machines or people and machines? Current research directions within British post-medieval and industrial archaeology' in A Horning and M Palmer (eds.) *Crossing paths or sharing tracks?* Society for Post-medieval Archaeology Monograph 5. (Woodbridge: Boydell and Brewer, 2009), pp. 31-40.

Newman, R, *The historical archaeology of Britain, c. 1540 - 1900.* (Stroud: Sutton, 2001).

Nielsen, R, 'Medieval warm period and the Little Ice Age myths'. 2005. Available at **http://home.iprimus.com.au/nielsens/**. Accessed 3 February 2009.

Nixon, ,, *Industrial archaeology of Derbyshire.* (Newton Abbot: David & Charles, 1969).

North, F J,, *Limestones.* (London: Thomas Murphy, 1930).

Oates, J A H, *Lime and limestone.* (Weinheim: Wiley-VCH, 1998).

Ogg, W G, 'The revival of liming' *Scottish Journal of Agriculture* 23 (1942), pp. 355-66.

Oosthuizen, S, 'Priceless landscape' *British Archaeology* 88 (2006), pp. 27-31.

Orange, H, 'Industrial archaeology: its place within the academic discipline, the public realm and the heritage industry' *Industrial Archaeology Review* 30 (2008), pp. 83-95.

Oschinsky, D, 'Medieval treatises on estate management' *Economic History Review.* Second Series, 8 (1956), pp. 296-309.

Outhwaite, R B, 'Progress and backwardness in English agriculture, 1500 - 1650' *Economic History Review* 39 (1986), pp. 1-18.

Overton, M, *Agricultural revolution in England: the transformation of the agrarian economy 1500 – 1850.* (Cambridge: Cambridge University Press, 1996).

Palmer, M, 'Post-medieval industrial landscapes: their interpretation and management' in D Hooke (ed.) *Landscape. The richest historical record.* Supplementary Series 1 (Westbury: Society for Landscape Studies, 2000), pp. 119-31.

Palmer, M, 'The archaeology of industrialization; introduction' in D Barker and D Cranstone (eds.) *The archaeology of industrialization.* (Leeds: Maney, 2004), pp. 1-4.

Palmer, M, 'Understanding the workplace: a research framework for industrial archaeology in Britain' *Industrial Archaeology Review* 27 (2005), pp. 9-17.

Palmer, M, 'Introduction: post-medieval landscapes since Hoskins – theory and practice' in P S Barnwell and M Palmer (eds.) *Post-medieval landscapes. Landscape history after Hoskins* 3. (Macclesfield: Windgather Press, 2007), pp. 1-5.

Palmer, M, 'Industrial archaeology and the archaeological community: fifty years on' *Industrial Archaeology Review* XXXII (2010), pp. 5-20.

Parker, D E, Legg, T P and Folland, C K, 'A new daily central England temperature series, 1772 – 1991' *International Journal of Climatology* 12 (1992), pp. 317-42.

Pearson, R W and Adams, F (eds.), *Soil acidity and liming.* (Madison, Wisc.: American Society of Agronomy, 1967).

Perry, P J, 'Where was the "Great Agricultural Depression"? A geography of agricultural bankruptcy in late Victorian England and Wales' *Agricultural History Review* 20 (1972), pp. 30-45.

Perry, P J, (ed.), *British agriculture 1875 - 1914.* (London: Methuen, 1973).

Phillips, A D M, 'The development of underdraining on a Yorkshire estate during the nineteenth century' *Yorkshire Archaeological Journal* 44 (1972), pp. 195-206.

Phillips, A D M and Clout H D, 'Underdraining in France during the second half of the nineteenth century' *Transactions of the Institute of British Geographers* 51 (1970), pp. 71-94.

Phillips, C B, *The gentry in Cumberland and Westmorland 1600 - 1665.* Unpublished PhD thesis, Lancaster University, 1973.

Phillips, C B and Smith, J H, *Lancashire and Cheshire from AD1540.* (London and New York: Longman, 1994).

Pickles, M F, 'Agrarian society and wealth in mid-Wharfedale, 1664 - 1743' *Yorkshire Archaeological Journal* 53 (1981), pp. 63-78.

Pickles, M and Bosworth, J, 'Farmhold structure in a district of piecemeal enclosure: the Manor of Askwith from 1596 to 1816' *Yorkshire Archaeological Journal* 63 (1991), pp. 109-26.

Platt, C, *The monastic grange in medieval England.* (London: Macmillan, 1969).

Porter, J, 'Waste land reclamation in the sixteenth and seventeenth centuries: the case of south-eastern Bowland, 1550 - 1630' *Transactions of the Historical Society of Lancashire and Cheshire* 127 (1977), pp. 1-23.

Post, J D, *The last great subsistence crisis in the western world.* (Baltimore and London: The Johns Hopkins University Press, 1977).

Prince, H C, 'The origin of pits and depressions in Norfolk' *Geography* (1964) 49, pp. 15-32.

Prince, H C, 'England circa 1800' in H C Darby (ed.) *A new historical geography of England after 1600* (Cambridge: Cambridge University Press, 1976), pp. 89-164.

Prothero, R E (Lord Ernle), *English farming past and present.* (London: Heinemann, 1961 edition). First published 1912.

Rackham, O, *The history of the countryside.* (London: Dent, 1986).

Raistrick, A, *The Pennine Dales.* (London: Eyre & Spottiswoode, 1968).

Raistrick, A, 'Industrial archaeology' in I G Simmons (ed.) *Yorkshire Dales* (London: HMSO, 1971), pp. 39-47.

Raistrick, A, *Industrial archaeology.* (St Albans: Paladin, 1973).

Rippon, S, *Historic landscape characterisation.* Practical Handbook 16. (York: Council for British Archaeology, 2004).

Rippon, S and Gardiner, M, 'Conclusions: the future of medieval landscape studies' in M Gardiner and S Rippon *Medieval landscapes. Landscape history after Hoskins.* (Macclesfield: Windgather Press, 2007), pp. 231-35.

Rix, M, 'Industrial archaeology' *The Amateur Historian* 2 (1955), pp. 225-29.

Rix, M, *Industrial archaeology.* (The Historical Association, 1967).

Robinson, G W, 'The use of lime' *Journal of the Royal Agricultural Society of England* 104 (1943), pp. 136-44.

Robinson, T, *Connemara. Listening to the wind.* (London: Penguin, 2007).

Robinson, D J and Cooke, R U, 'Lime-kilns in Surrey: a reconstruction of a rural industry' *Surrey Archaeological Collection* 59 (1962), pp. 19-26.

Romney, P (ed.), *The diary of Charles Fothergill. 1805.* Yorkshire Archaeological Society Record Series CXLII, (Leeds, 1984).

Rowe, D J (ed.), *General view of the agriculture of Northumberland, Cumberland and Westmorland* by J Bailey and G Culley. Facsimile edition, originally published 1805. (Newcastle upon Tyne: Frank Graham, 1972).

Rowley, T, *The English landscape in the twentieth century.* (London: Hambledon Continuum, 2006).

Russell, E J, 'Manures in December' *Journal of the Ministry of Agriculture* 27 (1920-21), pp. 876-82.

Ruston, A G, 'Lime burning on a Yorkshire farm' *Journal of the Ministry of Agriculture* 31 (1924), pp. 738-43.

Rynne, C, *Industrial Ireland 1750 - 1930. An archaeology.* (Cork: The Collins Press, 2006).

Salzman, L F, *Building in England down to 1540.* (Oxford: Clarendon Press, 1952).

Searle, A B, *Limestone and its products.* (London: Ernest Benn, 1935).

Searle, C E, ' "The odd corner of England": a study of a rural social formation in transition: Cumbria c. 1700 – 1914'. Unpublished PhD thesis. University of Essex, 1983.

Searle, C E, 'Customary tenants and the enclosure of the Cumbrian commons' *Northern History* 29 (1993), pp. 126-53.

Shannon, W D, *Approvement and improvement in early-modern England: enclosure in the lowland wastes of Lancashire c. 1500 – 1700*, unpublished PhD thesis, Lancaster University, 2009.

Sharpe France, R, 'A history of plague in Lancashire' *Transactions of the Historical Society of Lancashire and Cheshire* 90 (1939), pp. 1-175.

Sheeran, G, 'Patriotic views: aristocratic ideology and the eighteenth-century landscape' *Landscapes* 7 (2006), pp. 1-23.

Simmons, I G, *The moorlands of England and Wales. An environmental history 8000BC - AD2000.* (Edinburgh: Edinburgh University Press, 2003).

Skinner, B C, *The lime industry in the Lothians.* (Edinburgh: University of Edinburgh Press, 1969).

Smail, J, *Woollen manufacturing in Yorkshire. The Memorandum Book of John Brearley cloth frizzer at Wakefield 1758-1762.* Yorkshire Archaeological Society Record Series 155 (Woodbridge: Boydell Press, 2001).

Smith, A H, *Place-names of the West Riding of Yorkshire, Part II.* (Cambridge: Cambridge University Press, 1961).

Smith, R B, *Land and politics in the England of Henry VIII. The West Riding of Yorkshire: 1530-46.* (Oxford: Clarendon Press, 1970).

Smith, R S, 'Farming and the conservation of traditional meadowland in the Pennine Dales Environmentally Sensitive Area' in M B Usher and D B A Thompson (eds.) *Ecological change in the uplands* British Ecological Society, Special Publication 7. (Oxford: Blackwell Scientific Publications, 1988), pp. 183-99.

Smith, R S, Shiel, R S, Bardgett, R D, Millward, D, Corkhill, P, Evans, P, Quirk, H, Hobbs, P J and Kometa, S T, 'Long-term change in vegetation and soil microbial communities during the phased restoration of traditional grassland' *Journal of Applied Ecology* 45 (2008), pp. 670-79.

Smith, W J and K, *An architectural history of Towneley Hall, Burnley.* (Barrowford: Heritage Trust for the North West, 2004).

Smout, T C and Fenton, A, 'Scottish agriculture before the Improvers – an exploration' *Agricultural History Review* 13 (1965), pp. 73-93.

Smyth, W J, *Map-making, landscapes and memory. A geography of colonial and early modern Ireland c. 1530 - 1750.* (Cork: Cork University Press, 2006).

Spence, R T, 'Mining and smelting in Yorkshire by the Cliffords, Earls of Cumberland, in the Tudor and early Stuart period' *Yorkshire Archaeological Journal* 64 (1992), pp. 157-83.

SPMA, Available at **http://www.spma.org.uk/index.php**. Accessed 20 November 2008.

Stanier, P, *Stone quarry landscapes*. (Stroud: Tempus, 2000).

Steane, J M, *The archaeology of medieval England and Wales*. (London: Croom Helm, 1984).

Stratton, J M and Houghton Brown, J, *Agricultural records A.D. 220 - 1977*. (London: John Baker, 1978).

Sutton, K, 'Reclamation of wasteland during the eighteenth and nineteenth centuries' in H D Clout (ed.) *Themes in the historical geography of France*. (London: Academic Press, 1977), pp. 247-300.

Swain, J T, 'Industry and economy in North East Lancashire circa 1500 - 1640'. Unpublished PhD thesis, University of ambridge, 1983.

Swain, J T, *Industry before the industrial revolution. North-east Lancashire c. 1500 - 1640*. Third Series 32. (Manchester: Chetham Society, 1986).

Tarlow, S, *The archaeology of improvement in Britain, 1750 - 1850*. (Cambridge: Cambridge University Press, 2007).

Taylor, C C, 'The plus fours in the wardrobe: a personal view of landscape history' in D Hooke (ed.) *Landscape the richest historical record*. Supplementary Series 1 (Westbury: Society for Landscape Studies, 2000), pp. 157-62.

Taylor, C C, 'W. G. Hoskins and *The Making of the English landscape*' *The Local Historian* 35 (2) (2005), pp. 74-81.

Taylor, E G R, 'Camden's England' in H C Darby (ed.), *An historical geography of England before A.D. 1800*. (Cambridge: Cambridge University Press, 1951), pp. 354-86.

Taylor, H, 'The nineteenth century limekilns at Barnby Basin, South Yorkshire' *Yorkshire Archeological Journal* 81 (2009), pp. 311-27.

Taylor, T, *The buried soil*. (London: Fourth Estate, 2002).

Thirsk, J, 'Plough and pen: agricultural writers in the seventeenth century' in T H Aston, P R Cross, C Dyer and J Thirsk (eds.) *Social relations and ideas* (Cambridge: Cambridge University Press, 1983), pp. 295-318.

Thirsk, J, 'Making a fresh start: sixteenth-century agriculture and the classical inspiration' in M Leslie and T Raylor (eds.) *Culture and cultivation in early modern England. Writing and the land*. (Leicester: Leicester University Press, 1992), pp. 15-34.

Thirsk, J, *Alternative agriculture. A history*. (Oxford: Oxford University Press, 1997).

Thirsk, J and Cooper J P, *Seventeenth-century economic documents*. (Oxford: Clarendon Press, 1972).

Thompson, F M L, 'The second Agricultural Revolution, 1815 - 1880' *Economic History Review* 21 (1968), pp. 62-77.

Thorold Rogers, J E, *A history of agriculture and prices in England, vol. 4 1583 - 1702* and *vol. 7 1703 - 1793* (Oxford: Clarendon Press, 1963 edition). Originally published 1866-1902.

Thurston, J M, Williams, E D and Johnston, A E, 'Modern developments in an experiment on permanent grassland started in 1856: effects of fertilisers and lime on botanical composition and crop and soil analyses' *Annales Agronomiques* 27 (1976), pp. 1043-82.

Thwaite, H, *An edition of Abbotside wills and administrations, 1690 - 1760*. Unpublished M.Phil dissertation, University of Leeds, 1972.

Tilley, C, *A phenomenology of landscape. Places, paths and monuments*. (Oxford and Providence, R I: Berg, 1994).

Tomson, S J N and McIlwaine, J J, 'A Roman Limekiln in Womersley, North Yorkshire' *YAS Roman Antiquities Section Bulletin* 10 (1992-93), p. 17.

Toulmin Smith, L, *The itinerary of John Leland in or about the years 1535-1543*. (London: Centaur, 1964).

Turner, M, *English parliamentary enclosure*. (Folkestone: Dawson; Hamden, Conn.: Archon Books, 1980).

Turner, M, *Enclosures in Britain 1750 - 1830*. (London: Macmillan, 1984).

Turner, M E, Beckett, J V and Afton, B, 'Taking stock: farmers, farm records, and agricultural output in England, 1700 - 1850' *Agricultural History Review* 44 (1996), pp. 21-34.

Turner, M E, Beckett, J V and Afton, B, *Farm production in England 1700 - 1914*. (Oxford: Oxford University Press, 2001).

Tyson, B, 'Transportation and the supply of construction materials: an aspect of traditional building management' *Vernacular Architecture* 29 (1998), pp. 63-81.

Vale, B, *The Scropes of Bolton and of Masham, c. 1300 to c. 1450: a study of a northern noble family*. Unpublished D.Phil. thesis, University of York, 1987.

Vaughan, G, 'The collecting of Classical antiquities in England in the 18th century: a study of charles Towneley (1737 - 1805) and his circle'. Unpublished PhD thesis, University of Oxford, 1988.

Virgoe, J, 'John Holt and the *General view of the agriculture of the county of Lancaster*; an appraisal' *Transactions of the Historic Society of Lancashire and Cheshire* 154 (2005), pp. 93-116.

Wade Martins, S, *Farmers, landlords and landscapes*. (Macclesfield: Windgather Press, 2004).

Wade Martins, S, *Coke of Norfolk (1754 – 1842). A biography*. (Woodbridge: The Boydell Press, 2009).

Waites, B, *Moorland and vale-land farming in North-East Yorkshire. The monastic contribution in the thirteenth and fourteenth centuries*. Borthwick Papers 32 (York: Borthwick Institute of Historical Research, 1967).

Walton, J K, 'The strange decline of the Lakeland yeoman: some thoughts on sources, methods and definitions' *Tranasactions of the Cumberland and Westmorland Antiquarian and Archaeological Society* 86 (1986), pp. 221-33.

Walton, J R, 'Agriculture 1730 - 1900' in R A Dodgshon and R A Butlin (eds.) *An historical geography of England and Wales*. (London: Academic Press, 1978), pp. 239-65.

Ward, T, 'The Corn Laws and English wheat prices, 1815 - 1846' *Atlantic Economic Journal* 32 (2004), pp. 245-55.

White, R F, 'A Roman lime kiln near Cardington Mill, Bedford' *Bedfordshire Archaeological Journal* 12 (1977), pp. 23-26.

White, R, *Yorkshire Dales, landscpes through time.* (London: B T Batsford/English Heritage, 1997).

Whyte, I D, 'Written leases and their impact on Scottish agriculture in the seventeenth century' *Agricultural History Review* 27 (1979a), pp. 1-9.

Whyte, I D, *Agriculture and society in seventeenth-century Scotland.* (Edinburgh: John Donald, 1979b).

Whyte, I D, *Scotland before the Industrial Revolution.* (London and New York: Longman, 1995a).

Whyte, I D, *Climatic change and human society.* (London: Arnold, 1995b).

Whyte, I D, 'Patterns of parliamentary enclosure of waste in Cumbria: a case study from north Westmorland' *Landscape History* 22 (2000), pp. 77-89.

Whyte, I D, *Landscape and history since 1500.* (London: Reaktion Books, 2002).

Whyte, I D, *Transforming fell and valley.* (Lancaster: Centre for North-West Regional Studies, 2003a).

Whyte, I D, ' "Wild, barren and frightful" – parliamentary enclosure in an upland county: Westmorland 1767 – 1890' *Rural History* 14 (2003b), pp. 21-38.

Whyte, I D, 'Taming the fells: parliamentary enclosure and the landscape in northern England' *Landscapes* 6 (2005), pp. 46-61.

Whyte, I D, 'The costs of parliamentary enclosure in an upland setting: south and east Cumbria c. 1760 - 1860' *Northern History 43 (2006), pp. 97-115.*

Whyte, I D and K A, *The changing Scottish landscape 1500-1800.* (London and New York: Routledge, 1991).

Wigley, T M L, Lough, J M and Jones, P D, 'Spatial patterns of precipitation in England and Wales and a revised, homogeneous England and Wales precipitation series' *Journal of Climatology* 4 (1984), pp. 1-25.

Willan, T S and Crossley, E W, *Three seventeenth-century Yorkshire surveys.* Record Series CIV. (Leeds: Yorkshire Archaeological Society, 1941).

Williams, M, 'The enclosure and reclamation of waste land in England and Wales in the eighteenth and nineteenth centuries' *Transactions of the Institute of British Geographers* 51 (1970), pp. 55-69.

Williamson, T, 'Questions of preservation and destruction' in P Everson and T Williamson (eds.), *The archaeology of landscape* (Manchester: Manchester University Press, 1998).

Williamson, T, 'The rural landscape: 1500 - 1900, the neglected centuries' in D Hooke (ed.), *Landscape. The richest historical record.* Supplementary Series 1, (Westbury: Society for Landscape Studies, 2000a), pp. 109-17.

Williamson, T, 'Understanding enclosure' *Landscapes* 1 (2000b), pp. 56-79.

Williamson, T, *The transformation of rural England. Farming and the landscape 1700 - 1870.* (Exeter: University of Exeter Press, 2002).

Williamson, T, *Shaping medieval landscapes.* (Macclesfield: Windgather Press, 2003).

Williamson, T, 'HLC: some queries' *Landscapes* 8 (2007a), pp. 64-71.

Williamson, T, 'Archaeological perspectives on landed estates: research agendas' in J Finch and K Giles (eds.), *Estate landscapes: design, improvement and power in the post-medieval landscape.* (Woodbridge: Boydell and Brewer, 2007b).

Wilmot, S, *"The business of improvement": agriculture and scientific culture in Britain, c. 1770 - c. 1870.* Historical Geography Research Series. (Bristol: University of Bristol, 1990).

Wilson, K, *The history of Lothersdale.* (Lothersdale: Lothersdale Parish Council, 1972).

Winchester, A J L, *The harvest of the hills. Rural life in Northern England and the Scottish Borders,1400 – 1700.* (Edinburgh: Edinburgh University Press, 2000).

Winchester, A J L, *Thomas Denton. A perambulation of Cumberland 1687 - 1688.* Surtees Society CCVII. (Woodbridge: Boydell and Brewer, 2003).

Winchester, A J L, 'Regional identity in the Lake Counties: land tenure and the Cumbrian landscape' *Northern History* 42 (2005), pp. 29-48.

Winchester, A J L, *England's landscape. The North west.* (London: English Heritage/Collins, 2006).

Wingate, M, *Small-scale lime burning.* (London: Intermediate Technology Publications, 1985).

Wood, A, *Nineteenth century Britain 1815 - 1914.* (Harlow: Longman, 1982).

Woodward, D, ' "Gooding the earth": manuring practices in Britain 1500 - 1800' in S Forster and T C Smout (eds.) *The history of soils and field systems.* (Aberdeen: Scottish Cultural Press, 1994), pp. 101-10.

Wordie, J R, 'The chronology of English enclosures, 1500 - 1914' *Economic History Review* 36 (1983), pp. 483-505.

Wright, G N, 'Lime-kilns of the Pennine Dales' *The Yorkshire Ridings* 4 (1967), pp. 33-35.

Wright, G N, *The Yorkshire Dales.* (Newton Abbot: David & Charles, 1986).

Wrigley, E A, *Continuity, chance and change.* (Cambridge: Cambridge University Press, 1993).

Wrigley, E A and Schofield, R S, *The population history of England 1541 - 1871.* (London: Edward Arnold, 1981).

www.metoffice.com/research/hadleycentre/CRdata/Daily/HadCET_act.txt. Accessed 16 March 2006 and 26 February 2007.

www.whithorn.com/saint-ninian.htm. Accessed 10 August 2009.

Appendix

Summary of literature on lime kilns and lime burning: bibliographic references

Aldsworth, F, *Limeburning and the Amberley chalk pits: a history*. (Chichester: West Sussex County Council, 1979).

Barker, P F, *A survey of limekiln sites in south Cumbria and north Lancashire*.(Holme: Holme and District Local History Society, n.d [2008]).

Barnatt, J and Dickson, A, 'Survey and interpretation of a limekiln complex at Peak Forest, Derbyshire: and a review of limeburning in the north-west Peak' *Derbyshire Archaeological Journal* 124 (2004), pp. 141-215.

Bhanji, S, 'Limeburning in Exeter and Exe Estuary parishes: a documentary survey' *Report and Transactions of the Devonshire Association for the Advancement of Science* 131 (1999), pp. 165-97.

Bick, D, 'Lime-kilns on the Gloucestershire-Herefordshire border' *Industrial Archaeology Review* VII (1984), pp. 85-93.

Born, A, 'Limestone, limekilns and the limeburning industry north and west of Dartmoor' *Report and Transactions of the Devonshire Association for the Advancement of Science* 123 (1991), pp. 213-40.

Brock, J, 'Lime kilns around Richmond' *The Richmond Review* (1987), pp. 6-21.

Brown, J L, *The limekilns within the Pembrokeshire Coast National Park*. Unpublished report for the National Trust and the Pembrokeshire Coast National Park Authority.

Clarke, R J, 'The Closeburn Limeworks Scheme: a Dumfriesshire waterpower complex' *Industrial Archaeology Review* X (1987), pp. 5-22.

Cleasby, I, 'Limekilns in Sedbergh, Garsdale and Dent' Part 1 *The Sedbergh Historian* 3 (3) (1994), pp. 2-9.

Cleasby, I, 'Limekilns and limeburning in Sedbergh, Garsdale and Dent' Part 2 *The Sedbergh Historian* 3 (4) (1995), pp. 2-8.

Cleasby, I, 'Limekilns in Sedbergh, Garsdale and Dent *Current Archaeology* 145 (1995), pp. 16-20.

Daniel, P, 'Limeburning in Priddy' *Somerset Industrial Archaeological Society Bulletin* 86 (2001), pp. 8-15.

Davies, P B S, *Pembrokeshire limekilns*. (St. David's: Merrivale, 1997).

Donnachie, I, 'The lime industry in south-west Scotland' *Transactions of the Dumfriesshire and Galloway Natural History and Antiquarian Society* 48 (1971), pp. 146-52.

Evans, D, 'The technique of lime burning' *Devon and Cornwall Notes and Queries* 36 (1987), pp. 34-35.

Gibbons, P, 'Scottish limes. A preliminary review of information on limestone sources and lime production in Scotland during the 18th and 19th centuries' *Lime News* 1 (1993), pp. 5-18.

Green, H E, *The limestone mines of Walsall*. (Halesowen: The Black Country Society, 1977).

Groom, A R, 'A century of lime-burning at Buxton' *Lime News* 1 (1993), pp. 30-39.

Holt, M, 'Lime kilns in central Sussex' *Sussex Industrial History* 2 (1971), pp. 23-30.

Isham, K, *Lime kilns and limeburners in Cornwall*. (St. Austell: Cornish Hillside Publications, 2000).

Leach, J T, 'Burning lime in Derbyshire pye kilns' *Industrial Archaeology Review* XVII (1995), pp. 145-58.

Lach, J, 'Limeburning in Tiverton' *The Devon Historian* 60 (2000), pp. 8-14.

Leach, J, 'The Devon lime industry' *Southern History* 24 (2002), pp. 1-26.

Martins, C, 'The silent shore speaks' *Current Archaeology* 219 (2008), pp. 26-32.

Minchinton, W, *A limekiln miscellany: the South-West and South Wales*. (Exeter: Exeter Industrial Archaeology Group, 1984).

Pounds, N J G, 'Building and operating accounts of an eighteenth century lime-kiln' *Devon and Cornwall Notes and Queries* 32 (1971), pp. 237-40.

Robertson, A, *Limekilns of the North Pennines*. (Nenthead: North Pennines Heritage Trust, 1999).

Scottish Lime Centre Trust, *Charlestown Limeworks. Research and conservation*. (Edinburgh: Historic Scotland, 2006).

Skinner, B C, *The lime industry in the Lothians*. (Edinburgh: Edinburgh University Press, 1969).

Skinner, B C, 'The archaeology of the lime industry in Scotland' *Post-Medieval Archaeology* 9 (1975), pp. 225-30.

Toft, l A, 'Lime burning on the Gower Peninsula's limestone belt' *Industrial Archaeology Review* XI (1988), pp. 75-85.

Unwin, G, *Samuel Oldknow and the Arkwrights. The Industrial Revolution in Marple and Stockport*. (Manchester: Manchester University Press, 1968).

Wright, G N, 'Lime-kilns of the Pennine Dales' *The Yorkshire Ridings* 4 (1967), pp. 33-35.

www.ingramcontent.com/pod-product-compliance
Lightning Source LLC
LaVergne TN
LVHW071215210926
848482LV00015B/817
9781407307381